Spotlight on Numeracy

Spotlight on Numeracy

Harry Ogden and Rex Woodward

Pitman

PITMAN PUBLISHING LIMITED
128 Long Acre, London WC2E 9AN

PITMAN PUBLISHING INC
1020 Plain Street, Marshfield, Massachusetts 02050

Associated Companies
Pitman Publishing Pty Ltd, Melbourne
Pitman Publishing New Zealand Ltd, Wellington
Copp Clark Pitman, Toronto

© H. Ogden and R. Woodward 1985

First published in Great Britain 1985

British Library Cataloguing in Publication Data
   Ogden, Harry
      Spotlight on numeracy
      1. Mathematical—Problems, exercises, etc.
      I. Title II. Woodward, Rex
      510'.76   QA43

   ISBN 0–273–02016–1

All rights reserved. No part of this publication may be reproduced, stored in a retrieval system, or transmitted, in any form or by any means, electronic, mechanical, photocopying, recording and/or otherwise, without the prior written permission of the publishers. This book may not be lent, resold, hired out or otherwise disposed of by way of trade in any form of binding or cover other than that in which it is published, without the prior consent of the publishers. This book is sold subject to the Standard Conditions of Sale of Net Books and may not be resold in the UK below the net price.

East Devon College of
Further Education.

Typeset and printed in Great Britain at The Pitman Press, Bath

Contents

Preface ix

S.I. Units xi

1 Operations in Arithmetic 1
1.1 Addition 1
1.2 Subtraction 2
1.3 Multiplication 3
1.4 Division 4

    Exercises 1.A Core 4
    Exercises 1.B Business, Administration and Commerce 6
    Exercises 1.C Technical Services: Engineering and Construction 9
    Exercises 1.D General Manufacturing and Processing 12
    Exercises 1.E Services to People I: Community Services 14
    Exercises 1.F Services to People II: Food and Clothing 16

2 Operations with Fractions 21
2.1 Vulgar Fractions 21
2.2 Improper Fractions 21
2.3 Mixed Numbers 22
2.4 Equivalent Fractions 22
2.5 Cancelling 22
2.6 Addition and Subtraction of Fractions 23
2.7 Multiplication of Fractions 24
2.8 Division of Fractions 24
2.9 Fraction of a Quantity 24
2.10 One quantity as a Fraction of Another 24

    Exercises 2.A Core 25
    Exercises 2.B Business, Administration and Commerce 26
    Exercises 2.C Technical Services: Engineering and Construction 28
    Exercises 2.D General Manufacturing and Processing 30
    Exercises 2.E Services to People I: Community Services 31
    Exercises 2.F Services to People II: Food and Clothing 32

3 Operations with Decimals 35
3.1 Decimal Fractions 35
3.2 Addition of Decimals 35
3.3 Subtraction of Decimals 36
3.4 Multiplication of Decimals 36
3.5 Division of Decimals 37
3.6 Rounding-off 37

    Exercises 3.A Core 38
    Exercises 3.B Business, Administration and Commerce 40
    Exercises 3.C Technical Services: Engineering and Construction 41
    Exercises 3.D General Manufacturing and Processing 43
    Exercises 3.E Services to People I: Community Services 45
    Exercises 3.F Services to People II: Food and Clothing 46

4 Conversion of Fractions, Decimals and Percentages 48
4.1 Fractions to Decimals 48
4.2 Decimals to Fractions 48
4.3 Percentage 49
4.4 Fractions to Percentages 49
4.5 Percentages to Fractions 49
4.6 Decimals to Percentages 49
4.7 Percentages to Decimals 49
4.8 Percentage of a Quantity 50
4.9 Expressing a Quantity as a Percentage 50

    Exercises 4.A Core 50
    Exercises 4.B Business, Administration and Commerce 51
    Exercises 4.C Technical Services: Engineering and Construction 53
    Exercises 4.D General Manufacturing and Processing 55

Exercises 4.E Services to People I:
                Community Services 56
Exercises 4.F Services to People II: Food
                and Clothing 57

5 Estimates, Average, Ratio and Proportion 60

5.1 Estimates 60
5.2 Average 61
5.3 Ratio 61
5.4 Proportion 63

Exercises 5.A Core 63
Exercises 5.B Business, Administration
                and Commerce 65
Exercises 5.C Technical Services: Engin-
                eering and Construction 66
Exercises 5.D General Manufacturing and
                Processing 68
Exercises 5.E Services to People I:
                Community Services 70
Exercises 5.F Services to People II: Food
                and Clothing 72

6 Length, Area and Perimeter 74

6.1 Units of Length 74
6.2 Length Conversion Factors 74
6.3 Units of Area 75
6.4 Rectangle and Square 76
6.5 Parallelogram 77
6.6 Triangle 77
6.7 Circle 78

Exercises 6.A Core 79
Exercises 6.B Business, Administration
                and Commerce 83
Exercises 6.C Technical Services: Engin-
                eering and Construction 85
Exercises 6.D General Manufacturing and
                Processing 88
Exercises 6.E Services to People I:
                Community Services 91
Exercises 6.F Services to People II: Food
                and Clothing 94

7 Volume, Mass and Density 97

7.1 Units of Volume 97
7.2 Volume of a Prism 97
7.3 Volume of a Cylinder 99
7.4 Volume of a Pyramid and a Cone 100
7.5 Volume of a Sphere 100
7.6 Capacity of Containers 101
7.7 Mass 101
7.8 Density 102

Exercises 7.A Core 102
Exercises 7.B Business, Administration
                and Commerce 106
Exercises 7.C Technical Services:
                Engineering and
                Construction 107
Exercises 7.D General Manufacturing and
                Processing 110
Exercises 7.E Services to People I:
                Community Services 112
Exercises 7.F Services to People II: Food
                and Clothing 114

8 Angles and Shapes 116

8.1 Measurement of Angle 116
8.2 Types of Angles 117
8.3 The Right Angle 118
8.4 Angles in a Circle 118
8.5 Angles in a Semi-circle 119
8.6 Angles and Straight Lines 119
8.7 Angles in Triangles 120
8.8 Useful Angle Theorems 121
8.9 Properties of Shapes 122

Exercises 8.A Core 124
Exercises 8.B Business, Administration
                and Commerce 128
Exercises 8.C Technical Services:
                Engineering and
                Construction 130
Exercises 8.D General Manufacturing and
                Processing 133
Exercises 8.E Services to People I:
                Community Services 135
Exercises 8.F Services to People II: Food
                and Clothing 138

9 Practical Algebra 141

9.1 Constructing Simple Formulae 141
9.2 Like and Unlike Terms 142
9.3 Brackets 142
9.4 Substitution in Algebraic Expressions 142
9.5 Squares 143
9.6 Square Roots 143
9.7 Cubes 144
9.8 Simple Equations and Transposition 144
9.9 Transposition of Formulae 146

Exercises 9.A Core 146
Exercises 9.B Business, Administration
                and Commerce 149

Exercises 9.C Technical Services: Engineering and Construction 151
Exercises 9.D General Manufacturing and Processing 153
Exercises 9.E Services to People I: Community Services 155
Exercises 9.F Services to People II: Food and Clothing 156

10 Use of Tables, Graphs and Diagrams 158

10.1 Linear and Circular Scales 158
10.2 Tables 159
10.3 Conversion Tables 160
10.4 Tables of Squares and Square Roots 161
10.5 Applications of Squares and Square Roots 163
10.6 Presenting Information by Graphs 164
10.7 Procedure for Plotting Graphs 165
10.8 Conversion Graphs 166
10.9 Bar Charts and Histograms 167
10.10 Pie Charts 168

Exercises 10.A Core 169
Exercises 10.B Business, Administration and Commerce 173
Exercises 10.C Technical Services: Engineering and Construction 177
Exercises 10.D General Manufacturing and Processing 180
Exercises 10.E Services to People I: Community Services 184
Exercises 10.F Services to People II: Food and Clothing 187

11 Use of Calculators and Computers 191

11.1 Use of Calculators 191
11.2 Use of a Microcomputer 193

Exercises 11 198

Appendix 1 Project Work 202

Project 1 Furnishing a Bathroom 202
Project 2 Carpeting a Hotel Ballroom 202
Project 3 Design of a Cold Frame 202
Project 4 Building a Garage 203
Project 5 Concrete Base for a Garage 203
Project 6 Comparison of Duplicating Costs 203

Appendix 2 Computer Program 205

Answers 209

Preface

This book is written to support student-centred learning in Pre-Vocational courses for the 14 to 18 years age range in schools and colleges. The publication of *A Basis for Choice* and *ABC in Action* by the Further Education Curriculum Review and Development Unit (F.E.U.) has led to the introduction of new courses designed to meet the modern-day needs of a young person about to enter adult and working life. A common factor in these courses is the provision of opportunity for students to sample and explore a diverse range of vocational activities which will help them in making a final choice of their preferred future employment. An element of Work Experience may be included in the course, either to be gained in an employer's establishment or under simulated conditions within the institution.

Courses such as CGLI 365 Vocational Preparation (General) offer a framework which allows the development of full-time courses relevant to local employment needs and practices. The principles of student-centred learning are utilised to guide students through experience-related problems using a practical 'hands-on' approach which enables them to progress at their own pace drawing on the teacher as a resource when necessary. Students spend some of their time on a common core, which includes numeracy, and the remainder of the time exploring specific areas of employment in a range of vocational families. It is important that students see the relevance of the core subjects through their vocational application and become aware of the transferability of skills from one area to another.

The purpose of this book is to assist the vital integration of the numeracy core material with the work experience in the vocational areas. The authors recognise that teachers have a difficult organisational task in demonstrating the application of the principles and methods of numeracy to the practical situations which relate to the wide range of vocational options chosen by their students. Most teachers would readily admit that in some vocations the student's background knowledge could be greater than their own. For this reason the book is designed to provide a minimum base of instruction in mathematical method coupled with a large selection of job-related exercises.

Each chapter commences with brief statements of the theory involved and a number of worked examples to demonstrate the mathematical methods to be used. This is followed by a set of core exercises, Exercises A, which reinforce the basic principles and give practice in the mathematical methods. All students should attempt Exercises A. The sets of vocational exercises, Exercises B to F, give job-related applications of the basic principles and are selected by the students according to the vocational options they are following in the course. For instance, students on the CGLI 365 Vocational Preparation Course would probably require Exercises A plus a minimum of three sets of exercises chosen from Exercises B to F. The vocational exercises are designed to cover families of occupations generally as listed below.

VOCATIONAL EXERCISES	B	C	D	E	F
Family	Business administration and commerce	Technical services: Engineering and construction	General manufacturing and processing	Services to people I: Community services	Services to people II: Food and clothing
Occupations	Business studies Commerce Own business General office Law Banking Accounts Mortgage Wages Mail order	Mechanical engineering Electrical engineering Plant engineering Fabrication Construction Brickwork Joinery Plumbing Motor vehicle Painting and decorating	General manufacture Processing Chemicals Mining Plastics and rubber Farming Fisheries Market gardening	Community services Rates Roads Transport Distribution Carpets Playing fields Youth work Hairdressing Telephone Gas Hospitals	Food Clothing Catering Clothing manufacture Clothing sales Bakeries Snack bar Confectionery Wines and spirits Soft drinks Hotel trades

We acknowledge with gratitude the unfailing support and tolerance shown by our wives Lily and Margaret while this work was being written. Our thanks go also to Mrs Mary Jackson for typing the manuscript and to our ever helpful and encouraging editor Mr Brian Carvell.

We dedicate this book to the memory of Jack Slater, for many years our technician and mentor when we were young teachers.

1984   H. Ogden
                                        R. Woodward

The book includes rather more exercises than the minimum required for each topic. This is to allow for a wide ability range in the student group and to provide a modest lead-in to further vocational studies. It is hoped that the book will provide a useful source of vocationally orientated exercises for numeracy studies in the following courses:

CGLI 365 Vocational Preparation (General)
CGLI 364 Numeracy Level I and II
BEC General
RSA Arithmetic
Pitman Arithmetic
A.E.B. Test in Basic Arithmetic
Craft and Foundation courses in:
    Catering and food industries
    Hairdressing
    Construction
    Engineering and Agriculture
    Science industries
    Community care
    Commercial studies
    Distribution
    Information technology
Certificate of Pre-Vocational Education (CPVE)
Technical, Vocational Education Initiative (TVEI)
Youth Training programmes

S.I. Units

S.I. is the abbreviation used in all languages for the International System of Units (Système International des Unités). The system has six arbitrary basic units.

Basic Units

QUANTITY	NAME OF UNIT	UNIT/SYMBOL
Time	second	s
Length	metre	m
Mass	kilogramme	kg
Temperature	kelvin	K
Electric current	ampere	A
Luminous intensity	candela	cd

Selected Derived Units

QUANTITY	NAME OF UNIT	UNIT/SYMBOL
Force	newton	N
Energy, work	joule	J
Power	watt	W
Temperature	degree Celsius	°C
Electric charge	coulomb	C
Electric potential	volt	V
Electric resistance	ohm	Ω
Frequency	hertz	Hz
Torque	newton metre	Nm
Velocity	metre per second	m/s
Area	square metre	m^2
Volume	cubic metre	m^3

Common Prefixes for S.I. Units

PREFIX	SYMBOL	MULTIPLYING FACTOR
mega	M	1 000 000
kilo	k	1 000
centi	c	0.01
milli	m	0.001
micro	μ	0.000 001

Useful Conversion Factors

MULTIPLY	BY	TO GIVE
inches	25.4	millimetres
feet	0.3048	metres
yards	0.9144	metres
miles	1.61	kilometres
square inches	645.2	square millimetres
square feet	0.0929	square metres
square yards	0.8361	square metres
cubic inches	16 390	cubic millimetres
cubic feet	0.0283	cubic metres
cubic yards	0.765	cubic metres
cubic metres	1 000	litres
gallons (Imperial)	4.546	litres
pounds (mass)	0.454	kilogrammes
feet per minute	0.0051	metres per second
horsepower	0.7457	kilowatts

1 Operations in Arithmetic

Many of the everyday calculations that occur in working situations are solved by use of the familiar four rules of number—addition, subtraction, multiplication and division. In almost every occupation the ability to make quick and accurate calculation of simple quantities is an essential requirement of the job. The vital difference between problems met in working life and textbook exercises is the consequences that may arise from errors. Inaccurate calculation in working situations may cause workpieces or other products to be ruined with waste of expensive materials and many hours of skilled labour.

The purpose of this first chapter is to provide practice in simple calculations that may occur in many jobs. Remember that in working life your answers will rarely be checked by anyone else. You must develop a sense of responsibility for the result, and this will help you to gain complete confidence in your ability to calculate with speed and accuracy. When attempting the exercises in this chapter be aware that only the exactly correct answer is acceptable and the exercise must be repeated until this is achieved.

1.1 Addition

When adding quantities it is convenient to list them in columns, but care must be taken to ensure that numbers having the same place value appear in the same column.

Example 1.1 Add 73, 206 and 1 045.

```
    73
   206
  1045
  ————
  1324   (Ans.)
```

Example 1.2 Find the value of 2 763 + 194 + 1 309 + 65.

```
  2763
   194
  1309
    65
  ————
  4331   (Ans.)
```

Example 1.3 Find the total of 26, 594, 31, 8 244, 963 and 18.

```
    26
   594
    31
  8244
   963
    18
  ————
  9876   (Ans.)
```

Example 1.4 Find the sum of two hundred and nine, forty-nine and twelve thousand three hundred and twenty.

```
    209
     49
  12320
  —————
  12578   (Ans.)
```

Example 1.5 Find the value of 54p + 17p + 38p + £2.50.

```
   £
   0.54
   0.17
   0.38
   2.50
   ————
  £3.59   (Ans.)
```

Example 1.6 The amount of petrol sales on a garage forecourt over a period of 6 days are as shown below:

Mon	Tue	Wed	Thur	Fri	Sat
£816.74	£746.23	£501.19	£652.29	£937.15	£428.94

Find the total sales over the period.

```
  Mon    816.74
  Tue    746.23
  Wed    501.19
  Thur   652.29
  Fri    937.15
  Sat    428.94
         ———————
       £4082.54   (Ans.)
```

Example 1.7 The time taken on a train journey is 2 h 14 min. Calculate the arrival times of trains having the following departure times:

Depart 06.30 09.58 12.47 16.32 18.09 22.55

DEPART		ARRIVE
06.30	+ 2 hr 14 min	08.44
09.58	+ 2 hr 14 min	12.12
12.47	+ 2 hr 14 min	15.01
16.32	+ 2 hr 14 min	18.46
18.09	+ 2 hr 14 min	20.23
22.55	+ 2 hr 14 min	01.09 (*Ans.*)

Example 1.8 Find the total length of 3 metal rods having lengths of 25 mm, 18 mm and 37 mm.

```
mm
25
18
37
——
80 mm   (Ans.)
```

Example 1.9 Find the overall length of the workpiece shown in Fig. 1.1.

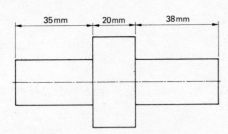

Fig. 1.1

Overall length = 35 mm + 20 mm + 38 mm

```
mm
35
20
38
——
93 mm   (Ans.)
```

1.2 Subtraction

When subtracting one quantity from another use the same column arrangement as for addition, again taking care that numbers having the same place value appear in the same column.

Example 1.10 Find the value of 14 070 − 9 865.

```
  14 070
−  9 865
———
   4 205   (Ans.)
```

Example 1.11 Find the difference between eighty-seven and twenty-nine.

```
  87
− 29
——
  58   (Ans.)
```

Example 1.12 Find the value of £19.65 − £8.78.

```
    £
  19.65
−  8.78
———
 £10.87   (Ans.)
```

Example 1.13 Calculate the dimension x on the shaft shown in Fig. 1.2.

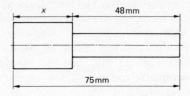

Fig. 1.2

x = 75 mm − 48 mm

```
  mm
  75
− 48
——
  27 mm   (Ans.)
```

Example 1.14 Calculate the dimension y on the part shown in Fig. 1.3.

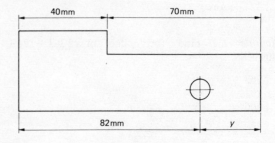

Fig. 1.3

Overall length = 40 mm + 70 mm

```
    mm
    40
    70
   ───
   110 mm   (Ans.)
```

y = overall length − 82 mm
y = 110 mm − 82 mm

```
    mm
   110
  − 82
   ───
    28 mm   (Ans.)
```

1.3 Multiplication

When two or more numbers are multiplied together the answer is called the **product** of the numbers.

Example 1.15   Find the product of 278 and 13.

```
     278
    × 13
    ────
     834
    2780
    ────
    3614   (Ans.)
```

Example 1.16   Find the value of 183 × 24 × 17.

```
     183
    × 24
    ────
     732
    3660
    ────
    4392
    × 17
    ─────
    30744
    43920
    ─────
    74664   (Ans.)
```

Example 1.17   Find the total mass of 25 sacks of flour each having a mass of 45 kg.

Total mass = 45 kg × 25

```
       kg
       45
      × 25
      ────
      225
      900
     ─────
     1125 kg   (Ans.)
```

Example 1.18   Find the total height of a pile of 16 washers each having a thickness of 4 mm.

Total height = 4 mm × 16

```
      mm
      16
    ×  4
    ────
      64 mm   (Ans.)
```

Example 1.19   Calculate the total cost of 13 m of curtain material at £3.49 per metre and 16 m of lining material at £1.63 per metre.

Cost of curtain material = £3.49 × 13

```
       £
     3.49
    × 13
    ─────
    10.47
    34.90
    ─────
    £45.37
```

Cost of lining material = £1.63 × 16

```
       £
     1.63
    × 16
    ─────
     9.78
    16.30
    ─────
    £26.08
```

Total cost = £45.37 + £26.08

```
       £
    45.37
    26.08
    ─────
    £71.45   (Ans.)
```

Example 1.20   Calculate the total cost of the following items of office supplies:

Two filing cabinets at £27.30 each
Four gross of files at £11.52 per gross
Ten reams of duplicating paper at £2.83 per ream
One pencil sharpener at £3.45
Six dozen pencils at £1.78 per dozen
Eight invoice books at 95p each

QUANTITY	ITEM	UNIT COST (£)	TOTAL COST (£)
2	Filing cabinet	27.30	54.60
4 gross	Files	11.52	46.08
10 reams	Duplicating paper	2.83	28.30
1	Pencil sharpener	3.45	3.45
6 dozen	Pencil	1.78	10.68
8	Invoice book	0.95	7.60
			£150.71 (Ans.)

1.4 Division

Example 1.21   Divide 4 085 by 19.

```
       215
  19 ) 4 085
       3 8
       ---
         28
         19
         --
         95
         95
         --
         —
```
215   (*Ans.*)

Example 1.22   Divide £78.82 by 14.

```
         5.63
  14 ) 78.82
       70
       --
        8.8
        8.4
        ---
         .42
         .42
         ---
          —
```
£5.63   (*Ans.*)

Example 1.23   If 9 kg of a chemical costs £43.83 find the cost per kilogramme.

Cost per kilogramme = £43.83 ÷ 9

```
       4.87
  9 ) 43.83
```
£4.87   (*Ans.*)

Example 1.24   How many pieces of length 28 mm can be sheared off a bar 3 052 mm long?

Number of pieces
= length of bar ÷ length of one piece
= 3 052 ÷ 28

```
         109
  28 ) 3 052
       2 8
       ---
         252
         252
         ---
          —
```
109 pieces   (*Ans.*)

Example 1.25   Calculate the dimension x on the part shown in Fig. 1.4.

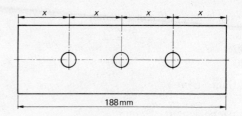

Fig. 1.4

Dimension x = 188 mm ÷ 4

```
       47
  4 ) 188
```
= 47 mm   (*Ans.*)

Exercises 1.A   Core

1.A1   Find the value of:
(a) 417 + 93 + 1 004
(b) 23 + 2 098 + 173
(c) 76 + 341 + 29
(d) 760 + 39 + 14 041 + 126
(e) 101 + 1 050 + 72 + 19
(f) 351 + 59 + 778 + 1 382

1.A2   Find the sum of:
(a) eighty-five and twenty-three
(b) seventeen and sixty-four
(c) ninety-six and thirty-three
(d) fifty-seven and forty-eight
(e) twenty-eight plus forty-seven

1.A3   Find the value of:
(a) £12.64 + £3.78
(b) £1.93 + 74p + 28p
(c) £109 + £14.40 + 65p
(d) £33.25 + £19.72 + £1.46
(e) 33p + 82p + £1.16 + 45p
(f) £6.99 + £44.13 + £2.57 + £17.22

1.A4   Add:
(a) 12 mm + 17 mm + 62 mm
(b) £12 507 + £393 + £1 709
(c) 14 m + 134 m + 78 m + 44 m
(d) 1 064 kg + 938 kg + 85 kg
(e) 1 h 30 min + 45 min + 2 h 23 min
(f) 2 weeks 3 days + 7 weeks 5 days + 13 days
(g) 34 litres + 117 litres + 95 litres
(h) 833 km + 2 094 km + 76 km + 577 km

1.A5   The estimated journey time of a motor coach operating a regular service is 4 h 45 min. Calculate the arrival times for each of the following departure times:

Depart: 08.15 09.30 11.17 14.40 17.38 20.55

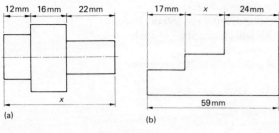
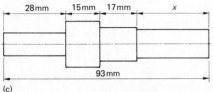
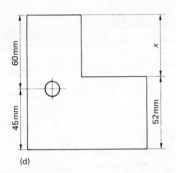
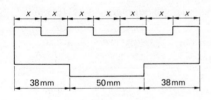
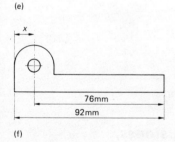

Fig. 1.5

1.A6 The takings for one day's sales in each department of a large store are shown below. Find the total takings of the store.

Clothing	£4 734.85	Food and drink	£1 927.68
Hardware	£ 827.74	Household goods	£1 085.93
Furniture	£2 840.20	Electrical goods	£3 728.47
Stationery	£ 714.09	Gardening	£ 934.56

1.A7 Find the value of:

(a) 847 − 398
(b) 2 014 − 766
(c) 12 362 − 10 949
(d) 39 405 − 17 081
(e) 4 070 − 1 985
(f) 106 219 − 38 447
(g) 10 000 − 2 186
(h) 73 852 − 40 509

1.A8 Find the value of:

(a) 183 m − 67 m
(b) 2 050 kg − 936 kg
(c) 418 litres − 139 litres
(d) 8 170 km − 3 095 km
(e) 704 g − 215 g
(f) 1 305 mm − 575 mm
(g) 12 h 28 min − 5 h 47 min
(h) 27 h 35 min − 16 h 56 min

1.A9 Find the value of:

(a) £101.30 − £41.75
(b) 83p − 29p
(c) £15 307 − £8 950.50
(d) £178.24 − £76.82
(e) £1 050 − £394.16
(f) £255.99 − £30.65
(g) £4.17 − 74p
(h) £30 719.44 − £3 635.78

1.A10 Find the dimension x on each of the parts shown in Fig. 1.5.

1.A11 Find the dimensions x and y on the part shown in Fig. 1.6.

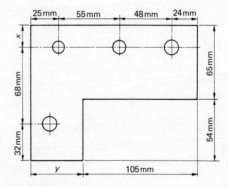

Fig. 1.6

1.A12 Find the **sum** and the **product** of:

(a) 86 and 13
(b) 24 and 49
(c) 120 and 62

Operations in Arithmetic 5

(d) 19 and 37
(e) twenty and twelve
(f) sixteen and thirty
(g) twenty-four and eight
(h) 17 and fifty

1.A13 Find the value of:

(a) 73 × 27
(b) 144 × 15
(c) 309 × 19
(d) 84 × 176
(e) 256 × 192
(f) 1 307 × 21
(g) 2 460 × 154
(h) 743 × 584
(i) 12 × 8 × 7
(j) 6 × 13 × 9
(k) 15 × 14 × 20
(l) 26 × 30 × 45
(m) 44 × 19 × 16
(n) 32 × 57 × 29

1.A14 Given that the area of a rectangle = length × breadth, complete the following table:

LENGTH (mm)	BREADTH (mm)	AREA (mm^2)
12	30	
26	54	
108	43	
14	65	
32	28	
320	170	
97	63	

1.A15 Find the value of:

(a) £44 × 25
(b) £13.50 × 9
(c) 64p × 24
(d) £10.65 × 13
(e) £32.40 × 56
(f) £1.75 × 144
(g) £890.65 × 21
(h) £5.95 × 360

1.A16 Calculate:

(a) 195 m × 30
(b) 450 g × 13
(c) 56 km × 39
(d) 3 h 15 min × 7
(e) 225 mm × 25
(f) 133 kg × 82
(g) 21 h 48 min × 12
(h) 37 min × 42

1.A17 Find the total cost of the following items of hardware:

10 hinges at £1.65 each
5 dozen screws at 18p per dozen
3 wood chisels at £3.16 each
1 saw at £2.95
4 gross nails at 70p per gross

1.A18 Find the value of:

(a) 1 288 ÷ 23
(b) 425 ÷ 17
(c) 10 380 ÷ 12
(d) 3 738 ÷ 42
(e) 2 774 ÷ 73
(f) 3 760 ÷ 16
(g) 24 622 ÷ 13
(h) 11 913 ÷ 57

1.A19 Calculate:

(a) £32.04 ÷ 9
(b) 85p ÷ 17
(c) £111.84 ÷ 12
(d) 8 188 m ÷ 23
(e) 2 408 mm ÷ 56
(f) £786.90 ÷ 43
(g) £3 799.60 ÷ 35
(h) 41 416 kg ÷ 167

1.A20 Complete the following table. (All length dimensions are in millimetres.)

Number of pieces × length per piece = total length

	NUMBER OF PIECES	LENGTH PER PIECE	TOTAL LENGTH
(a)	8	63	
(b)		27	270
(c)	13		1 274
(d)	34		1 802
(e)	48	105	
(f)		18	2 412
(g)	27	93	
(h)		55	2 255

Exercises 1.B Business, Administration and Commerce

1.B1 A shopkeeper presents his accounts to customers as below. What is the total of each account?

(a) 18p   (b) 21p   (c) 29p   (d) 27p   (e) 94p
    20p   12p   36p   29p   67p
    23p   17p   41p   34p   24p
    ___   ___   ___   ___   ___

    ___   ___   ___   ___   ___

1.B2 A single man draws a State retirement pension of £36 per week. His need to budget weekly spending is important and his standard outgoings are: rent £19, heating and lighting £8 and food £6.50. How much per week is left for clothes and other expenses?

1.B3 From Monday to Friday a workman buys a newspaper on his way to work at a cost of 18p per day. On Saturday and Sunday he has papers delivered to his house which cost 45p on Saturday and 66p on Sunday. What is his weekly paper bill?

1.B4 (a) A milkman delivers 2 pints of milk each day of the week to a house. The cost of each pint is 20p. What would be the weekly milk bill?
(b) What would be the cost of milk for 1 year (52 weeks)?

1.B5 (a) A vendor sells shirts on a market stall and the prices range from £2 to £4. He makes the following sales on one afternoon: 52 shirts at £2, 28 shirts at £3 and 17 shirts at £4. How much money has he collected at the end of the afternoon?
(b) A market stall has overheads as follows:
Rent of the stall = £20 per day
Staff wages = £35 per day
Transport of goods = £15 per day
Refreshments for staff = £2.50 per day
Lighting and heating = £6.50 per day
(i) What does it cost the stall-holder in overheads per day?
(ii) What would it cost per week if he operated the stall on four days per week?

1.B6 When opening a bank account a young man makes a deposit of £1 200. Over a period of 5 weeks he withdraws the following amounts:

1st week	2 cheques at £40 each
2nd week	1 cheque at £30
3rd week	4 cheques at £22 each
4th week	3 cheques at £12 each
5th week	7 cheques at £21 each

How much has he left in his account at the end of the fifth week?

1.B7 An office boy is sent to the local take-away for morning snacks for the office staff. The orders are:

Marjorie	Flask of coffee, chocolate biscuit, buttered roll
Joan	Mug of coffee, two buttered rolls, cream cake
Susan	Chocolate biscuit, meat pie
George	Mug of tea, meat pie, sausage roll, cream bun
Jack	Carton of soup, two sausage rolls, chocolate biscuit

The price list at the take-away shows:

Sausage rolls	15p each
Meat pies	28p each
Chocolate biscuits	4p each
Cartons of soup	25p each
Buttered rolls	12p each
Flasks of coffee	48p each
Mugs of coffee	19p each
Mugs of tea	15p each
Cream cakes	23p each

(a) How much money will the office boy require for the total order?
(b) How much will each person have to pay?

1.B8 (a) A ream of paper contains 500 sheets and is used in a duplicating machine at the rate of four sheets per second. How long will it take, in minutes and seconds, to print the whole ream?
(b) Shorthand is taken down from dictation at 120 words per minute. If the letter dictated contains 3 600 words, how long will it take to write the shorthand?
(c) To transcribe a report from shorthand, a secretary needs 10 min for every 100 words. How long will it take, in hours and minutes, to transcribe a report of 4 000 words?
(d) A typical telephone call connection takes an operator 20 seconds. She makes these connections at the rate of 300 per day, between carrying out other office duties. How much time does she spend on telephone work alone?

1.B9 Estimated outgoings for a semi-detached three-bedroomed house are:

Rates £289 per year
Gas, paid four times per year, at £66 per quarter
Electricity, paid four times per year, at £71 per quarter
Telephone, paid four times per year, at £43 per quarter
Insurance £266 per year
Maintenance and decorating costs £233 per year

(a) How much per week must the householder put aside to meet these costs in a full year?
(b) How much more does the householder pay for electricity than for gas in a full year?

1.B10 A newsagent prepares weekly accounts for his regular customers in the form shown below.

Item 1	18p × 5	0.90
Item 2	14p × 5	0.70
Item 3	26p × 2	0.52
Item 4	35p × 1	0.35
		£2.47

Where

Item 1 = evening newspapers
Item 2 = morning newspapers
Item 3 = Sunday newspapers
Item 4 = magazines

Using the same form prepare separate accounts for the following customers:

CUSTOMER	ITEM 1	ITEM 2	ITEM 3	ITEM 4
(a)	18p × 5	16p × 5	26p × 2	35p × 2
(b)		21p × 5	26p × 1	35p × 1
(c)	17p × 5	16p × 5		60p × 1
(d)	18p × 5		26p × 2	
(e)		14p × 5	26p × 3	45p × 1
(f)	18p × 5	14p × 5		35p × 2
(g)		21p × 5	26p × 2	15p × 1
(h)	17p × 5		26p × 1	55p × 1

1.B11 The table shows the number of items of mail received by a mail-order firm over a period of 3 months. Copy the table and complete the totals across and down.

MONTH	ORDERS	PAYMENTS	RETURNED GOODS	TOTALS
Jan	7 531	5 049	133	
Feb	4 966	8 184	49	
Mar	6 038	6 275	76	
Totals				

1.B12 The table shows the sales in 1 week of a garage. Copy the table and complete the totals across and down.

DAY	PETROL (£)	OIL (£)	TYRES (£)	ACCESS-ORIES (£)	TOTALS (£)
Mon	897.62	78.45	101.30	121.75	
Tues	605.19	39.16	24.55	208.19	
Wed	650.33	64.52	176.80	73.70	
Thur	702.21	48.90	50.50	169.23	
Fri	1 052.86	87.16	238.15	104.88	
Sat	473.41	16.84	68.50	115.72	
Totals (£)					

1.B13 A workman's gross weekly wage and the amount of income tax paid is shown in the table for a period of 6 weeks. Calculate:

(a) his total gross wage for the 6 weeks;
(b) the amount of tax paid in the same period.

WEEK	GROSS WAGE (£)	INCOME TAX (£)
1	102.56	14.17
2	97.50	13.80
3	98.86	13.95
4	109.35	15.63
5	102.24	14.27
6	105.17	14.82

1.B14 Find the total amount to be paid for each of the following invoices:

(a) 8 reams of typing paper at £4.17 per ream
12 notepads at 78p each
4 boxes of staples at £1.05 per box
16 ring files at 98p each

(b) 17 gal of petrol at £1.83 per gallon
6 litres of antifreeze at 84p per litre
4 tyres at £23.95 each
1 car radio at £46.35

(c) 15 sheets of hardboard at £1.63 per sheet
19 m of planking at 36p per metre
6 bags of cement at £1.58 per bag

(d) 3 doz 16p postage stamps
1 000 envelopes at 53p per 100
8 rolls of tape at 72p per roll
14 boxes of carbons at £2.64 per box

(e) 25 kg of potatoes at 11p per kilogramme
24 tins of soup at 33p each
16 kg of flour at 39p per kilogramme
30 loaves at 37p each
30 kg of sugar at 42p per kilogramme

(f) 12 pairs of jeans at £12.50 each
10 pairs shoes at £13.95 each
14 skirts at £12.75 each
12 scarves at £2.36 each
48 pairs tights at 99p each

1.B15 The table shows the number of incoming and outgoing telephone calls during a working week at an estate agent's office.

DAY	INCOMING CALLS	OUTGOING CALLS
Mon	83	31
Tue	105	62
Wed	91	55
Thur	73	48
Fri	86	51
Sat	124	27

(a) Find the number of incoming calls received in the week.
(b) Find the number of outgoing calls made in the week.

(c) Find the difference between the number of calls received and the number of calls made in the week.
(d) On which day was there least difference between the number of calls received and the number of calls made?
(e) If the average cost of an outgoing call is 13p what is the total cost for the week?

1.B16 (a) A manufacturer's catalogue has 32 pages and there are 28 items on each page. Calculate the total number of items in the catalogue.
(b) If the printing cost is 8p per page, find the cost of producing 500 catalogues.

1.B17 A retail shop employs five assistants who all receive the same weekly wage. If the total wage bill for a full year is £20 410, what is the weekly wage of one assistant?

1.B18 (a) A small business rents a computer at a cost of £1 500 per year. What is the cost per working day of hiring the computer if the business operates a 5-day week for 48 weeks in the year?
(b) A photographer buys a cine camera on a hire purchase agreement. The cost of the camera is £276 and he pays a deposit of £30. An interest charge of £51 is added to the balance which is to be repaid by 12 equal monthly payments. Calculate the amount of each payment.

1.B19 A shop rents video films at the following rates:

Monday to Friday
1 tape for 1 night   £1.00
2 tapes for 1 night   £1.80
2 tapes for 2 nights £3.50

Saturday and Sunday
1 tape for 1 night   £1.50
2 tapes for 2 nights £5.00

How much would be charged for each of the following orders:
(a) 1 tape for Tuesday, 2 tapes for Thursday and Friday?
(b) 2 tapes for Wednesday and Thursday, 1 tape for Sunday?
(c) 3 tapes for Tuesday and Wednesday, 2 tapes for Saturday and Sunday?
(d) 2 tapes for Wednesday, 4 tapes for Saturday and Sunday?
(e) 1 tape for Monday, 2 tapes for Wednesday, 1 tape for Sunday?

1.B20 Details of the transactions in a building society account are given below:

		£
4 March	Balance	234.28
21 March	Paid in	50.00
11 April	Withdrawn	85.00
24 May	Withdrawn (cheque)	107.67
30 June	Interest added	21.34
25 July	Paid in	135.00
8 Aug	Paid in (cheque)	76.80
24 Aug	Withdrawn	180.00

(a) What amount was paid into the account, including interest, between 4 March and 24 August?
(b) How much was in the account on 1 July?
(c) What was the final balance?

Exercises 1.C Technical Services: Engineering and Construction

1.C1 Pieces of timber of various lengths are cut into three sections as shown below. Find the original length of each piece. (Neglect the width of saw cut.)

(a) 25 mm	(b) 122 mm	(c) 134 mm	(d) 41 mm
28 mm	215 mm	137 mm	48 mm
33 mm	112 mm	239 mm	53 mm

1.C2 If a piece of timber 600 mm long was supplied in each case for Exercise 1.C1, what would be left after cutting the three pieces? (Neglect the width of saw cut.)

1.C3 When laying felt on the roof of a garage, a workman hammered in felt nails at the rate of one every 7 seconds. If 400 nails were needed, how long, in minutes and seconds, did it take to fasten down the felt?

1.C4 A tank which contains 750 litres of water is to be emptied through a tap which passes 24 litres per minute. How much time, in minutes and seconds, is needed to empty the tank?

1.C5 Calculate the dimension x for each of the components shown in Fig. 1.7.

1.C6 If it takes 2 min for a fitter to file the burrs from a machined surface, how much time, in hours and minutes, does he require to debur a batch of 136 components?

1.C7 (a) A plumber required six pieces of copper pipe of the following lengths:

116 mm   127 mm   133 mm   147 mm   262 mm
312 mm

What was the total length of pipe required?
(b) If the plumber repeated the same job at seven houses, how much pipe would be required for all the work?

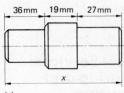
(a)

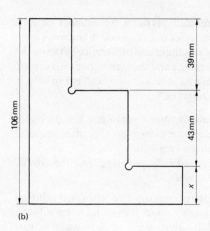
(b)

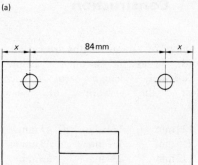
(c)

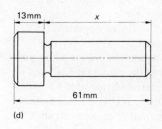
(d)

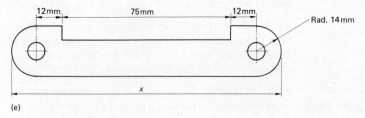
(e)

Fig. 1.7

1.C8 Twenty-five pieces of metal each 23 mm long are to be cut from a bar. Allowing 3 mm for the width of each saw cut, what total length of bar is required?

1.C9 (a) A strip of metal is to have nine holes drilled along its length. The holes are equally spaced along the centre line at a centre distance of 85 mm. Calculate the distance between the centres of the first and the last hole.
(b) A casting having a mass of 150 kg is made from an alloy which contains 45 kg of copper for every 5 kg of tin. What mass of tin will be contained in the casting?
(c) Concrete is mixed from sand, cement and stone chippings. For every 10 kg of sand there is 1 kg cement and 15 kg stone chippings. How much cement would be needed to mix 312 kg of concrete?

1.C10 The teeth on the blade of a band saw pass through a piece of timber at the rate of 760 teeth per minute. If it takes 13 min to cut through the timber, how many teeth will have passed through the wood?

1.C11 The table shows the number of bags of cement sold each day at a builders' yard over a period of 10 weeks.
(a) Complete the table to show the number of bags sold each week.
(b) How many bags of cement were sold in the 10-week period?

WEEK	MON	TUE	WED	THUR	FRI	SAT	TOTAL
1	26	27	30	163	421	210	
2	28	35	97	175	270	113	
3	34	33	107	210	327	92	
4	42	34	167	193	302	115	
5	103	33	142	93	260	187	
6	67	61	102	180	292	281	
7	71	107	63	190	315	210	
8	78	93	111	210	256	191	
9	82	163	171	82	302	126	
10	90	96	66	166	217	132	

1.C12 If the bags of cement in Exercise 1.C11 were sold at a price of £1.90 each, calculate:

(*a*) the takings in each of the 10 weeks;
(*b*) the total takings over the 10-week period.

1.C13 (*a*) A cast iron bush has an external diameter of 95 mm and an internal diameter of 73 mm. Calculate the wall thickness of the bush.
(*b*) A scribing block has a rectangular base 85 mm long and 55 mm wide. Calculate the area of the base. (Area = length × width.)
(*c*) A pile of 26 identical steel washers has a height of 78 mm. Calculate the thickness of one washer.

1.C14 (*a*) The diameter of a circular bar is reduced in one cut in a lathe operation. The original diameter of the bar is 103 mm and the depth of cut is 7 mm. What is the finished diameter of the bar?
(*b*) During a sliding operation on a centre lathe the cutting tool moves a distance of 120 mm along the length of the bar in 1 min. Calculate the distance moved by the tool (i) in 1 second, (ii) in 2 min 15 seconds and (iii) in 3 min 28 seconds.

1.C15 Calculate the dimensions *x* and *y* on the components shown in Fig. 1.8.

1.C16 (*a*) Fourteen revolutions of a nut cause it to advance 56 mm along the length of a thread. What is the pitch of the thread?
(*b*) Calculate the volume of the rectangular block of metal shown in Fig. 1.9. (Volume = length × width × height.)

1.C17 The table shows the power consumed in kilowatts by the main electric motors of the machine tools in a workshop.

Four lathes	2 kW each
Two milling machines	4 kW each
Three drilling machines	1 kW each
One mechanical saw	2 kW
Two shaping machines	3 kW each

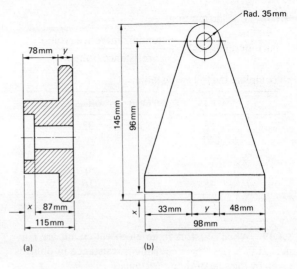

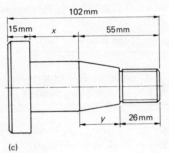

Fig. 1.8

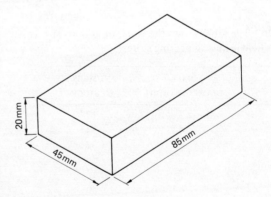

Fig. 1.9

Find:
(*a*) the total power consumed by all the machine tools;
(*b*) the power consumed when only the lathes and the shaping machines are operated;
(*c*) the number of lathes that would consume the same power as the milling machines and the shaping machines operating at the same time.

Operations in Arithmetic 11

1.C18 Use the relations

$$\text{current (amperes)} = \frac{\text{potential difference (volts)}}{\text{resistance (ohms)}}$$

to complete the following table:

VOLTS	AMPERES	OHMS
88		22
240	2	
240		48
	2	6
48		16

1.C19 When resistors in an electrical circuit are connected in series, their equivalent resistance in ohms is given by the sum of their resistances

$$R = R_1 + R_2 + R_3$$

Use this expression to complete the following table:

EQUIVALENT RESISTANCE R (ohms)	RESISTOR R_1 (ohms)	RESISTOR R_2 (ohms)	RESISTOR R_3 (ohms)
	14	7	12
10		16	40
	400	120	200
60	15		20
800		200	150

1.C20 The table shows the stock of various size twist drills held in an engineering stores.

DRILL DIAMETER (mm)	NUMBER IN STOCK
3	17
4	34
5	38
6	25
7	13
8	36
9	11
10	42
11	14
12	53

Find:
(a) the total number of drills in stock;
(b) the number of drills of less than 8 mm diameter;
(c) the number of drills of greater than 5 mm diameter;
(d) the number of drills of greater than 5 mm and less than 8 mm diameter.

Exercises 1.D General Manufacturing and Processing

1.D1 A gardener buys fifteen daffodil bulbs at 32p for ten, two bags of peat at 135p each, two small shrubs at 163p each. How much did he spend?

1.D2 A landscape gardener requires to line both sides of a straight path 52 m long with rose bushes. The spacing between the bushes is to be 1 m and the path must start and end with a bush.

(a) Calculate the number of rose bushes required.
(b) What length of path could be lined on both sides with 72 bushes?

1.D3 The daily output for one working week of a factory producing plastic containers is shown below.

Mon	2 943
Tue	2 778
Wed	3 015
Thur	3 132
Fri	2 844

(a) Calculate the total output for the week.
(b) What would the total output have been if the factory had been closed on Friday?

1.D4 (a) A chemist sells 21 boxes of pills from his stock of 72 boxes containing 140 pills each. How many pills are left in stock?
(b) A chemist is paid £1.20 for each prescription filled. He fills 45 prescriptions on Monday, 32 on Tuesday, 36 on Thursday and 75 on Friday. How much money does the chemist collect for prescriptions filled on the 4 days?

1.D5 A factory packs in boxes containing 120 pills.

(a) How many pills are required to fill 850 boxes?
(b) How many boxes are required to pack 78 360 pills?

1.D6 A miner receives a basic wage of £110.70 per week and a bonus which depends on his weekly output. In a 6-week period the bonus amounted to:

1st week	£18.40
2nd week	£15.75
3rd week	£13.24
4th week	£19.67
5th week	£20.08
6th week	£17.55

Calculate:
(a) his total wage for each of the six weeks;
(b) the amount of his wages for the 6-week period.

1.D7 The local farm shop sells produce at the following prices:

Potatoes	12p per kilogramme
Cabbages	30p
Cauliflowers	40p
Turnips	25p
Carrots	20p per kilogramme
Onions	32p per kilogramme

What would it cost to purchase:

(a) a cabbage, a cauliflower and 2 turnips?
(b) 1 kg onions, 2 kg carrots and a cabbage?
(c) 4 kg potatoes, 3 cauliflowers and a turnip?
(d) 2 kg carrots, a cabbage and 3 kg potatoes?
(e) 1 kg onions, 4 kg potatoes and 3 turnips?

1.D8 (a) When growing lettuces in a greenhouse, a gardener successfully grows 17 out of every 20 planted. He plants 500 in one patch. How many lettuces does he grow?
(b) If the gardener makes a profit of 2p on each lettuce how much profit does he make?

1.D9 (a) One face at a coal-mine produces 150 tonnes per shift. Six men operate the machinery to produce the coal. How much coal does each man produce?
(b) Tubs of coal pass a check point in a mine at the rate of four per minute. Each tub holds half a tonne. How many tonnes pass the check point in 1 h 16 min?

1.D10 (a) A water pump clears water from a mine shaft at the rate of 22 litres per minute. The pump runs for 4 h. How much water does it clear?
(b) How long must the pump run to clear 1 232 litres of water?

1.D11 (a) Six lorries which can each carry a mass of 7 tonnes are used to remove a mass of 3 906 tonnes from an open-cast mine. How many round trips does each lorry make?
(b) The distance from the open-cast mine to the tipping area is 43 km. What distance would each lorry travel when loaded to remove the total mass? (Neglect the return journeys.)

1.D12 The table shows the main dimensions of four deep-sea trawlers.

VESSEL	DISPLACE-MENT (t)	LENGTH (m)	BEAM (m)	DRAUGHT (m)
Sapphire Queen	568	47	8	$3\frac{1}{2}$
Amber Lady	607	48	9	$4\frac{1}{4}$
Redwood Rose	452	42	8	$4\frac{1}{4}$
Coral Lancer	393	39	7	$3\frac{1}{2}$

(a) Calculate:
 (i) the total displacement of the four vessels;
 (ii) the combined length of *Amber Lady* and *Coral Lancer*;
 (iii) the difference in displacement of *Sapphire Queen* and *Redwood Rose*.
(b) Given that

$$\frac{\text{average}}{\text{displacement}} = \frac{\text{sum of the four displacements}}{4}$$

calculate the average displacement of the four vessels.
(c) Calculate for the four vessels (i) the average length, (ii) the average beam and (iii) the average draught.

1.D13 (a) A field is rectangular in shape and has a length of 48 m and a width of 33 m. Calculate the total length of fencing required to enclose the field completely.
(b) Calculate the total cost of the fencing at £1.25 per metre length.

1.D14 A small processing plant employs the following workers:

One foreman at £130.45 per week
Six process operators at £93.17 per week
Four manual workers at £85.78 per week

Calculate the weekly bill of the plant.

1.D15 A market gardener employs part-time packers at a rate of £1.68 per hour.

(a) What would be the earnings of a part-time worker for 17 h?
(b) If the hourly rate were increased to £1.85 how much extra would the worker be paid for the 17 h?

1.D16 The lighting cost of a small factory for a period of 13 weeks is £390. The factory operates a 5-day week and employs 26 workers. Calculate:

(a) the lighting cost per working day;
(b) the lighting cost per employee for the 13-week period.

1.D17 A factory manufactures boxed games, each of which contains the following materials:

Paper	54 g
Wood	98 g
Plastic	30 g
Metal	15 g

Calculate:

(a) the total mass of material in each game;
(b) the mass of each material contained in a batch of 120 games. (Give the answers in grammes.)

1.D18 A processing operation uses 950 litres of water per hour. Calculate the water consumption of the operation in litres for:

(a) a working shift of 7 h;
(b) a working week of 35 h;
(c) three working shifts.

1.D19 In one day a tanker made the following deliveries of heating oil to small factories on an industrial estate:

Brown's Furniture Works	5 750 litres
Ace Chemicals	3 660 litres
Zero Paints	6 130 litres
White Lion Plastics	4 075 litres
Evans Soap Works	2 988 litres
Juniper Matches	7 105 litres
Thames Meat Products	3 550 litres

Calculate:
(a) the total number of litres of oil delivered in the day;
(b) the difference in litres between the largest and the smallest deliveries.

1.D20 Figure 1.10 shows the floor plan of a small plastics factory.

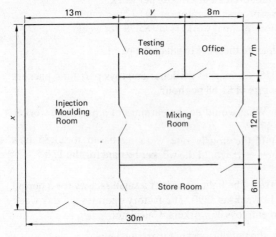

Fig. 1.10

(a) Find:
(i) the length x of the Injection Moulding Room;
(ii) the length y of the Testing Room.
(b) Given that

area of room (m²) = length (m) × width (m)

Calculate the area of (i) the Office, (ii) the Store Room, (iii) the Mixing Room, (iv) the Testing Room and (v) the Injection Moulding Room.
(c) What is the difference in area between the largest and the smallest rooms?

Exercises 1.E Services to People 1: Community Services

1.E1 Carpet of the same width is to be laid in four houses on the stairs, in the hall and along the landing. The lengths required are:

	HOUSE 1	HOUSE 2	HOUSE 3	HOUSE 4
Stairs	18 m	21 m	16 m	31 m
Hall	21 m	26 m	17 m	42 m
Landing	17 m	19 m	18 m	28 m

(a) How many metres of carpet are required for each house?
(b) What total length of carpet is required for all four houses?
(c) It takes eight tacks per metre to hold a carpet down. How many tacks are required for a carpet 27 m long?

1.E2 A Scout troop is asked to deliver leaflets through the doors of houses on an estate. The estate has 26 streets. Seven streets have 28 houses, eleven have 35 houses, three have 72 houses and five have 37 houses. If the average distance between each house is 15 m:
(a) what will be the total distance walked by the Scout troop to deliver all the leaflets;
(b) what will be the total number of leaflets delivered?

1.E3 Bus fares on a local route are:

Main Street to Market Square	21p
Market Square to Bus Station	37p
Bus Station to Main Street	18p

The bus carries 26 passengers from Market Square to the Bus Station, 31 passengers from Main Street to Market Square and 47 passengers from the Bus Station to Main Street. How much was collected in fares?

1.E4 (a) The total fares collected on a coach journey amounted to £78.72. There were 32 people on the coach. How much did each pay for the journey?
(b) The journey was 41 k long. How much per kilometre did each person pay?

1.E5 (a) Road tax on a car is £85 and third party insurance is £113 per year. Garaging costs are £62 per year. How much does it cost per week to tax, insure and garage the car?
(b) A car travelled 65 000 km on a set of four tyres. The tyres were worn out after 2 years' motoring and the car was used on 5 days each week.
(i) How many running days did the car have in the 2-year period?
(ii) How many kilometres per day did it travel?

14 Spotlight on Numeracy

1.E6 For one revolution of a bus tyre the bus travels a distance of 3 m. How far does the bus travel for 363 revolutions of its wheels?

1.E7 A lorry travels 19 km for every 5 litres of diesel oil used. How far will it travel on four full tanks of diesel oil, if the fuel tank holds 165 litres?

1.E8 A single tree needs 11 m^2 of parkland for successful growth.
(a) If 5 665 m^2 of parkland are available, how many trees could be planted?
(b) What area of ground would be required to plant 325 trees?

1.E9 Two people on a 15-day walking holiday travelled a total distance of 312 km. If they rested on 2 days of the holiday, how far did they walk per day?

1.E10 In 1 year (52 weeks) a man spent 5 weeks in hospital and 3 weeks in a convalescent home. He took 2 weeks annual holiday and 5 working days off for other statutory holidays. If he worked on 5 days per week, how many days did he work in the year?

1.E11 The gas bill of a large house for a quarter of a year (13 weeks) was £117.
(a) What was the cost per week?
(b) If gas was used at the same rate, what would be the cost for a full year?

1.E12 The amount of rates to be paid each year on a property is given by

Rates to be paid = rateable value × rate in the £

Use this information to complete the following table:

PROPERTY	RATEABLE VALUE (£)	RATE IN THE £ (£)	RATES TO BE PAID (£)
Terraced house	200	1.50	
Shop	700	1.50	
Small factory	1 500	1.20	
Detached house	400	1.50	

1.E13 The rateable values of properties in a small town are shown in the table below.

USE OF PREMISES	RATEABLE VALUE (£)
Domestic	894 317
Business	406 539
Industrial	178 795

(a) What is the total rateable value of all properties?
(b) What is the difference between the rateable values of the domestic properties and the industrial properties?

1.E14 The expenses of running a youth club for a full year are shown below.

Youth Leader's salary	£2 689
Hire of premises	£1 092
Heating and lighting	£ 985.16
Equipment	£ 642.70
Cleaning	£ 806
Repairs	£ 374.29

Calculate:
(a) the total cost for the full year;
(b) the cost of cleaning and repairs for the full year;
(c) the cost per week of hiring the premises.

1.E15 (a) The time taken on a train journey is 3 h 42 min. Calculate the arrival times of trains having the following departure times:

07.48 10.26 13.04 17.18 22.30

(b) The return rail fare for the journey is £43.35 for an adult and £26.42 for a child. Calculate the total fare to be charged for two adults and three children.
(c) If the same journey by coach takes 5 h 34 min, how much time is saved by taking the train?
(d) The cost of coach travel is £21.50 for an adult and £16.85 for a child. How much money is saved by two adults and three children taking the coach?

1.E16 Calculate the total weekly wage bill for the following hourly paid staff of a parks maintenance department:

4 head gardeners working 40 h each at £3 per hour
8 gardeners working 40 h at £2 per hour
6 manual workers working 50 h at £1.50 per hour
2 drivers working 44 h at £1.50 per hour

1.E17 Figure 1.11 shows the floor plan of a suite of offices.

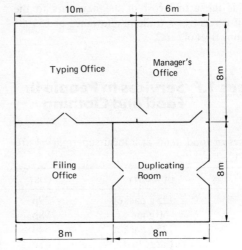

Fig. 1.11

Given that

$$\text{area of room (m}^2) = \text{length (m)} \times \text{width (m)}$$

calculate the area of carpeting required for each of the following: (a) the Manager's Office, (b) the Filing Office, (c) the Duplicating Room and (d) the Typing Office.

1.E18 To carpet the office suite shown in Exercise 1.E17, three grades of carpet are to be used.

Imperial at £5 per square metre
Empire at £3 per square metre
Tufteck at £2 per square metre

(a) Calculate the total cost of carpeting the office suite using the following grades of carpet:

Manager's Office	Imperial
Filing Office	Empire
Duplicating Room	Tufteck
Typing Office	Empire

(b) What would be the total cost of carpeting all the offices in (i) Imperial, (ii) Empire and (iii) Tufteck.

1.E19 The weekly rates at a private nursing home are:

Short-stay patients	£94
Long-stay patients	£82
Children under sixteen	£51

Calculate the income in a week when 13 short-stay patients, 8 long-stay patients and 2 children were resident at the nursing home.

1.E20 A self-drive car-hire company is replacing its fleet of hire cars by purchasing the following new cars:

Four Ford Escorts at £4 484 each
Four Talbot Horizons at £4 677 each
Two Vauxhall Cavaliers at £5 153 each
Eight Austin Metros at £3 604 each

(a) Calculate the total cost of the new cars.
(b) What is the actual cost of the new cars to the company if it receives a trade-in allowance of £53 085 for its existing fleet of cars?

Exercises 1.F Services to People II: Food and Clothing

1.F1 Prices of food items at a local supermarket are listed below.

ITEM	QUANTITY	COST
Tea	125 g packet	29p
Coffee	50 g jar	98p
Cocoa	125 g packet	54p
Bread	Large loaf	48p
Bread	Small brown loaf	29p
Currant teacake	Packet of 4	30p
Milk	½ litre bottle	24p
Cheese	125 g	80p
Cornflakes	250 g box	45p
Flour	450 g packet	33p
Eggs	½ doz	42p
Currants	250 g packet	47p
Sugar	1 kg bag	46p
Beetroot	Small jar	27p
Vinegar	30 cl bottle	25p
Sardines	100 g tin	51p
Salmon	125 g tin	£1.20
Caster sugar	250 g bag	36p
Corned beef	250 g tin	£1.03
Soap powder	900 g packet	91p
Bacon	250 g packet	65p
Cream	250 g jar	49p

Use the price list to make out bills giving the total cost for each of the customers (a) to (h).

(a) 1 packet tea
     2 bottles milk
     125 g cheese
     1 small brown loaf
(b) 2 large loaves
     4 currant teacakes
     1 box cornflakes
     6 eggs
     1 bag sugar
(c) 1 tin salmon
     1 jar coffee
     1 packet flour
     2 jars cream
     1 doz eggs
(d) 2 packets cocoa
     3 bottles milk
     1 packet currants
     1 bottle vinegar
     1 tin sardines
     6 eggs
(e) 2 large loaves
     1 jar beetroot
     1 packet bacon
     2 boxes cornflakes
     1 box soap powder
(f) 2 packets tea
     2 bottles milk
     250 g cheese
     1 bag caster sugar
     1 tin corned beef
(g) 3 small brown loaves
     2 doz eggs
     1 packet currants
     1 tin salmon
     1 bottle vinegar

(h) 3 bottles milk
2 large loaves
1 small brown loaf
2 packets cornflakes
1 bag sugar
2 doz eggs
2 packets cocoa
8 currant teacakes

1.F2 Customers (a) to (g) in Exercise 1.F1 paid their bills with £5 notes. Customer (h) paid with a £10 note. How much change did each customer receive?

1.F3 The following prices were displayed in the window of a butcher's shop:

Whole leg of lamb	£1.09 per lb
Whole shoulder of lamb	64p per lb
Lamb chops	89p per lb
Rolled shoulder of pork	89p per lb
Whole leg of pork	79p per lb
Braising steak	£1.38 per lb
Minced steak	£1.20 per lb
Rump steak	£2.88 per lb
Fillet steak	£3.00 per lb
Pork sausages	68p per lb
Beef sausages	63p per lb
Kippers	65p per pair

Use this price list to make out bills giving the total cost for each of the customers (a) to (f).

(a) 2 lb lamb chops
1 lb braising steak
½ lb fillet steak
(b) 3 lb whole leg of lamb
½ lb pork sausages
2 pair kippers
(c) 4 lb whole leg of pork
1 lb minced steak
1 lb beef sausages
½ lb rump steak
(d) 2 lb rolled shoulder of pork
1 lb braising steak
1 pair kippers
(e) 3 lb whole shoulder of lamb
1 lb fillet steak
1 lb pork sausages
2 lb lamb chops
(f) ½ lb rump steak
½ lb fillet steak
1 lb beef sausages
1 lb lamb chops
1 pair kippers

1.F4 If all the customers in Exercise 1.F3 paid their bills with £20 notes, what amount of change would each receive?

1.F5 (a) A jar of jam has a mass of 460 g. If the mass of the empty jar is 36 g, what is the mass of the jam?
(b) If 55 tins of biscuits cost £165, what is the cost of 1 tin?
(c) A bakery produces 1 835 loaves on one night shift. 980 are white loaves and the rest are brown. How many brown loaves are made?

1.F6 (a) While delivering pies to a shop, a tray holding 36 pies is dropped and 21 are damaged. How many pies are left whole?
(b) On opening his shop, a shopkeeper has sandwiches made up as follows:

46 beef   48 ham   72 chicken

He sells 61 chicken sandwiches, 42 ham and 30 beef. How many sandwiches has he left?
(c) In a café a scone with butter costs 34p and a cup of coffee 24p. Four people order the following:

Coffee and scone
Coffee
Coffee
Coffee and scone

What is the total amount of the bill?
(d) If 13 kg of potatoes costs 286p, what is the cost per kilogramme?

1.F7 The following items appeared on the stock list of a grocery shop:

8 doz tins soup
1 gross tins beans
½ gross tins peas
6 doz tins fruit
54 tins tomatoes

Calculate the total number of tins on the list.

1.F8 A greengrocer's delivery van was loaded with the following items:

250 kg potatoes
40 kg apples
15 kg turnips
35 kg cabbages
12 kg tomatoes
16 kg beans

(a) What was the total mass of the load?
(b) After several stops the driver had delivered:

80 kg potatoes
12 kg apples
7 kg turnips
8 kg cabbages
4 kg tomatoes
6 kg beans

What was the remaining mass of the load?

Operations in Arithmetic

1.F9 A cosmetics shop was offering a discount of £2.75 if the following make-up items were bought as a package:

Kiku perfume	£4.99
Sheer lip glaze	£1.55
No. 7 eye shadow	£1.60
Velvet face powder	£1.10
Yardley lipbrush	£1.99

What was the discount price of the package?

1.F10 The following items were on special offer at a jewellery store:

Silver pendant reduced from £9.40 to £4.99
Silver rope chain reduced from £22.38 to £11.99
Gold hoop earrings reduced from £6.75 to £3.30
Sapphire stud earrings reduced from £16.50 to £8.49
Daisy chain necklet reduced from £29.22 to £15.75
Gold identity bracelet reduced from £47.08 to £22.99
Gold charm bracelet reduced from £115.48 to £74.99
Gold cross reduced from £48.50 to £27.95
Serpentine chain reduced from £21.17 to £9.99
Heart pendant reduced from £52.23 to £29.69

Calculate the total amount of money saved by each of the customers, (a) to (f), who bought the following items:

(a) Sapphire stud earrings, serpentine chain;
(b) Gold charm bracelet, gold hoop earrings;
(c) Gold cross, silver rope chain;
(d) Silver pendant;
(e) Daisy chain necklet, heart pendant, gold identity bracelet;
(f) Silver rope chain, sapphire stud earrings.

1.F11 (a) A waitress earns £56 in 4 days. How much can she earn in (i) 1 day and (ii) 5 days?
(b) The sales of a snack bar in 1 month amounted to £3 453. The cost of food was £1 250, staff wages were £680 and overheads £810. What profit was made for the month?

1.F12 The sales of an ice-cream vendor for 1 week are shown below.

DAY	CORNETS	WAFERS	TUBS	TOTALS
Mon	407	518	106	
Tue	358	524	92	
Wed	619	584	212	
Thur	314	282	84	
Fri	596	412	179	
Sat	704	847	316	
Sun	688	838	353	
Totals				

Copy the table and complete the totals across and down.

1.F13 The price list at a hairdressing salon is shown below.

Full perm	£12.50
Shampoo and set	£5.00
Crimping	£4.50
Cut	£3.00
Conditioner	£1.00

The table shows the number of treatments carried out on 6 days of 1 week.

DAY	FULL PERM	SHAMPOO AND SET	CRIMPING	CUT	CONDITIONER
Mon	6	8	4	4	5
Tue	4	9	2	3	4
Wed	5	6	2	2	3
Thur	3	10	6	6	5
Fri	8	14	6	8	8
Sat	8	12	8	5	9
Totals					
Takings (£)					

(a) Copy the table and complete, using the price list to find the takings for each treatment in the week.
(b) What were the total takings for the week?

1.F14 The prices at a men's hairdressers are:

Cut	£1.50
Shave	£1.00
Cut and shave	£2.00
Cut and shampoo	£2.50
Pensioner's cut	£1.00

The services given on each day of 1 week were:

Monday   Cut 35, shave 12, cut and shave 6, cut and shampoo 3, pensioner's cut 14
Tuesday   Cut 29, shave 14, cut and shave 7, cut and shampoo 6, pensioner's cut 12
Wednesday   Cut 41, shave 16, cut and shave 8, cut and shampoo 10, pensioner's cut 18
Thursday   Cut 17, shave 5, cut and shave 2, cut and shampoo 2, pensioner's cut 0
Friday   Cut 58, shave 21, cut and shave 19, cut and shampoo 12, pensioner's cut 0
Saturday   Cut 63, shave 28, cut and shave 31, cut and shampoo 18, pensioner's cut 0

(a) Use this information to complete the following table:

DAY	CUT	SHAVE	CUT AND SHAVE	CUT AND SHAMPOO	PENSIONER'S CUT
Mon					
Tue					
Wed					

18 Spotlight on Numeracy

Thur
Fri
Sat

Totals

(b) Use the totals and the price list to calculate the takings for the week.
(c) On which day did the hairdresser close for a half-day?
(d) On which days was the reduced price for pensioners not available?

1.F15 (a) A boutique has the following items in stock:

Dresses	338
Tops	216
Trousers	178
Jeans	124
Blouses	248
Skirts	146
Suits	73
Other items	417

What is the total number of items in stock?

(b) Find the total amount to be charged to each of customers (i) to (iv) making the following purchases:

(i)	Dress	£9.20
	Blouse	£6.99
	Tights	56p
(ii)	Trousers	£8.95
	Scarf	£2.99
	Top	£7.58
(iii)	Jeans	£15.35
	Top	£8.75
	Skirt	£10.00
(iv)	Suit	£39.50
	Coat	£27.85
	Dress	£10.37

1.F16 (a) The shoe department of a men's outfitters requires to keep a stock of 50 pairs of each common size of men's shoes. The present stock is:

SIZE	STOCK
7	31
8	27
9	19
10	38

How many pairs of shoes must be ordered to bring the stock to the required level?
(b) The length of cloth needed to make a dress is 3 m, and the length required to make a pair of trousers is 2 m. What length of cloth is required to make 14 dresses and 15 pairs of trousers?

1.F17 The table shows the bookings at a restaurant for 6 days of a week.

DAY	LUNCH	DINNER
Mon	104	72
Tue	138	64
Wed	98	89
Thur	123	47
Fri	132	58
Sat	110	93

(a) What was the total number of meals served in the week?
(b) Which was the most popular day for lunch?
(c) Which was the most popular day for dinner?
(d) How many more lunches were served than dinners?

1.F18 A party of three ordered the following courses in a restaurant:

(i)	Soup	60p
	Rump steak	£4.50
	Sweet	95p
	Coffee	60p
(ii)	Melon	85p
	Chicken	£3.90
	Cheese and biscuits	65p
	Coffee	60p
(iii)	Prawn cocktail	£1.20
	Trout	£4.15
	Sweet	85p
	Coffee	60p

If the party also ordered a bottle of wine at £5.30, what was the total amount of the bill?

1.F19 A clothing manufacturer produces 38 garments at a total cost of £532. He sells all the garments to a retailer at a price of £16.99 each.
(a) What was the manufacturer's cost per garment?
(b) How much was the total profit made by the manufacturer?

1.F20 The following lists of ingredients are for recipes to provide portions for four people. Rewrite the lists to provide sufficient ingredients for portions for 10 people.

(a) Beef Goulash
   660 g stewing steak
   220 g onions
   72 g dripping
   2 litres stock
   2 green peppers
   240 g tomatoes
   1 clove of garlic
   440 g potatoes
      Add seasoning

Operations in Arithmetic 19

(b) Chilli Con Carne
    1 onion
    1 clove garlic
    24 g dripping
    440 g minced beef
    1 tspn flour
    220 g tomatoes
    1 tspn chilli powder
    220 g baked beans
    Add seasoning

(c) Apple Crunch
    440 g apples
    64 g butter
    2 tbspn granulated sugar
    1 lemon (grated rind)
    4 slices brown bread
    4 tbspn desiccated coconut
    4 tbspn demerara sugar

2 Operations and Fractions

Fractions are used to represent parts or portions of whole values. Figure 2.1 shows a sheet of hardboard which is equally divided into two parts, A and B. Each half of the hardboard is a fraction of the whole sheet and can be shown as

[Part A | Part B]

Fig. 2.1

part A = $\frac{1}{2}$ of the whole sheet
part B = $\frac{1}{2}$ of the whole sheet
part A + part B = whole sheet
$\therefore \frac{1}{2} + \frac{1}{2} = 1$

In Fig. 2.2. the same sheet of hardboard is divided into four equal parts. Each part can be shown as $\frac{1}{4}$ of the whole sheet. The whole of the sheet is the sum of the four parts.

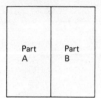

Fig. 2.2

$\therefore \frac{1}{4} + \frac{1}{4} + \frac{1}{4} + \frac{1}{4} = 1$

By comparing Fig. 2.1 and Fig. 2.2 is can be seen that two of the shaded parts could together be equal to part A.

$\therefore \frac{1}{4} + \frac{1}{4} = \frac{1}{2}$

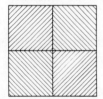

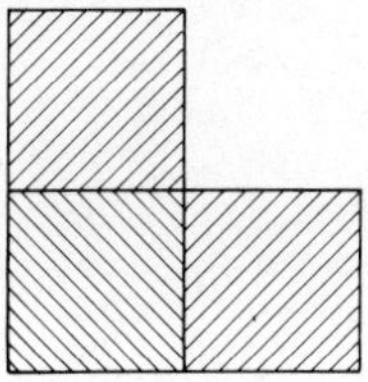

Fig. 2.3

If a piece was cut so that it contained three of the shaded parts, as shown in Fig. 2.3, then the piece could be represented as

$\frac{1}{4} + \frac{1}{4} + \frac{1}{4} = \frac{3}{4}$ of the whole sheet

2.1 Vulgar Fractions

Fractions which have a value of less than 1 are known as **vulgar fractions**. $\frac{1}{4}, \frac{1}{2}, \frac{3}{4}, \frac{7}{8}, \frac{17}{19}$ are all vulgar fractions. The number above the dividing line in the fraction is called the **numerator**. The number below the dividing line is called the **denominator**. In vulgar fractions the numerator is always less than the denominator.

When fractions have the same number for their denominator they are said to have a **common denominator**. These fractions can be added together by adding their numerators.

$$\frac{1}{3} + \frac{1}{3} = \frac{1+1}{3} = \frac{2}{3}$$

$$\frac{1}{7} + \frac{3}{7} = \frac{1+3}{7} = \frac{4}{7}$$

$$\frac{1}{17} + \frac{5}{17} + \frac{3}{17} = \frac{1+5+3}{17} = \frac{9}{17}$$

Subtraction can be done in the same way.

$$\frac{8}{15} - \frac{2}{15} = \frac{8-2}{15} = \frac{6}{15}$$

2.2 Improper Fractions

When two vulgar fractions are added together the numerator of the answer may be greater than the denominator. This type of fraction is called an **improper fraction**.

$$\frac{3}{5} + \frac{4}{5} = \frac{7}{5}$$

$\frac{19}{8}, \frac{5}{4}, \frac{3}{2}, \frac{21}{15}$ are all improper fractions.

2.3 Mixed Numbers

Quantities which contain both a whole number and a vulgar fraction are said to be **mixed numbers**. $1\frac{1}{2}, 4\frac{3}{5}, 11\frac{2}{9}$ are all mixed numbers. Mixed numbers are converted to improper fractions by multiplying the whole number by the denominator and then adding the numerator of the fraction.

Example 2.1 Convert $4\frac{3}{5}$ to an improper fraction.

$$4\frac{3}{5} = \frac{(4 \times 5) + 3}{5} = \frac{20 + 3}{5} = \frac{23}{5} \quad (Ans.)$$

Improper fractions can be converted to mixed numbers by dividing the numerator by the denominator to find the value of the whole number. The remainder becomes the numerator of the fraction part of the mixed number.

Example 2.2 Convert $\frac{26}{3}$ to a mixed number.

$$\frac{26}{3} = 8 + \frac{2}{3} = 8\frac{2}{3} \quad (Ans.)$$

2.4 Equivalent Fractions

Figure 2.4(a) shows a sheet of hardboard divided into two equal parts, and Fig. 2.4(b) shows the same size of sheet divided into six equal parts. It can be seen that the shaded part of the two sheets are equal.

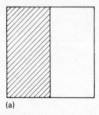

(a)

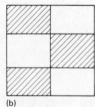

(b)

Fig. 2.4

$$\frac{1}{2} = \frac{3}{6}$$

These fractions are equal in value and are said to be **equivalent fractions**.

An equivalent fraction for any vulgar fraction can be found by multiplying the numerator and the denominator by the same number.

$$\frac{4}{7} = \frac{4 \times 3}{7 \times 3} = \frac{12}{21}$$

$\frac{4}{7}$ and $\frac{12}{21}$ are equivalent fractions.

Example 2.3 Find three equivalent fractions to $\frac{3}{8}$.

$$\frac{3}{8} = \frac{3 \times 2}{8 \times 2} = \frac{6}{16}$$

$$\frac{3}{8} = \frac{3 \times 5}{8 \times 5} = \frac{15}{40}$$

$$\frac{3}{8} = \frac{3 \times 10}{8 \times 10} = \frac{30}{80} \quad (Ans.)$$

2.5 Cancelling

Equivalent fractions can be found for some fractions by dividing the numerator and denominator by the same number. This process is known as **cancelling**.

$$\frac{35}{45} = \frac{35 \div 5}{45 \div 5} = \frac{7}{9}$$

Cancelling is usually done to reduce a fraction to its simplest form, i.e. its lowest terms.

Example 2.4 Cancel each of the following fractions to its lowest terms.

(a) $\frac{12}{45}$ (b) $\frac{117}{153}$ (c) $\frac{195}{315}$

(a) $\frac{12}{45} = \frac{12 \div 3}{45 \div 3} = \frac{4}{15}$

(b) $\frac{117}{153} = \frac{117 \div 9}{153 \div 9} = \frac{13}{17}$

(c) The simplification can be done in two stages:

$$\frac{195}{315} = \frac{195 \div 5}{315 \div 5} = \frac{39}{63}$$

$$= \frac{39 \div 3}{63 \div 3} = \frac{13}{21} \quad (Ans.)$$

Example 2.5 Simplify $\frac{378}{672}$.

$$\frac{378}{672} = \frac{378 \div 2}{672 \div 2} = \frac{189}{336}$$

$$= \frac{189 \div 3}{336 \div 3} = \frac{63}{112} = \frac{63 \div 7}{112 \div 7} = \frac{9}{16} \quad (Ans.)$$

2.6 Addition and Subtraction of Fractions

In Section 2.1 we saw that fractions having common denominators can be added together by adding their numerators. Thus, to add fractions, we need to bring them to a common denominator. This can be done by replacing the original fractions with equivalent fractions having a common denominator.

Example 2.6   Find the value of $\frac{5}{12} + \frac{3}{8}$.

Replacing $\frac{5}{12}$ by an equivalent fraction

$$\frac{5}{12} = \frac{5 \times 2}{12 \times 2} = \frac{10}{24}$$

Replacing $\frac{3}{8}$ by an equivalent fraction

$$\frac{3}{8} = \frac{3 \times 3}{8 \times 3} = \frac{9}{24}$$

$$\therefore \frac{5}{12} + \frac{3}{8} = \frac{10}{24} + \frac{9}{24} = \frac{19}{24} \quad (Ans.)$$

It can be seen from the last example that the lowest common denominator is the smallest number that both denominators will exactly divide into. This fact can be used to give a general method of adding fractions.

Example 2.7 $\frac{3}{4} + \frac{1}{6}$.

Step 1. The lowest common denominator is 12, because it is the smallest number that both 4 and 6 will divide into without a remainder.

Step 2. Each denominator is now divided into the lowest common denominator and the result is multiplied by its numerator. A new fraction is now created using the lowest common denominator. Thus,

$$\frac{3}{4} + \frac{1}{6} = \frac{(3 \times 3) + (2 \times 1)}{12}$$

$$= \frac{9 + 2}{12}$$

$$= \frac{11}{12} \quad (Ans.)$$

Example 2.8 $\frac{5}{6} + \frac{7}{8}$.

$$\frac{5}{6} + \frac{7}{8} = \frac{(4 \times 5) + (3 \times 7)}{24}$$

$$= \frac{20 + 21}{24}$$

$$= \frac{41}{24} = 1\frac{17}{24} \quad (Ans.)$$

Example 2.9 $\frac{5}{6} + \frac{2}{3} + \frac{7}{9}$.

$$\frac{5}{6} + \frac{2}{3} + \frac{7}{9} = \frac{(3 \times 5) + (6 \times 2) + (2 \times 7)}{18}$$

$$= \frac{41}{18} = 2\frac{5}{18} \quad (Ans.)$$

Subtractions of fractions is done using the same method.

Example 2.10 $\frac{11}{25} - \frac{2}{5}$.

$$\frac{11}{25} - \frac{2}{5} = \frac{(1 \times 11) - (5 \times 2)}{25}$$

$$= \frac{11 - 10}{25} = \frac{1}{25} \quad (Ans.)$$

When mixed numbers are involved they may be first converted to improper fractions.

Example 2.11 $1\frac{3}{5} - \frac{2}{3}$.

$$1\frac{3}{5} = \frac{8}{5}$$

$$\frac{8}{5} - \frac{2}{3} = \frac{(3 \times 8) - (5 \times 2)}{15}$$

$$= \frac{24 - 10}{15} = \frac{14}{15} \quad (Ans.)$$

It is sometimes easier to treat the whole number part of mixed numbers as a separate calculation.

Example 2.12   Find the value of $3\frac{1}{2} + 1\frac{3}{4} + 1\frac{5}{8}$.

$$3\frac{1}{2} + 1\frac{3}{4} + 1\frac{5}{8} = (3 + 1 + 1) + \frac{1}{2} + \frac{3}{4} + \frac{5}{8}$$

$$= 5 + \frac{1}{2} + \frac{3}{4} + \frac{5}{8}$$

$$= 5 + \frac{(4 \times 1) + (2 \times 3) + (1 \times 5)}{8}$$

$$= 5 + \frac{4 + 6 + 5}{8}$$

$$= 5 + \frac{15}{8} = 6\frac{7}{8} \quad (Ans.)$$

2.7 Multiplication of Fractions

Fractions are multiplied by multiplying the numerators together and the denominators together. The answer should then be reduced to its lowest terms by cancelling.

Example 2.13 Simplify $\frac{2}{5} \times \frac{3}{4}$.

$$\frac{2}{5} \times \frac{3}{4} = \frac{6}{20}$$

$$= \frac{3}{10} \quad (Ans.)$$

Example 2.14 $\frac{5}{9} \times \frac{3}{8} \times \frac{3}{10}$.

$$\frac{5}{9} \times \frac{3}{8} \times \frac{3}{10} = \frac{45}{720}$$

$$= \frac{5}{80}$$

$$= \frac{1}{16} \quad (Ans.)$$

An alternative method for this example is to cancel the fractions before multiplication

$$\frac{\cancel{5}^1}{\cancel{9}_1} \times \frac{\cancel{3}^1}{8} \times \frac{\cancel{3}^1}{\cancel{10}_2} = \frac{1}{16} \quad (Ans.)$$

Example 2.15 Simplify $1\frac{2}{3} \times 3\frac{1}{2}$.

$$1\frac{2}{3} = \frac{5}{3}, \quad 3\frac{1}{2} = \frac{7}{2}$$

$$\frac{5}{3} \times \frac{7}{2} = \frac{35}{6} = 5\frac{5}{6} \quad (Ans.)$$

2.8 Division of Fractions

When dividing by a fraction simply invert the divisor and then multiply.

Example 2.16 Simplify $\frac{2}{3} \div \frac{8}{9}$.

$$\frac{2}{3} \div \frac{8}{9} = \frac{2}{3} \times \frac{9}{8}$$

$$= \frac{18}{24} = \frac{3}{4} \quad (Ans.)$$

Example 2.17 Find the value of $1\frac{5}{6} \div 1\frac{1}{4}$.

$$1\frac{5}{6} = \frac{11}{6}, \quad 1\frac{1}{4} = \frac{5}{4}$$

$$\frac{11}{6} \div \frac{5}{4} = \frac{11}{6} \times \frac{4}{5} = \frac{44}{30}$$

$$= \frac{22}{15} = 1\frac{7}{15} \quad (Ans.)$$

2.9 Fraction of a Quantity

To find any fraction of a quantity simply multiply the quantity by the fraction, e.g.

$$\tfrac{3}{4} \text{ of } £160 = \frac{3}{4} \times 160 = £120$$

Example 2.18 Find the value of:

(a) $\tfrac{2}{3}$ of 480;
(b) $\tfrac{5}{9}$ of 162 kg;
(c) $\tfrac{6}{15}$ of 210 m.

(a) $\tfrac{2}{3} \times 480 = 320$ (*Ans.*)
(b) $\tfrac{5}{9} \times 162 = 90$ kg (*Ans.*)
(c) $\tfrac{6}{15} \times 210 = 84$ (*Ans.*)

Example 2.19 If $\tfrac{3}{20}$ of a batch of 600 electrical components were found to be faulty, how many were satisfactory?

Number of faulty components $= \dfrac{3}{20} \times 600 = 90$

Number of satisfactory components $= 600 - 90$
$= 510 \quad (Ans.)$

2.10 One Quantity as a Fraction of Another

Often it is useful to state one quantity as a fraction of another, e.g.:

- the number of women workers as a fraction of the total labour force of a factory;
- the number of faulty electric lamps as a fraction of the total number of lamps produced;
- the number of visitors in a particular week as a fraction of the total number of visitors in a year at a holiday resort.

To express one quantity as a fraction of another, the first quantity is made the numerator and the second is made the denominator. The fraction should then be simplified by cancelling to its lowest terms.

Example 2.20 Express 85 as a fraction of 500.

$$\frac{85}{500} = \frac{17}{100} \quad (Ans.)$$

Example 2.2 A warehouse is illuminated by the following electric lamps:

150 100 W   95 60 W   55 40 W

What fraction of the total number of lamps are (a) 100 W, (b) 60 W and (c) 40 W?

Total number of lamps = 150 + 95 + 55 = 300

(a) $\frac{150}{300} = \frac{1}{2}$ (Ans.)

(b) $\frac{95}{300} = \frac{19}{60}$ (Ans.)

(c) $\frac{55}{300} = \frac{11}{60}$ (Ans.)

Exercises 2.A Core

2.A1 A circle is divided into 24 equal parts. How many parts are there in:

(a) half of the circle;
(b) a quarter of the circle;
(c) one-third of the circle;
(d) two-thirds of the circle;
(e) three-quarters of the circle?

2.A2 Add:

(a) $\frac{2}{7}+\frac{3}{7}$ (b) $\frac{4}{19}+\frac{2}{19}$ (c) $\frac{5}{13}+\frac{3}{13}$

(d) $\frac{2}{5}+\frac{3}{5}$ (e) $\frac{9}{21}+\frac{4}{21}+\frac{6}{21}$ (f) $\frac{1}{3}+1\frac{1}{3}$

(g) $\frac{3}{4}+\frac{1}{4}+1\frac{1}{4}$ (h) $\frac{3}{16}+\frac{7}{16}+1\frac{1}{16}$

2.A3 Find the value of:

(a) $\frac{9}{16}-\frac{5}{16}$ (b) $\frac{27}{32}-\frac{11}{32}$ (c) $1\frac{1}{4}-\frac{3}{4}$ (d) $\frac{7}{10}-\frac{3}{10}$

(e) $\frac{59}{64}-\frac{27}{64}$ (f) $1\frac{1}{5}-\frac{4}{5}$ (g) $2\frac{1}{3}-1\frac{2}{3}$ (h) $3\frac{1}{4}-1\frac{3}{4}$

2.A4 Change to improper fractions:

(a) $1\frac{4}{7}$ (b) $2\frac{2}{5}$ (c) $5\frac{3}{4}$ (d) $12\frac{1}{2}$ (e) $3\frac{7}{9}$ (f) $5\frac{4}{11}$ (g) $1\frac{13}{55}$

(h) $10\frac{9}{16}$

2.A5 Change to mixed numbers:

(a) $\frac{11}{8}$ (b) $\frac{7}{4}$ (c) $\frac{13}{5}$ (d) $\frac{17}{6}$ (e) $\frac{29}{12}$

(f) $\frac{31}{8}$ (g) $\frac{122}{53}$ (h) $\frac{131}{9}$

2.A6 Cancel to the lowest terms:

(a) $\frac{16}{64}$ (b) $\frac{18}{24}$ (c) $\frac{108}{208}$ (d) $\frac{72}{360}$ (e) $\frac{15}{75}$

(f) $\frac{196}{240}$ (g) $\frac{66}{242}$ (h) $\frac{39}{130}$

2.A7 Give **two** equivalent fractions for each of the following:

(a) $\frac{7}{15}$ (b) $\frac{2}{5}$ (c) $\frac{12}{36}$ (d) $\frac{4}{9}$ (e) $\frac{50}{150}$

(f) $\frac{17}{40}$ (g) $\frac{5}{14}$ (h) $\frac{24}{36}$

2.A8 Find the value of:

(a) $\frac{3}{8}+\frac{1}{2}$ (b) $\frac{9}{16}+\frac{3}{8}$ (c) $\frac{3}{10}+\frac{9}{15}$

(d) $\frac{1}{3}+\frac{1}{4}+\frac{1}{6}$ (e) $\frac{5}{12}+\frac{2}{3}+\frac{1}{6}$ (f) $\frac{1}{2}+\frac{2}{3}+\frac{1}{4}$

(g) $\frac{5}{6}+\frac{3}{8}+\frac{2}{3}$ (h) $\frac{5}{6}+\frac{2}{3}+\frac{4}{9}$

2.A9 Simplify:

(a) $1\frac{3}{4}+2\frac{1}{2}$ (b) $1\frac{7}{8}+1\frac{3}{16}$ (c) $2\frac{2}{5}+3\frac{1}{3}$ (d) $4\frac{5}{16}+1\frac{3}{8}$

(e) $3\frac{1}{2}+2\frac{3}{4}+4\frac{1}{8}$ (f) $1\frac{5}{6}+2\frac{8}{9}+3\frac{2}{3}$ (g) $6\frac{1}{3}+4\frac{3}{4}+7\frac{1}{2}$

(h) $3\frac{7}{12}+5\frac{5}{6}+9\frac{2}{3}$

2.A10 Calculate:

(a) $\frac{11}{12}-\frac{3}{8}$ (b) $\frac{4}{5}-\frac{1}{4}$ (c) $\frac{13}{15}-\frac{2}{5}$ (d) $\frac{5}{6}-\frac{5}{8}$ (e) $1\frac{3}{5}-\frac{2}{3}$

(f) $2\frac{3}{4}-1\frac{5}{16}$ (g) $9\frac{1}{3}-4\frac{3}{4}$ (h) $3\frac{3}{10}-1\frac{11}{15}$

2.A11 Find the value of:

(a) $1\frac{1}{2}+3\frac{1}{4}-2\frac{3}{4}$ (b) $1\frac{5}{8}+2\frac{1}{6}-1\frac{2}{3}$ (c) $5\frac{1}{2}+4\frac{1}{3}-1\frac{5}{6}$

(d) $10\frac{1}{4}+1\frac{1}{3}-2\frac{1}{2}$ (e) $4\frac{5}{6}-2\frac{2}{9}+\frac{2}{3}$ (f) $3\frac{6}{10}+4\frac{2}{5}-1\frac{13}{15}$

Operations with Fractions 25

(g) $7\frac{1}{4} - 3\frac{5}{16} + 1\frac{9}{32}$ (h) $4\frac{4}{21} + 1\frac{6}{7} - 3\frac{2}{3}$

2.A12 Calculate:

(a) $\frac{1}{2} \times \frac{3}{4}$ (b) $\frac{2}{3} \times \frac{4}{5}$ (c) $\frac{5}{9} \times \frac{3}{8}$ (d) $\frac{10}{17} \times \frac{2}{5}$ (e) $\frac{3}{10} \times \frac{4}{5}$

(f) $\frac{7}{8} \times \frac{9}{13}$ (g) $\frac{1}{2} \times \frac{2}{3} \times \frac{3}{4}$ (h) $\frac{4}{7} \times \frac{3}{4} \times \frac{5}{12}$ (i) $\frac{3}{16} \times \frac{4}{5} \times \frac{2}{9}$

(j) $\frac{8}{15} \times \frac{2}{3} \times \frac{5}{16}$ (k) $\frac{7}{24} \times \frac{3}{4} \times \frac{12}{21}$ (l) $\frac{12}{45} \times \frac{9}{15} \times \frac{3}{4}$

2.A13 Calculate:

(a) $\frac{3}{8} \div \frac{1}{4}$ (b) $\frac{1}{2} \div \frac{7}{8}$ (c) $\frac{4}{9} \div \frac{2}{3}$ (d) $\frac{8}{15} \div \frac{4}{5}$ (e) $\frac{7}{20} \div \frac{3}{10}$

(f) $\frac{15}{28} \div \frac{3}{7}$ (g) $1\frac{1}{2} \div \frac{1}{3}$ (h) $3\frac{1}{4} \div \frac{2}{5}$ (i) $2\frac{2}{5} \div \frac{3}{4}$ (j) $9\frac{1}{6} \div \frac{11}{12}$

(k) $4\frac{1}{8} \div 2\frac{1}{2}$ (l) $2\frac{5}{8} \div 3\frac{1}{2}$

2.A14 Find:

(a) $\frac{1}{2}$ of 340 (b) $\frac{2}{3}$ of 90 (c) $\frac{5}{12}$ of 840
(d) $\frac{6}{7}$ of 98 (e) $\frac{3}{10}$ of 1 200 (f) $\frac{4}{5}$ of 1 650
(g) $\frac{3}{8}$ of 5 000 (h) $\frac{9}{15}$ of 225 (i) $\frac{7}{20}$ of 10 000
(j) $\frac{13}{15}$ of 600

2.A15 Find:

(a) $\frac{3}{7}$ of 280 m (b) $\frac{5}{8}$ of 64 cm (c) $\frac{3}{5}$ of 200 kg
(d) $\frac{17}{20}$ of 1 000 litres (e) $\frac{7}{25}$ of 800 g (f) $\frac{11}{12}$ of £72
(g) $\frac{4}{15}$ of 135 mm (h) $\frac{13}{30}$ of 900 km (i) $\frac{6}{7}$ of 630 m
(j) $\frac{4}{15}$ of 270 kg

2.A16

(a) Express 36 as a fraction of 84.
(b) Express 14 as a fraction of 49.
(c) Express 8 as a fraction of 64.
(d) Express 25 as a fraction of 80.
(e) Express 32 as a fraction of 160.
(f) Express 18 as a fraction of 108.
(g) Express 20 as a fraction of 1 000.
(h) Express 54 as a fraction of 144.

2.A17 Given that the area of a rectangular room = length × width, calculate the area of the following rooms in square metres:

ROOM	LENGTH (m)	WIDTH (m)
(a)	$3\frac{1}{2}$	$2\frac{1}{2}$
(b)	$4\frac{1}{4}$	$2\frac{3}{4}$
(c)	$3\frac{3}{4}$	$1\frac{1}{2}$
(d)	$6\frac{1}{2}$	4
(e)	8	$3\frac{1}{2}$
(f)	$7\frac{1}{2}$	$3\frac{3}{4}$
(g)	$10\frac{1}{4}$	$5\frac{1}{2}$
(h)	$6\frac{3}{4}$	$4\frac{1}{4}$

2.A18 (a) Three pieces, each 200 mm long, are sheared off a bar of length 900 mm.
 (i) What fraction of the original bar is each piece?
 (ii) What fraction of the original bar is left after the three pieces have been removed?
(b) Contracts totalling £120 000 were awarded between three companies. Brown Steel Ltd received one-half of the total, Jones and Sons received one-third and New Enterprises received one-sixth. What was the value of the contracts received by each company?

2.A19 Calculate:

(a) $\frac{2}{3}$ of 69p (b) $\frac{3}{4}$ of 80p (c) $\frac{9}{10}$ of £1.50 (d) $\frac{3}{5}$ of £3.55
(e) $\frac{5}{6}$ of £12.60 (f) $\frac{7}{12}$ of £14.40 (g) $\frac{1}{3}$ of £18.75
(h) $\frac{4}{5}$ of £75.50 (i) $\frac{7}{9}$ of £1.08 (j) $\frac{5}{8}$ of £25
(k) $\frac{17}{100}$ of £14 (l) $\frac{3}{20}$ of £44.60 (m) $\frac{3}{15}$ of 75p
(n) $\frac{8}{17}$ of £68 (o) $\frac{5}{11}$ of £16.50 (p) $\frac{8}{19}$ of £3.80
(q) $\frac{7}{24}$ of £1 200 (r) $\frac{5}{6}$ of £7.92 (s) $\frac{4}{7}$ of £24.78
(t) $\frac{2}{3}$ of £28.71

2.A20 Express:

(a) $\frac{1}{4}$ yd in inches (1 yd = 36 in);
(b) $\frac{5}{8}$ lb in ounces (1 lb = 16 oz);
(c) $\frac{7}{12}$ ft in inches (1 ft = 12 in);
(d) $\frac{3}{7}$ stone in pounds (1 stone = 14 lb);
(e) $2\frac{3}{4}$ gal in pints (1 gal = 8 pints);
(f) $\frac{3}{4}$ gal in quarts (1 quarts = 2 pints);
(g) $\frac{5}{7}$ cwt in pounds (1 cwt = 112 lb);
(h) $\frac{3}{8}$ mile in yards (1 mile = 1 760 yd);
(i) $\frac{4}{5}$ ton in pounds (1 ton = 2 240 lb);
(j) 1 ft 6 in in yards;
(k) $\frac{3}{4}$ mile in feet;
(l) $\frac{1}{2}$ stone in ounces;
(m) $3\frac{1}{2}$ quarts in pints;
(n) $1\frac{4}{5}$ ton in hundredweights (1 tons = 20 cwt);
(o) $7\frac{3}{4}$ cwt in stones (1 cwt = 8 stone).

Exercises 2.B Business, Administration and Commerce

2.B1 A market stall has overheads as follows:
Rent of the stall £25 per day
Staff wages £45 per day
Transport £16 per day
Lighting and refreshments
 for staff £9 per day
What fraction of the total overheads is taken in (a) rent, (b) wages, (c) transport, (d) lighting, etc.?

2.B2 From his gross earnings of £120, a man paid £30 in tax, £8 in hire purchase payments and £4 to a holiday fund. What fraction of his gross earnings went on (a) tax, (b) hire purchase and (c) the holiday fund?

2.B3 Four items must be ordered to maintain the stock level of a shop. Item (a) 30, item (b) 45, item (c) 60 and item (*d*) 15. What fraction of the total number to be ordered is each item?

2.B4   A family spent £2.30 at a sweet shop. 82p was spent on chocolates, 60p on lollipops, 35p on small items and 53p on mixtures. What fraction of the total cost was spent on (*a*) chocolates, (*b*) lollipops, (*c*) small items and (*d*) mixtures?

2.B5   (*a*) A woman went shopping with £43.68 in her purse. She spent one-half of the money on food and a further £5.20 on household items. How much money had she left?
(*b*) A girl received £12 for part-time working. She spent £1.80 on travelling and £1.20 on lunch. What fraction of her total wage was left?

2.B6   (*a*) From 500 reams of paper, one-fifth was used for note making and a quarter of the remainder was used for duplicating. How many reams were left?
(*b*) A spirit duplicating machine was used to produce spirit copies on sheets of paper from a master. For every seven satisfactory copies produced, one copy was spoiled in the printing process. If 210 satisfactory copies were made:

(i) How many sheets were spoiled?
(ii) What was the total number of sheets used?
(iii) What fraction of the total sheets were spoiled?

2.B7   (*a*) The Inland Revenue allowed a business man to depreciate the original cost of his car by one-quarter in the first year and by a further one-fifth of the original cost in the second year. If the purchase price was £6 000, what would be the value of the car after 2 years?
(*b*) When using his car, the business man is allowed to claim travelling expenses at the rate of $8\frac{1}{4}$p per kilometre. How much would be his total claim for a year in which he travelled 24 000 km on business?

2.B8   (*a*) An office worker is paid £32 for overtime working. The amount was reduced by three-tenths for income tax. How much did he receive?
(*b*) A local government officer is required to make pension contributions amounting to seven-hundredths of his gross salary. If his salary is £8 500 per annum, how much does he pay in superannuation contribution in the year?

2.B9   A mail-order firm receives 480 telephone calls in 1 day. A quarter of the calls are inquiries, five-eighths are orders and the remainder are from suppliers. How many calls were: (*a*) inquiries, (*b*) orders and (*c*) from suppliers?

2.B10   A general office worker works $7\frac{1}{2}$ h in 1 day. She spends one-fifth of this time on clerical work, two-fifths on typing and the remainder on duplicating. How much time, in hours and minutes, does she spend on (*a*) clerical work, (*b*) typing and (*c*) duplicating?

2.B11   A mail-order firm printed 3 600 catalogues which were distributed as follows:

1 400 to agents
  900 to shops
1 300 to individual customers

What fraction of the total number of catalogues was sent to (*a*) agents, (*b*) shops and (*c*) individual customers?

2.B12   A service engineer is paid at the rate of 'time and a third' for week-day overtime and 'time and a half' for week-end overtime. If his basic rate for normal working hours is £2.40 per hour, calculate his total earnings for the following week:

	NORMAL HOURS	OVERTIME HOURS
Monday	$8\frac{1}{2}$	2
Tuesday	$8\frac{1}{2}$	
Wednesday	$8\frac{1}{2}$	2
Thursday	$8\frac{1}{2}$	$1\frac{1}{2}$
Friday	8	
Saturday		4
Sunday		$2\frac{1}{2}$

2.B13   Calculate the floor area of a rectangular office of length $7\frac{1}{4}$ m and width $5\frac{3}{4}$ m. (Area = length × width.)

2.B14   For a spring sale, a department store reduces the prices of certain lines by one-third. The list shows the normal prices of goods. Calculate the sale prices.

Ladies' tights	99p
Gent's socks	£1.20
Jeans	£12.60
Cassettes	£3.99
Paperbacks	£1.50
22 in television	£390
Writing pads	63p
Handbags	£7.62
Ladies' shoes	£10.92
Gent's shoes	£17.16
Electric irons	£24.30

2.B15   The total cost of manufacturing a range of office typewriters is made up of:

$\frac{1}{3}$ material costs
$\frac{5}{9}$ labour costs
$\frac{1}{9}$ overheads

Operations with Fractions   27

Use this information to complete the following table:

TYPEWRITER MODEL	TOTAL MANUFAC- TURING COST (£)	MATERIAL COST (£)	LABOUR COST (£)	OVER- HEADS (£)
Word King Standard	74.79			
Word King Super	98.46			
Word Ace Manual	163.98			
Word Ace Electric	215.10			

2.B16 (a) What is the total number of items in the following stationery order?

$\frac{3}{4}$ doz memo pads
$3\frac{1}{2}$ doz pencils
$2\frac{3}{4}$ doz writing pads
$12\frac{1}{2}$ doz A4 envelopes
$4\frac{1}{4}$ gross adhesive labels

(b) A mail-order firm offers a discount of one-eighth on the list price of articles to its agents. How much would an agent pay for goods having a total list price of £61.52?

2.B17 A furniture shop requires a deposit of $\frac{3}{20}$ of the selling price of all items sold on hire purchase. Calculate the amount of deposit for each of the following items:

	SELLING PRICE
Bookcase	£70.80
Dining table	£107.60
Settee	£153.20
3-piece suite	£369.40
Double bed	£133.20
Dressing table	£96.00
Wardrobe	£174.60
Sideboard	£188.20
Hi-fi cabinet	£75.00
Writing desk	£132.80

2.B18 A publishing firm employs 225 workers on its staff. One-third are aged under 25 years and one-fifteenth are over 60 years. Calculate:
(a) the fraction of the staff who are aged between 25 and 60 years;
(b) the number of workers in each of the three age groups.

2.B19 A consignment of 120 cassette tapes was sent in three parcels. The first parcel contained 35 tapes, the second parcel three-eighths of the order and the remaining tapes were sent in the third parcel. Find:
(a) what fraction of the order was in the first parcel;
(b) how many tapes were in the second parcel;
(c) how many tapes were in the third parcel.

2.B20 A bank has 2 000 customers who hold the following types of account: (a) $\frac{9}{20}$ hold current accounts, (b) $\frac{3}{10}$ hold deposit accounts and (c) $\frac{1}{4}$ hold both current and deposit accounts. Calculate the number of customers in each category.

Exercises 2.C Technical Services: Engineering and Construction

2.C1 It takes $5\frac{1}{2}$ h to sand down and varnish a table top. One-third of the time is spent applying the varnish.
(a) What fraction of the total time is spent on sanding?
(b) How much time is spent on sanding?

2.C2 A car tyre has 26 lb/in^2 pressure when blown up correctly. A puncture allows air to leak out. What will the pressure, in pounds per square inch, be when the pressure has been reduced by (a) one-quarter, (b) five-eigths and (c) three-quarters?

2.C3 The fuel tank of a light van holds 54 litres and the van travels 8 km on 1 litre of fuel. What fraction of a full tank of fuel would be consumed in each of the following journeys: (a) 208 km, (b) 256 km and (c) 336 km?

2.C4 Pieces of bar were parted off on a capstan lathe to the following lengths: $\frac{3}{4}$ in, $\frac{1}{2}$ in, $\frac{5}{8}$ in, $\frac{7}{16}$ in and $\frac{11}{16}$ in. What length of bar in inches would be required if the width of the parting tool was $\frac{3}{16}$ in?

2.C5 (a) From a packet of 200 welding rods, one-fifth of the rods had been used. How many rods were left?
(b) A paint can contained 5 litres of paint when it was despatched from the factory. During transit five-eighths of the paint had leaked out. How much paint was in the can on arrival?

2.C6 (a) The oil sump on a car holds 6 litres when showing full on the dipstick. The engine burns half a litre per 1 000 km. After a journey of 3 000 km what fraction of the oil has been consumed?
(b) A car makes a journey of 240 miles. Half the journey is on motorways travelling at 70 miles per hour. A quarter of the journey is on dual carriageways, travelling at 50 miles per hour and the remainder is in urban built-

up areas travelling at 30 miles per hour. The petrol consumption of the car is:

20 miles per gallon at 70 miles per hour
30 miles per gallon at 50 miles per hour
40 miles per gallon at 30 miles per hour

Estimate how much petrol was used on the journey.

2.C7 The gable end of a house required 5 500 bricks when being built. Two thousand of these bricks started to flake after 12 years. What fraction of the bricks remained in good condition?

2.C8 To decorate a room, paint was required in the following quantities:

5 litres emulsion
$1\frac{1}{2}$ litres white gloss
$\frac{3}{4}$ litre blue gloss
$1\frac{3}{4}$ litres undercoat

What fraction of the total paint used was: (a) emulsion paint, (b) white gloss, (c) blue gloss and (d) undercoat?

2.C9 A brick wall containing 400 bricks is demolished and, along with building rubble, is loaded into a skip. The mass of the skip is 900 kg and the total mass of the skip and its contents is 3 400 kg. If the mass of one brick is $\frac{7}{8}$ kg, what fraction of the total contents of the skip is the building rubble?

2.C10 (a) A blacksmith, when forging a bar of steel 13 cm long, draws it out to $20\frac{1}{2}$ cm long. By what fraction of the original size has the metal been lengthened?
(b) In a turning operation the cutting tool travels along the work at the rate of $\frac{1}{32}$ in per revolution of the work piece. The workpiece rotates at 256 r/min. How far will the cutting tool have travelled in half a minute?

2.C11 Calculate the volume of concrete required to cast a rectangular machine foundation $2\frac{1}{2}$ m long, $1\frac{3}{4}$ m wide and $\frac{3}{4}$ m deep. (Volume = length × width × depth.)

2.C12 A model of a light aircraft is to be built to a scale of $\frac{1}{25}$ of full size. Calculate the dimensions of the model for each of the following dimensions of the aircraft:

Wingspan	1 000 cm
Length	900 cm
Height	325 cm
Propeller diameter	200 cm
Undercarriage wheel diameter	75 cm

2.C13 A collant tank contains 96 litres of cutting fluid. The cutting fluid is made up of $\frac{3}{48}$ soluble oil and the remainder water.

(a) What fraction of the cutting fluid is water?

(b) How many litres of soluble oil are contained in the tank?

2.C14 In a batch of 1 200 turned components $\frac{7}{50}$ are found to be oversize and $\frac{3}{40}$ undersize.

(a) What fraction of the batch are correct size?
(b) How many components are (i) oversize, (ii) undersize and (iii) correct size?

2.C15 (a) A manufactured component consists, by volume, of $\frac{3}{8}$ ferrous metal, $\frac{5}{12}$ non-ferrous metal and the remainder is plastic material. Determine what fraction of the volume of the component is plastic material.
(b) If the volume of the component is $72\,000$ mm^3, calculate the volume of (i) the ferrous metal, (ii) the non-ferrous metal and (iii) the plastic material.

2.C16 Calculate the distance moved by a nut along a thread having a pitch of 4 mm when the nut is turned through (a) $2\frac{1}{2}$ revolutions, (b) $3\frac{1}{4}$ revolutions and (c) $12\frac{3}{4}$ revolutions.

2.C17 (a) A batch of 600 electrical resistors were checked and $\frac{3}{50}$ of the batch were found to be faulty. Calculate:
 (i) the number of faulty resistors;
 (ii) the number of satisfactory resistors;
 (iii) the fraction of the batch that was satisfactory.
(b) In a factory employing 400 workers, $\frac{2}{5}$ are machinists, $\frac{3}{10}$ are assembly workers and the remainder are unskilled workers. Find the number of (i) machinists, (ii) assembly workers and (iii) unskilled workers.

2.C18 An engineering works employs the following skilled workers:

135 fitters
225 machinists
 50 welders
 25 electricians
 15 maintenance workers

State the number of workers employed in each trade as a fraction of the total number of skilled workers.

2.C19 The total machining time to manufacture an engineering component is made up of:

Turning 12 min
Milling 18 min
Shaping 36 min
Drilling   6 min

Express the time taken by each operation as a fraction of the total machining time.

2.C20 The total resistance R of two electrical resistors R_1 and R_2 when connected in series is given by

$$R = R_1 + R_2$$

Use this formula to complete the following table.

R_1 (ohms)	R_2 (ohms)	R (ohms)
$3\frac{1}{2}$	5	
$12\frac{1}{2}$	$6\frac{1}{4}$	
$2\frac{1}{5}$	$3\frac{2}{5}$	
$1\frac{3}{4}$	$2\frac{1}{2}$	
$4\frac{3}{5}$	$7\frac{1}{2}$	

Exercises 2.D General Manufacturing and Processing

2.D1 A bottle holds 5 litres of a chemical. What fraction will be left in each case when the following quantities are taken out?
(a) $1\frac{1}{4}$ litres (b) $\frac{5}{8}$ litre (c) $3\frac{3}{4}$ litres (d) $3\frac{1}{8}$ litres
(e) $4\frac{3}{8}$ litres (f) $2\frac{3}{5}$ litres (g) $1\frac{7}{10}$ litres (h) $3\frac{3}{10}$ litres

2.D2 A manufacturing plant operates on a 7-day-week production system. The plant closes for 9 public holidays per year, and 56 other days for maintenance and annual holidays. What fraction of the time does the plant operate in a 365-day year?

2.D3 Labels on tins are attached by a machine at the rate of 240 per minute, but due to a machine fault 32 in every 240 labels are attached upside-down. What fraction of the labels are the right way up after 10 min of machine running time?

2.D4 (a) My garden was $2\frac{1}{3}$ ha in area until I bought another $\frac{5}{6}$ ha from a neighbour. How much land have I now?
(b) A young parks worker was paid on Friday and spent $\frac{5}{8}$ of his wage on Saturday and a further $\frac{5}{16}$ on Monday. What fraction of his wage was left for the rest of the week?

2.D5 (a) A garden is $1\frac{2}{3}$ ha in area. Two-thirds of this is planted with vegetables. What area is left unplanted?
(b) An orchard has an area of $4\frac{1}{2}$ ha, half of which is planted with apple trees. If half of the trees are pruned, what area has pruned trees?

2.D6 The total livestock of a farm is 715. There are 310 sheep, 250 cows, 120 bullocks and 35 horses. What fraction of the total stock are (a) cows, (b) bullocks, (c) horses and (d) sheep?

2.D7 In the London marathon a runner covered the 26 miles in 6 h 14 min. It took the runner 12 min each for the first 8 miles, 14 min each for the next 12 miles, 68 min for the next 4 miles and 42 min for the last 2 miles. What fractions of the full time did it take for (a) the first 8 miles, (b) the next 12 miles, (c) the next 4 miles, (d) the last 2 miles and (e) the runner to reach the half-way stage?

2.D8 A coal grader selects the sizes of coal pieces and directs them into different containers. After a shift, the coal produced has been graded as follows: $\frac{1}{3}$ to grade 1, $\frac{1}{4}$ to grade 2, $\frac{1}{5}$ to grade 3. What fraction would be left for grade 4?

2.D9 (a) A trawler landed 18 t of fish. 4 t of the fish were used for cat food, 9 t were frozen and the rest was kept fresh. What fraction of the total landed was kept fresh?
(b) The contents of a trawl was found to be $\frac{2}{5}$ cod, $\frac{1}{8}$ sprats, $\frac{1}{4}$ mackerel and the rest were flat fish. What fraction of the total catch was flat fish?

2.D10 A market gardener employs two full-time gardeners, each working 45 h per week, and six part-time packers, each working 10 h per week. What fraction of the total working hours is provided by (a) full-time workers and (b) part-time workers?

2.D11 The total income of a mixed farm may be divided under the following headings:

$\frac{3}{8}$ dairy products
$\frac{1}{4}$ meat products
$\frac{1}{12}$ garden produce
$\frac{7}{24}$ grain products

If the total income of the farm for a full year is £32 280, how much income was earned under each heading?

2.D12 A trawler steams for $3\frac{1}{2}$ h at $5\frac{1}{2}$ knots, followed by $7\frac{1}{2}$ h at $6\frac{1}{2}$ knots. How many nautical miles were covered in the 11 h period? (1 knot = 1 nautical mile per hour.)

2.D13 A manufacturing company pays overtime at the rate of 'time and a half'. Calculate the total overtime bill in 1 week for the following workers:

	BASIC RATE	OVERTIME
2 foremen	£3.20	4 h each
20 process workers	£2.60	$6\frac{1}{2}$ h each
4 maintenance fitters	£2.90	$10\frac{1}{2}$ h each
3 drivers	£2.80	12 h each

2.D14 The weekly production of a plastics factory is 18 000 units made up of:

Two-fifths injection mouldings
One-third flash mouldings
One-sixth laminates
One-tenth other items

Calculate the number of items produced of each type.

2.D15 The annual production of a tyre manufacturer is divided into the following categories:

One-third car (radial)
One-twelfth car (crossply)
One-quarter lorry tyres
One-eighth agricultural tyres
One-sixth aircraft tyres
Remainder fork lift truck tyres

What fraction of the total production is for (a) fork lift trucks and (b) cars?

2.D16 Three boreholes, A, B and C, are drilled in a mining exploration. Borehole A is 200 m deep. Borehole B is four-fifths of the depth of borehole A. Borehole C is seven-eighths of the depth of borehole B. Calculate the depth of (a) borehole B and (b) borehole C.

2.D17 A miner earns a basic wage of £2.40 per hour. He is paid 'time and a quarter' for overtime working and receives a bonus payment of one-tenth of his total wage. How much does he earn in a week when he works 42 h normal working and 6 h overtime?

2.D18 A farmer's working day is spent as follows:

(a) tending livestock 3 h
(b) milking 2 h 30 min
(c) feeding livestock 1 h 30 min
(d) field maintenance 3 h 30 min
(e) fencing 1 h 30 min

What fraction of his working day does he spend on each occupation?

2.D19 The production cost of a manufactured item is £6.40. This cost is made up of:

Materials one-quarter
Labour five-eighths
Packaging and overheads one-eighth

Find the cost per item of (a) materials, (b) labour and (c) packaging and overheads.

2.D20 A farmer buys a piece of land 45 m long and $30\frac{1}{2}$ m wide at a cost of £2.50 per square metre. How much does he pay for the land? (Area (m^2) = length (m) × width (m)).

Exercises 2.E Services to People I: Community Services

2.E1 An illness lasting 86 days required 48 days convalescing, followed by 14 days holiday. What fraction of the total time was spent on convalescing?

2.E2 A social worker visited seven houses in a cul-de-sac on an estate. The estate had 1 155 houses and the cul-de-sac had 42. What fraction of houses did the social worker visit on (a) the estate and (b) the cul-de-sac.

2.E3 The area of floor space in a house is 130 m^2. A quarter is carpeted, three-eighths is carpet tiled and the rest is lino tiled. What area is lino tiled?

2.E4 The total cost of running a car is £260 per year. The cost is made up of £85 road tax, £115 insurance and £60 garage fees. What fraction of the total cost is (a) insurance, (b) road tax and (c) garage fees?

2.E5 Newspapers are delivered to the houses on an estate. One-fifth of the houses take newspaper A, one-quarter take newspaper B, three-eighths take newspaper C and one-tenth take newspaper D. What fraction of the houses on the estate do not have a newspaper delivered?

2.E6 (a) A highway worker is on site for 9 h in a day. He has three-quarters of an hour for lunch break, and 15 min in the morning and afternoon for tea breaks. His work is interrupted by traffic for a total of $2\frac{1}{4}$ h. What fraction of the total time spent on site did he work?
(b) An ambulance driver works a 5-day week for 48 weeks in 1 year. In a year of 365 days what fraction of the year is he on duty?

2.E7 The rates collected by a local council are spent in the following manner: one-quarter on fire service, one-sixth on libraries, one-tenth on waste disposal, one-thirtieth on planning, one-thirtieth on public protection and the rest on other services. What fraction is spent on other services?

2.E8 Education takes $\frac{14}{25}$ of the total rates of a town, and social services take $\frac{3}{25}$. The police force and highway maintenance both take $\frac{1}{10}$. What fraction is left for other services?

2.E9 Thirty-five houses in a street each have a dustbin. Seven dustbins are $\frac{1}{4}$ full, 12 are $\frac{1}{2}$ full, 6 are $\frac{5}{8}$ full and 10 are $\frac{3}{4}$ full. If all the bins had been in a line, and each filled in turn, how many full bins would there be?

2.E10 (a) Goods from a store may be wrapped in either paper bags or plastic bags. If $\frac{7}{16}$ of the purchases in a day are wrapped in plastic bags, what fraction of the total purchases was wrapped in paper bags.
(b) If the total sales of the store were 5 120 for the same day, how many purchases were wrapped in paper bags?

2.E11 (a) Two pieces of carpet of length $4\frac{7}{8}$ m and $3\frac{3}{16}$ m are joined together to form one piece. What is the length of the new piece?
(b) On inspection a roll of carpet was found to be

stained, and a piece measuring $2\frac{7}{8}$ m was removed. The remaining carpet was cut into two pieces of $12\frac{13}{16}$ m and $13\frac{11}{32}$ m length. What was the original length of the roll of carpet?

2.E12 To reach a camping site, a holiday maker travelled the first 120 km by bus, 75 km by train and the rest of the journey on foot. The total distance was 200 km. What fraction of the total distance was (a) the train journey, (b) the bus journey and (c) the walk?

2.E13 Of the young people who attend the local youth club, $\frac{3}{16}$ play snooker, $\frac{5}{12}$ play football, $\frac{3}{8}$ are in the Scouts or Guides, the rest belong to a drama group. If the drama group is seven strong, how many are in (a) the snooker group, (b) the football group, (c) the Scouts and Guides group?

2.E14 The bookings at a hairdressing salon for one day are (a) 6 perms, (b) 8 rinse and sets, (c) 4 cut and sets and (d) 2 tint and sets. Determine what fraction each treatment is of the total bookings.

2.E15 A holidaymaker breaks his leg while abroad and receives hospital treatment before returning home by special flight and ambulance. His insurance company agrees to pay two-thirds of the medical costs and three-quarters of the transport costs. How much of the total expense must be met by the holidaymaker if the medical costs are £414 and the transport costs are £668?

2.E16 (a) A telephone switchboard accepts 360 incoming calls in one day. The operator takes approximately three-quarters of a minute to answer each call and route it to the required department. How much time does the operator spend on incoming calls?
(b) In the same day the operator places the following outgoing calls: 46 local, 32 trunk and 12 overseas. What fraction of the total outgoing calls were trunk calls?

2.E17 A nurse earns a gross wage of £96 per week. Income tax, £10.25, National Insurance, £6.44, and pension contribution, £7.31, are deducted from her pay. What fraction of her gross wage is her take-home pay?

2.E18 (a) A double-decker bus has seats for 32 passengers downstairs and 35 people upstairs. On a journey seven-eighths of the seats downstairs, and four-sevenths of the seats upstairs are occupied. How many passengers are on the bus?
(b) On the return journey there are 19 passengers upstairs and 23 downstairs. What fraction of the total number of seats are unoccupied?

2.E19 Figure 2.5 shows the floor plan of a dentist's surgery. Calculate the total area of carpet in square metres required to carpet all the rooms.

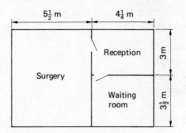

Fig. 2.5

2.E20 In a multi-storey car park there are spaces for 1 040 cars. The table shows the number of cars in the car park at six times during the day.

TIME	NO. OF CARS
09.00	640
11.00	880
13.00	960
15.00	720
17.00	480
19.00	800

Rewrite the table to show what fraction of the car park is occupied at each time of the day.

Exercises 2.F Technical Services II: Food and Clothing

2.F1 A roll of cloth is 10 m^2, and it takes a piece $2\frac{3}{4} \text{ m}^2$ to make a dress. After cutting the dress to the pattern $\frac{1}{8} \text{ m}^2$ is wasted. What area of cloth is wasted if 3 dresses are made from the roll?

2.F2 There are 91 m of cotton thread on a reel which is used on a sewing machine to join short lengths of material together. Each run uses 9 m of thread. What fraction of the thread is left after 9 runs?

2.F3 The same size of men's shoes are marked size 8 on the British market, size 9 on the American market and size $41\frac{1}{2}$ on the Continent. A market stall has 52 pairs of these shoes marked 8, 5 pairs marked 9 and 90 pairs marked $41\frac{1}{2}$. What fraction of the total number of shoes are (a) British size, (b) American size and (c) Continental size?

2.F4 A hotel manager receives his salary every month in amounts equal to one-twelfth of his annual salary. In his first year as manager his salary was £6 360 p.a. This was increased by two-fifths in his second year. Calculate his monthly salary (a) in the first year and (b) in the second year.

2.F5 Figure 2.6 shows the floor plan of a small canteen. Calculate the floor area of each room. (Area of room (m²) = length (m) × width (m).

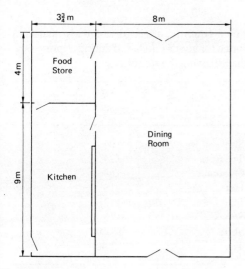

Fig. 2.6

2.F6 The following costs are involved in producing a restaurant meal for 50 people: food £55, wages £75, overheads £45. If the restaurant charged £3.50 per head, what fraction of the total income would be the cost of (*a*) food, (*b*) wages and (*c*) overheads?

2.F7 A restaurant is required to prepare 384 portions of turkey for a Christmas function. Twenty-four portions can be obtained from one bird.
(*a*) What fraction of the total number of portions is supplied from one bird?
(*b*) What fraction of the total number of portions is supplied from four birds?

2.F8 (*a*) A butcher delivers 32 kg of meat to a canteen. The meat produces 96 portions at $\frac{1}{4}$ kg. What fraction of the meat was not used?
(*b*) If steak costs £2.80 per pound, find the cost of (i) $\frac{3}{4}$ lb, (ii) $2\frac{1}{2}$ lb and (iii) 4 oz. (1 lb = 16 oz.)

2.F9 A sewing machine was priced by supplier A at £480. The same model was offered by supplier B at £416. What fraction of supplier A's price is saved by purchasing from supplier B?

2.F10 A jar of pickles has a mass of 500 g. The empty jar has a mass of 100 g, the remainder being pickles and vinegar. The mass of vinegar is 150 g. What fraction of the total mass is (*a*) pickles, (*b*) vinegar and (*c*) the jar?

2.F11 A bakery produces 750 scones, 1 000 barm cakes and 200 loaves. What fraction of the total number of items is (*a*) the scones, (*b*) the barm cakes and (*c*) the loaves?

2.F12 A café owner prepares the following sandwiches: 75 beef, 65 ham, 35 cheese, 70 chicken, 55 salad. What fraction of the total number of sandwiches has each filling?

2.F13 (*a*) Five drums of cooking oil are found to be $\frac{1}{3}$, $\frac{1}{2}$, $\frac{2}{5}$, $\frac{3}{4}$ and $\frac{7}{8}$ full. The capacity of each drum is 5 litres. They are all poured into a deep fat fryer which holds 25 litres. What fraction of the fryer will be filled?
(*b*) Four 1-gal jugs of milk are found to be $\frac{1}{2}$, $\frac{1}{4}$, $\frac{3}{8}$ and $\frac{5}{8}$ full. If the contents of all the jugs were poured into a 4-gal drum:
  (i) What fraction of the drum would be full?
  (ii) How much milk would the drum contain, in gallons and pints?
  (iii) How many more quarts of milk would it take to fill the drum?
(1 gallon = 8 pints; 1 quart = 2 pints.)

2.F14 A fish and chip shop sold 132 fish and 420 bags of chips on Friday evening. A fish costs three times as much as a bag of chips and the total takings were £163.20.
(*a*) How much does a bag of chips cost?
(*b*) What fraction of the monies taken was from fish sales?
(*c*) How much does one fish cost?

2.F15 (*a*) The daily receipts of a snack bar were: Monday £130, Tuesday £162, Wednesday £210, Thursday £96, Friday £180, Saturday £222. What fraction of the total receipts was taken on each day?
(*b*) If the profit from sales was one-quarter of the receipts, what profit was made on each day?
(*c*) What were the total profits and the total receipts for the week?

2.F16 (*a*) A list of ingredients for a cake is shown below. (Four eggs and flour would also be required.)

100 g raisins	100 g sultanas
75 g glace cherries	225 g butter
125 g mixed peel	250 g brown sugar
425 g currants	100 g ground almonds

What fraction of the total mass is each ingredient?
(*b*) A list of ingredients for rhubarb tart is shown below. (Milk and sugar would also be required.)

$1\frac{1}{4}$ lb cooked rhubarb
8 oz plain flour
4 oz self-raising flour
4 oz margarine
2 oz lard

What fraction of the total mass is each ingredient?

2.F17 A girl makes the following purchases at a boutique:

    Skirt   £16
    Top   £12
    Coat   £24
    Scarf   £2
    Shoes   £10

What fraction of the total cost did she pay for each item?

2.F18 A tailor offers a customer a choice of three materials when making a skirt:

    Material A   £6 per metre
    Material B   £4.50 per metre
    Material C   £3 per metre

If $2\frac{1}{2}$ m of material are required to make the skirt, what would be the cost of material for each choice?

2.F19 The table shows the number of lunches served in a works' canteen over a working week. Rewrite the table to show what fraction of the total number of lunches for the week is served each day.

    Monday   368
    Tuesday   412
    Wednesday   380
    Thursday   320
    Friday   440

2.F20 A men's outfitter offers a discount of a quarter of the listed price for goods purchased during a spring sale. Rewrite the list giving the sale prices.

	LIST PRICE
Suit	£64
Trousers	£24
Overcoat	£68
Hat	£8
Jeans	£12
Shoes	£16
Shirt	£8.80
Tie	£2.40

34   Spotlight on Numeracy

3 Operations with Decimals

3.1 Decimal Fractions

The normal system of numbers that we use for everyday calculation is called the **denary** system, each digit in the number has a unit of ten times that of its neighbour to the right, e.g. 7 438:

the 8 represents 8×1 = 8
the 3 represents 3×10 = 30
the 4 represents 4×100 = 400
the 7 represents $7 \times 1\,000$ = 7 000
                                    7 438

The **decimal** system extends the number to include values less than 1. The same rule applies so that each digit in a **decimal fraction** has a unit ten times that of its neighbour to the right, e.g. 0.692 5:

the 6 represents $6 \times \frac{1}{10} = \frac{6}{10}$ = 0.6
the 9 represents $9 \times \frac{1}{100} = \frac{9}{100}$ = 0.09
the 2 represents $2 \times \frac{1}{1000} = \frac{2}{1000}$ = 0.002
the 5 represents $5 \times \frac{1}{10000} = \frac{5}{10000}$ = 0.000 5
                                                              0.692 5

Each digit in a decimal number can be given a place value, e.g. 432.786 5:

Place value 100 10 1 $\frac{1}{10}$ $\frac{1}{100}$ $\frac{1}{1000}$ $\frac{1}{10000}$
             4   3   2.7 8   6   5

Example 3.1 Write as decimal fractions (a) $\frac{3}{10}$, (b) $\frac{9}{100}$ and (c) $\frac{7}{1000}$.

(a) 0.3 (*Ans.*)
(b) 0.09 (*Ans.*)
(c) 0.007 (*Ans.*)

Example 3.2 Write as decimal fractions (a) $\frac{7}{10} + \frac{3}{100}$ and (b) $\frac{9}{10} + \frac{7}{100} + \frac{3}{1000} + \frac{1}{10000}$.

(a) $\frac{7}{10}$ = 0.7
    $\frac{3}{100}$ = 0.03
              0.73 (*Ans.*)

(b) $\frac{9}{10}$ = 0.9
    $\frac{7}{100}$ = 0.07
    $\frac{3}{1000}$ = 0.003
    $\frac{1}{10000}$ = 0.000 1
                 0.973 1 (*Ans.*)

Example 3.3 Write as decimal fractions (a) $\frac{37}{100}$ and (b) $\frac{133}{1000}$.

(a) $\frac{37}{100} = 37 \times 0.01 = 0.37$ (*Ans.*)
(b) $\frac{133}{1000} = 133 \times 0.001 = 0.133$ (*Ans.*)

3.2 Addition of Decimals

When adding decimal quantities the numbers must appear in columns with the decimal points occurring directly below each other. In this way all the figures in the same column will have the same place value.

Example 3.4 $4.716 + 2.98 + 3.057$.

    4.716
    2.98
    3.057
   10.753 (*Ans.*)

Example 3.5 Find the overall length L of the workpiece shown in Fig. 3.1.

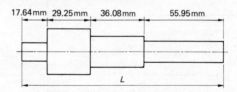

Fig. 3.1

The overall length L mm is given by $(17.64 + 29.25 + 36.08 + 55.95)$ mm

    17.64
    29.25
    36.08
    55.95
   138.92 $L = 138.92$ mm (*Ans.*)

Example 3.6 The power used in kilowatts by four machines in a workshop is shown below.

Power saw   0.73 kW
Drilling machine 0.84 kW
Lathe   2.07 kW
Grinding machine 1.7 kW

Find the total power consumption.

Total power consumption = (0.73 + 0.84 + 2.07 + 1.7) kW

```
 0.73
 0.84
 2.07
 1.7
─────
 5.34    Total power consumption = 5.34 kW
                                        (Ans.)
```

3.3 Subtraction of Decimals

When subtracting decimals the same column arrangement is used, with the decimal points appearing directly below each other.

Example 3.7   Find the value of 27.096 − 9.848.

```
 27.096
−  9.848
───────
 17.248   (Ans.)
```

Example 3.8   Find the dimension L on the shaft shown in Fig. 3.2.

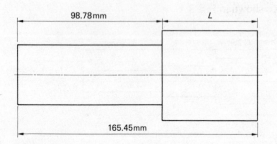

Fig. 3.2

Dimension L is given by (165.45 − 98.78) mm

```
 165.45
−  98.78
───────
  66.67    L = 66.67 mm   (Ans.)
```

Example 3.9   A shaping machine removes a cut of 6.75 mm from a block of metal 53.88 mm thick. What is the final thickness of the block?

Final thickness = (53.88 − 6.75) mm

```
 53.88
−  6.75
──────
 47.13    Final thickness = 47.13 mm   (Ans.)
```

3.4 Multiplication of Decimals

The number of decimal places in a decimal quantity is the number of figures to the right of the decimal point. Hence, 3.182 has 3 decimal places, 1.09 has 2 decimal places and 19.7136 has 4 decimal places. When two decimal quantities are multiplied together the number of decimal places in the answer must be the same as the total number of decimal places in both quantities.

Example 3.10   3.46 × 1.6.

Step (1)   Disregard the decimal points and multiply the two numbers

```
   346
 ×  16
 ─────
  2076
  3460
 ─────
  5536
```

Step (2)   Count the total number of decimal places in both numbers; 3.46 has 2 decimal places and 1.6 has 1 decimal place

Total number of places = 2 + 1 = 3

Step (3)   The answer must now be given the same number of decimal places (i.e. 3), then,

5.536   (Ans.)

Example 3.11   Calculate the floor area of a room 4.73 m long and 2.86 m wide.

Area (m²) = length (m) × width (m)
Area = 4.73 × 2.86

Disregard the decimal points

```
    473
 ×  286
 ──────
   2838
  37840
  94600
 ──────
 135278
```

Count the total decimal places in both numbers

2 + 2 = 4 decimal places

Give the answer the same number of decimal places

13.527 8 m² (*Ans.*)

Example 3.12 Calculate the volume of a block of metal of length 8.5 cm, width 5.3 cm and thickness 1.37 cm.

Volume (cm³) = length (cm) × width (cm) × thickness (cm)

Volume = 8.5 × 5.3 × 1.37 cm³
85 × 53 × 137 = 617185
Total number of decimal places = 1 + 1 + 2 = 4
Volume = 61.718 5 cm³ (*Ans.*)

3.5 Division of Decimals

Division is made easier by converting the decimal quantity of the divisor to a whole number by multiplying by 10, 100, etc. The dividend (the number to be divided) must be multiplied by the same value so that the ratio of the two numbers remains the same.

Example 3.13 Find the value of 39.78 ÷ 2.34.

Step (1) Convert the divisor to a whole number by multiplying by 100

2.34 × 100 = 234

Step (2) Multiply the dividend by the same value

39.78 × 100 = 3 978

Step (3) Find the answer by long division

```
         17
    234)3978
        234
        ---
        1638
        1638
        ----
          —
```

39.78 ÷ 2.34 = 17 (*Ans.*)

Example 3.14 Convert a dimension of 439.42 mm to inches (1 in = 25.4 mm).

Dimension in inches = 439.42 ÷ 25.4
                    = 4 394.2 ÷ 254

```
          17.3
    254)4394.2
        254
        ---
        1854
        1778
        ----
         762
         762
         ---
           —
```

Dimension = 17.3 in (*Ans.*)

3.6 Rounding-off

When carrying out calculations with decimal numbers it is often required to state the answer correct to a given number of decimal places. For instance, in the machining of metals it is pointless to give a calculated dimension as say 15.496 3 mm when the metric micrometer used to measure the work will only measure directly to 0.01 mm. In this case the measurement would be rounded-off for practical purposes to 15.50 mm.

Two rules are commonly used to round-off decimal numbers in order to reduce the number of decimal places:

1. If the first figure after the required number of decimal places is 5 or greater, then add 1 to the previous figure and discard the figure.
2. If the first figure after the required number of decimal places is less than 5 then discard it.

Thus, 2.167 correct to 2 decimal places becomes 2.17 and 2.164 correct to 2 decimal places becomes 2.16.

Example 3.15 Show the value 8.476 3 correct to (*a*) 3 decimal places, (*b*) 2 decimal places and (*c*) 1 decimal place.

(*a*) 8.476 3 is 8.476 correct to 3 d.p. (*Ans.*)
(*b*) 8.476 3 is 8.48 correct to 2 d.p. (*Ans.*)
(*c*) 8.476 3 is 8.5 correct to 1 d.p. (*Ans.*)

Rounding-off may also be done to give a value correct to a number of **significant figures**. Usually the same two rules are used.

Example 3.16 Round-off the following values to 3 significant figures: (*a*) 11.57, (*b*) 2.042, (*c*) 0.076 38, (*d*) 0.092 13, (*e*) 5 039 and (*f*) 107 348.

(*a*) 11.57 is 11.6 to 3 s.f. (*Ans.*)
(*b*) 2.042 is 2.04 to 3 s.f. (*Ans.*)
(*c*) 0.076 38 is 0.076 4 to 3 s.f. (*Ans.*)
(*d*) 0.092 13 is 0.092 1 to 3 s.f. (*Ans.*)
(*e*) 5 039 is 5 040 to 3 s.f. (*Ans.*)
(*f*) 107 348 is 107 000 to 3 s.f. (*Ans.*)

Example 3.17 The table below shows the number of plastic components produced by an injection moulding machine over a period of 8 h continuous operation.

1	2	3	4	5	6	7	8
1 859	1 904	1 876	1 883	1 924	1 876	1 915	1 866

Rewrite the table (*a*) correct to 3 significant figures and (*b*) correct to 2 significant figures.

(*a*)

1	2	3	4
1 860	1 900	1 880	1 880

5	6	7	8
1 920	1 880	1 920	1 870	(*Ans.*)

(b)

1	2	3	4
1 900	1 900	1 900	1 900
5	6	7	8
1 900	1 900	1 900	1 900 (*Ans*.)

This example shows that the production rate that appeared in the first table as an irregular performance shows a steady production rate of about 1 900 per hour when written correct to 2 significant figures. Rounding-off by significant figures is often used in industry to show meaningful trends from large quantities of measured data.

Exercise 3.A Core

3.A1 Add:
(*a*) $1.2 + 3.4$ (*b*) $2.1 + 16.9$ (*c*) $8.5 + 7.6$
(*d*) $12.3 + 1.7 + 2.6$ (*e*) $13.4 + 26.8 + 40.5$
(*f*) $18.2 + 9.7 + 11.3$ (*g*) $26.3 + 17.9 + 12.8$
(*h*) $7.7 + 102.4 + 19.5$ (*i*) $31.8 + 6.3 + 1.8$
(*j*) $3.2 + 7.6 + 9.2 + 6.4$ (*k*) $1.3 + 2.9 + 1.8 + 3.6$
(*l*) $17.9 + 9.4 + 22.6$ (*m*) $41.4 + 63.8 + 80.9$
(*n*) $56.7 + 55.8 + 57.4$

3.A2 Add:
(*a*) $1.25 + 3.69$ (*b*) $19.43 + 7.88$ (*c*) $1.09 + 2.74$
(*d*) $1.34 + 2.06 + 1.44$ (*e*) $5.6 + 1.39$
(*f*) $3.2 + 1.55 + 4.07$ (*g*) $17.3 + 1.642$
(*h*) $2.518 + 1.695 + 3.041$ (*i*) $1.5 + 22.62 + 3.853$
(*j*) $7.619 + 12.8 + 13.07$
(*k*) $1.076\,5 + 2.914\,6 + 4.309\,8$
(*l*) $2.062\,5 + 1.409 + 8.6$
(*m*) $19.383 + 19.209 + 19.514$
(*n*) $33.007 + 106.75 + 2.094$

3.A3 Find the value of:
(*a*) $2.768 - 1.345$ (*b*) $4.092 - 2.719$
(*c*) $17.64 - 13.05$ (*d*) $38.21 - 17.64$
(*e*) $8.642 - 3.804$ (*f*) $2.034\,4 - 1.692\,7$
(*g*) $103.8 - 76.94$ (*h*) $251 - 39.363$
(*i*) $1.89 + 2.62 - 1.97$ (*j*) $18.05 + 21.7 - 32.485$
(*k*) $43.8 + 19.76 - 52.09$ (*l*) $3.6 + 7.3 - 5.4 + 1.62$
(*m*) $109.742 + 83.615 - 151.482$
(*n*) $15.6 + 37.094 + 18.22 - 51.913$

3.A4 Find the overall length *L* in each of the following (all dimensions are in millimetres):

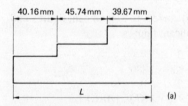

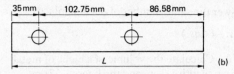

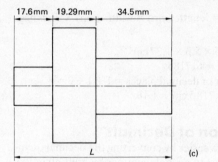

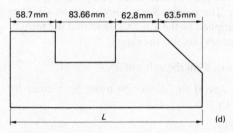

Fig. 3.3

3.A5 Find the dimension *x* in each of the following (all dimensions are in millimetres):

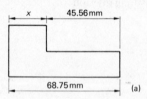

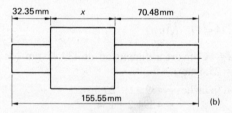

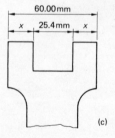
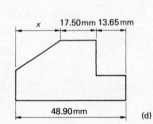

Fig. 3.4

3.A6 Find the overall dimensions x and y of each of the following floor plans (all dimensions are in metres):

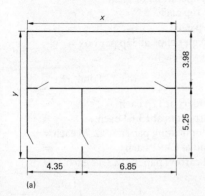
(a)

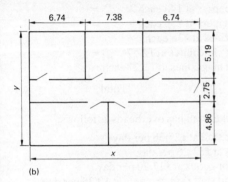
(b)

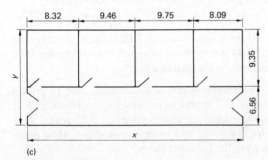
(c)

Fig. 3.5

3.A7 Multiply:
(a) 1.5×6 (b) 2.75×4 (c) 13.68×3
(d) 0.85×11 (e) 3.6×13 (f) 1.98×20
(g) 14.64×12 (h) 31.39×8 (i) 107.612×5
(j) 0.094×14 (k) 2.716×29 (l) 0.085×40
(m) 7.171×33 (n) 19.642×17

3.A8 Multiply:
(a) 1.3×1.2 (b) 0.7×1.5 (c) 2.9×0.6
(d) 10.7×1.4 (e) 3.26×1.1 (f) 1.09×2.4
(g) 3.76×1.8 (h) 6.73×4.25 (i) 9.04×5.07
(j) 1.918×2.74 (k) 5.307×9.61
(l) 48.36×12.25 (m) $1.75 \times 3.2 \times 1.6$
(n) $4.41 \times 2.09 \times 1.33$

3.A9 Divide:
(a) $127.8 \div 9$ (b) $102 \div 12$ (c) $125.6 \div 8$
(d) $134.5 \div 5$ (e) $31.71 \div 7$ (f) $5.55 \div 3$
(g) $347.6 \div 11$ (h) $10.92 \div 13$ (i) $0.222 \div 6$
(j) $39.74 \div 20$ (k) $17.81 \div 100$ (l) $352.05 \div 15$
(m) $162.24 \div 26$ (n) $595.84 \div 32$

3.A10 Divide:
(a) $12.8 \div 0.4$ (b) $14.4 \div 1.2$ (c) $110.7 \div 0.9$
(d) $674.4 \div 1.2$ (e) $51.42 \div 1.4$ (f) $6.4 \div 0.08$
(g) $356.5 \div 15.5$ (h) $669.6 \div 37.2$
(i) $83.111 \div 21.7$ (j) $174.455 \div 1.85$
(k) $7.4256 \div 2.08$ (l) $986.7 \div 50.6$
(m) $4.3757 \div 3.29$ (n) $27.7498 \div 16.42$

3.A11 Given that the area of a rectangle is length × breadth, complete the following table:

LENGTH (m)	BREADTH (m)	AREA (m^2)
13	1.8	
24	0.9	
17	2.6	
1.8	1.4	
4.9	3.2	
13.6	1.5	
9.2	8.7	

3.A12 Given that the volume of a block is length × width × thickness, complete the following table:

LENGTH (cm)	WIDTH (cm)	THICKNESS (cm)	VOLUME (cm^3)
10	8	1.5	
12	1.2	1.1	
13.5	1.4	1.2	
20	3.5	1.4	
40	9.8	0.5	

3.A13 Round-off to 3 decimal places:
(a) 1.0446 (b) 3.7984 (c) 12.6109 (d) 8.0404
(e) 1.6995

3.A14 Round-off:
(a) 28.716 to 2 d.p. (b) 1.744 to 2 d.p.
(c) 5.86 to 1 d.p. (d) 0.94318 to 4 d.p.
(e) 7.649 to 2 d.p. (f) 1.3997 to 3 d.p.
(g) 18.06 to 1 d.p. (h) 11.097 to 2 d.p.

3.A15 Round-off to 3 significant figures:
(a) 3158 (b) 2.076 (c) 12.93 (d) 20576
(e) 0.08425

3.A16 Find the value of each of the following and give the answer correct to 2 decimal places:

(a) 13.8×1.76 (b) $20 \div 6$ (c) 1.4857×3

Operations with Decimals 39

3.A17 Find the value of 28.43 × 1.65 correct to 3 significant figures.

3.A18 Divide 59.028 9 by 7 and give the answer correct to (a) 3 decimal places and (b) 3 significant figures.

3.A19 Round-off:

(a) 2.793 8 to 3 d.p.   (b) 0.051 to 2 d.p.
(c) 8.167 16 to 4 d.p. (d) 39.776 to 4 s.f.
(e) 0.045 47 to 3 s.f.

3.A20 Given that 1 in = 25.4 mm convert the following dimensions to millimetres:

(a) 10 in (b) 1.5 in (c) 9.8 in (d) 2.6 in (e) 0.9 in

Exercise 3.B Business, Administration and Commerce

3.B1 Withdrawals were made from bank balances by a number of customers as shown below. What would be the remaining balance in each case?

(a) £17.64 withdraw £4.36
(b) £21.62 withdraw £8.31
(c) £76.92 withdraw £16.16
(d) £20.42 withdraw £12.65

3.B2 On a particular day the exchange rate between the pound and the American dollar was £1 = $1.20.

(a) Convert to dollars (i) £400, (ii) £1.20 and (iii) £1 000.
(b) Convert to pounds (i) $9.36, (ii) $15.60 and (iii) $50.

3.B3 Details from a bank statement are shown in the table below. Complete the table to show the balance at each date.

DATE	DETAILS	DEBITS (£)	CREDITS (£)	BALANCE (£)
16 Jul	Balance forward			273.02
21 Jul	Credit transfer		103.71	
2 Aug	Cheque D. Bloggs		79.93	
5 Aug	Cheque 123456	312.26		
	Budget Account	60.00		
12 Aug	Employer		431.67	
	Cheque 123457	103.86		

3.B4 A dining-room table and four chairs cost £373.75. The table alone cost £119.15, what is the cost of each chair?

3.B5 A bank cashier earns £722.68 per month. The following deductions are made from his salary: £216.80 income tax, £43.36 superannuation and £68.29 for other stoppages. How much is his take-home pay?

3.B6 Find the total of each of the accounts listed below

(a) 5 reams of paper at £1.62 each   =
    12 pencils at 16p each   =
    6 pens at 29p each   =
    4 boxes of paper clips at 33p per box =
    2 rubbers at 17p each   =
                                  Total £ ___

(b) 24 manilla folders at 17p each   =
    2 pencil sharpeners at £3.63 each   =
    4 reams of duplicating paper at £2.76 each =
    2 cans of fluid at £4.91 each   =
    3 packets of carbon paper at £3.12 each =
                                   Total £ ___

(c) 6 newspapers at 14p each   =
    2 newspapers at 25p each   =
    4 magazines at 43p each   =
    1 box of chocolates at £2.57   =
    2 parish magazines at £1.27 each =
                                Total £ ___

3.B7 A market stall has overheads as follows:
Rent of the stall at £27.36 per day
Staff wages at £37.26 per day
Transport of goods at £17.20 per day
Lighting and staff refreshments at £9.35 per day

What is the total cost per day of running the stall?

3.B8 If the stall in Exercise 3.B7 is open on 4 days per week for 44 weeks in the year, what would be the total cost of running the stall in a year?

3.B9 Seventeen reams of paper were used to print a newsletter. 7.7 reams were used for information, 4.25 reams were used on advertisements, 3.82 reams were used for the cover and special inset pages. How many reams were left for other items?

3.B10 (a) A pile of paper 3.5 cm high contains 490 sheets. How many sheets would be in a pile 1.75 cm high?
(b) A pile of paper 0.4 cm high contains 56 sheets. How many sheets would need to be added to make the pile 1.2 cm high?

3.B11 (a) Repairs to a typewriter over a period of 4 years were: 1st year £12.06, 2nd year £27.32, 3rd year £31.79, 4th year £62.63. What is the total repair bill for the 4-year period?
(b) If the original typewriter cost £319.43, what will be the total cost, including repairs, over the 4 years?

3.B12 (a) A shopper spent £47.36 on food, £26.36 on kitchen needs and £2.36 on other goods. What total amount did she spend?

(b) An office worker enters a café with £5.30 in his wallet. If he spends £1.87 on the meal, how much money has he left?

3.B13 A will includes a provision for £5 763 to be divided equally between four relatives of the deceased. How much does each receive?

3.B14 An insurance broker buys a small computer system at the following cost:

128K desk computer	£1 755.60
Twin disc drive	£438.70
Daisywheel printer	£823.48
Software	£279.44

(a) What was the total cost of the system?
(b) What was the difference in cost between the computer and the printer?

3.B15 A businessman changes £600 to dollars when the exchange rate is £1 = $1.60. He changes the dollars back to pounds when the exchange rate is £1 = $1.50.

(a) How many dollars did he receive for his £600?
(b) How much did he receive when he changed the dollars back to pounds?
(c) How much profit did he make on the transaction?

3.B16 A building society is prepared to loan 0.95 of the cost of a new house and 0.90 of the cost of an older house. How much will the society lend on (a) a new house costing £26 000, (b) an older house costing £18 000 and (c) a new house costing £43 000?

3.B17 The cost of buying goods under a deferred payments scheme is 1.27 times the cost of the goods. What would be the total cost under this scheme of buying goods valued at (a) £650, (b) £470 and (c) £330. Give the answers to the nearest pound.

3.B18 An advertising leaflet measures 8 in by 6 in. Given that 1 in = 2.54 cm, find the size of the leaflet in centimetres.

3.B19 A commercial traveller claims expenses for the use of his car at the rate of 13.4 p per mile. How much expenses can he claim for a week in which he records the following mileages?

Monday	105 miles
Tuesday	98 miles
Wednesday	136 miles
Thursday	49 miles
Friday	73 miles

Give the answer to 2 decimal places of a pound.

3.B20 The table shows the daily attendance at an Office Machinery Exhibition:

Monday	3 094
Tuesday	2 731
Wednesday	4 066
Thursday	2 449
Friday	5 174

Rewrite the table giving the attendance correct to 2 significant figures.

Exercise 3.C Technical Services: Engineering and Construction

3.C1 (a) The following voltage drops are measured across six electrical resistors connected in a series circuit:

17.65   21.11   24.92   4.31   10.37   12.12

What is the total voltage drop?
(b) A dimension of 3.19 in is to be converted to millimetres, give the answer correct to 2 d.p. (1 in = 25.4 mm.)

3.C2 (a) A firm buys 5-litre cans of paint in bulk and pays £217.40 for 500 litres. What is the cost of one 5-litre can?
(b) A roll of wall-paper has an area of 7.8 m². If the pattern size makes it necessary to waste 0.65 of the roll, what area of paper is used?

3.C3 A service engineer bought petrol for his car at three garages, they charged, 46 p per litre, 45 p per litre and 44 p per litre. He bought 16 litres from the first, 13 litres from the second and 21 litres from the third. What was the total amount he spent on petrol?

3.C4 (a) A spring stretches 11.23 cm for every kilogramme suspended from it. If 23 kg is loaded on the spring how far will it stretch?
(b) A builder uses 5.2 m² of plywood in every kitchen on an estate which has 92 houses. How much plywood does he use?

3.C5 Calculate dimensions x and y in Fig. 3.6.

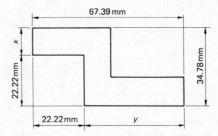

Fig. 3.6

3.C6 (a) A journey from Abbotsville to Swantown is 279.34 miles. 168.3 miles of the journey is on motorways and the rest on dual carriageways. What distance is travelled on dual carriageways?

(b) A metal chain is 15.66 m long. How many pieces 0.33 m long can be cut from the chain and what will be the length of the remaining piece?

3.C7 A length of PVC-covered cable has a mass of 112 kg. The mass of plastic is four times greater than the mass of copper. What mass of the cable is (a) copper and (b) PVC?

3.C8 A milling cutter feeds into the work at 12 mm per revolution. The workpiece is 250 mm long. How many revolutions must the cutter make to machine the full length of the workpiece. (Ignore lead in.)

3.C9 A truck on a journey from Preston to Sunderland covers 167 miles. Five-eighths of the distance is on motorways and the remainder on ordinary roads. What distance was travelled on (a) motorway and (b) ordinary roads?

3.C10 (a) A dimension on a drawing reads 0.625 mm. A template is to be made 25 times full size. What will be the template dimension?
(b) If the pitch (distance between threads) of a screw is 1.5 mm, how many revolutions of a nut will be required to advance it a distance of 35 mm along the screw?

3.C11 Six pieces of wood are sawn from a plank having a length of 2500 mm. If the pieces are 123.5 mm, 234.5 mm, 136.25 mm, 144.4 mm, 246.8 mm and 432.12 mm in length, what will be the length of plank remaining?

3.C12 Use the relationship

$$\text{current (amperes)} = \frac{\text{potential difference (volts)}}{\text{resistance (ohms)}}$$

to complete the following table:

VOLTS	AMPERES	OHMS
12		10
24		7.5
6		12
240		50
18		8
7.5		25
110		40
36		30

3.C13 The fuel tank of a van holds 54 litres and the van travels 9.2 km for each litre of fuel used. If the van started out with a full tank, what amount of fuel will be left after a journey of (a) 208 km, (b) 256 km and (c) 336 km?

3.C14 (a) Pieces of bar are parted-off to the following lengths 0.75 in, 0.5 in, 0.625 in, 0.4375 in and 0.6875 in. What is minimum length of bar required? (Neglect width of parting tool.)
(b) During transit 0.625 litres of paint spilled from a 5-litre container. How much paint was left?
(c) To paint a room 5.5 litres of paint was used, consisting of 1.25 litres gloss, 1.73 litres undercoat and the remainder emulsion paint. How much emulsion paint was used?

3.C15 Find the dimension x in each of the components shown in Fig. 3.7. (All dimensions in millimetres.)

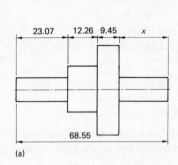
(a)

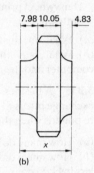
(b)

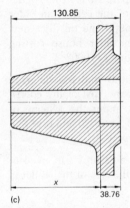
(c)

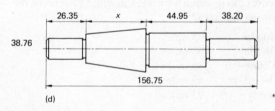
(d)

Fig. 3.7

3.C16 Find the dimensions x and y on the components shown in Fig. 3.8. (All dimensions are in millimetres.)

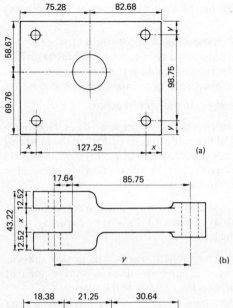

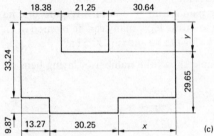

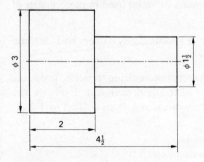

Fig. 3.8

3.C17 (*a*) Find the height of a pile of seven spacing washers each 3.75 mm thick.
(*b*) What would be the height of the pile if three of the washers were removed?
(*c*) The diameter of a circular bar is reduced in one cut in a lathe operation. The original diameter of the bar was 85.56 mm and the depth of cut was 2.37 mm. What is the finished diameter of the bar?

3.C18 Figure 3.9 shows a component dimensioned in inches. Make a sketch of the component and show all the dimensions in millimetres. (1 in = 25.4 mm.)

Fig. 3.9

3.C19 During a sliding operation on a centre lathe (Fig. 3.10) the cutting tool moves a distance of 0.32 mm per revolution of the spindle. If the spindle speed is 289 r/min, how far does the tool move along the work (*a*) in 1 min and (*b*) in 1.8 min?

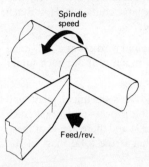

Fig. 3.10

3.C20 Figure 3.11 shows a block of metal of width 170 mm being machined on its upper surface in a shaping operation. If the table feed is 2.5 mm per stroke, calculate the number of strokes required to machine the full width of the surface.

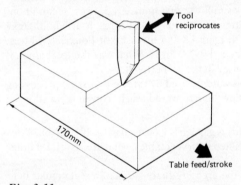

Fig. 3.11

Exercise 3.D General Manufacturing and Processing

3.D1 A forester is able to plant 465 trees on 1.5 ha of land. How many could he plant on (*a*) 5.3 ha and (*b*) 7.72 ha?

3.D2 A milk tanker can carry 27.5 m³ of milk. If, during a day, the Milk Marketing Board processes 2 600 m³ of milk, how many tanker loads would be required to distribute the milk?

3.D3 1 cm³ of a metal has a mass of 2.7 g, what is the mass of a piece of this metal which has a volume of 0.84 cm³?

3.D4 A soap factory produces detergent at the rate of 27.6 m³ each minute. How much is produced in (a) 12 min and (b) 32.5 min?

3.D5 Liquid is poured from bottles containing 0.56 litres, 0.81 litres, 1.37 litres, 1.21 litres, 2.12 litres, 3.65 litres and 2.67 litres into one large container. How much liquid will be in the container?

3.D6 A team of 9 underground workers at a coal-mine produce 391 t of coal, how much coal did each miner produce?

3.D7 Containers are filled at a chemical plant and transported by railway to the processing works. Each container holds a mass of 9.39 t and there are 43 of these containers in the train. What is the total mass of the chemical?

3.D8 A tank containing a fluid is to be pumped dry. Three pumps are to be used, each pumping at the rate of 21.37 litres per minute. The capacity of the tank is 8 000 litres but 1 325 litres of fluid have already been removed. How long will it take the three pumps to empty the tank? (Give the answer in minutes correct to 1 d.p.)

3.D9 (a) The water input into a chemical plant is metered as follows: first hour 964.3 litres, second hour 1 052.2 litres, third hour 1 443.9 litres, fourth hour 866.7 litres, fifth hour 1 530.6 litres, sixth hour 986.4 litres, seventh hour 1 087.9 litres. What was the total amount of water used in the 7 h period?
(b) If this water is used by three equal-sized processing plants, how much would each consume in the 7 h period?

3.D10 (a) A garden is 1.66 ha in area. If 1.106 ha of this area is planted with vegetables, what area is left for other plants?
(b) Find the area in square yards of a rectangular field measuring 83.5 yd by 62.3 yd. (Area of rectangle = length × breadth.)

3.D11 A marathon runner covered 8 miles in 96.32 min, 14 miles in 168.79 min, 4 miles in 69.17 min and the last 2 miles in 49.1 min. What was his time for the complete event?

3.D12 A coal grader deposits 1.736 t in the grade 1 hopper, 2.961 t in the grade 2 hopper, 0.72 t in the grade 3 hopper and 7.321 t in the grade 4 hopper. What is the total mass of coal deposited in all the hoppers?

3.D13 Prices of produce at a local farm shop are listed below:

Potatoes	13.6 p per kilogramme
Cabbages	14 p each
Turnips	23.4 p per kilogramme
Carrots	20.7 p per kilogramme
Onions	33.3 p per kilogramme

What would be the total cost of each of the following orders,
(a) 2 kg of onions, 3 kg of carrots and a cabbage?
(b) 5 kg of potatoes, 2 kg of turnips and 2 cabbages?
(c) 1 kg of turnips, 12 cabbages and 2 kg of carrots?
(d) 2 doz cabbages, 15 kg potatoes and 4 kg onions?
Give the answers to the nearest penny.

3.D14 Bottles of detergent contain 5 litres. How much will be left in the bottle in each case when the following quantities are used: (a) 1.25 litres, (b) 0.625 litres, (c) 3.175 litres and (d) 3.125 litres? (Give the answers in litres correct to 1 d.p.)

3.D15 From his weekly wage a deckhand spends 0.65 of it on Friday, a further 0.27 of it on Saturday. How much had he left if his wage was £140?

3.D16 (a) An allotment was 3.33 ha in size and was increased by the purchase of an additional area of 0.615 ha. What is the new size of the allotment?
(b) A grower plants fruit trees which require 0.35 ha each in which to grow. How many can be planted in an area of (i) 7.2 ha, (ii) 15 ha and (iii) 30.5 ha?

3.D17 The table shows the number of laying hens on six farms:

Lower Halpton	2 075
Green Ridge	1 839
Tithebarn	1 962
Six Oaks	1 551
Crossdale	2 125
Fir View	1 674

(a) Rewrite the table giving the number of hens correct to 2 significant figures.
(b) Use the values in your table to estimate the number of eggs produced at each farm, assuming every 10 hens produce 6 eggs.

3.D18 A manufacturer produces a piece of farming equipment having a total mass of 80 kg. 0.763 of the total mass is metal, 0.148 is wood and the remainder is plastic.
(a) Calculate the mass of plastic used in each piece of equipment.
(b) Calculate the mass of metal used in three pieces of equipment.
(Give the answer in kilogrammes correct to 1 decimal place.)

3.D19 A chemical manufacturing process produces 14.56 gal per hour of a liquid chemical. Estimate the production of the chemical in gallons correct to 1 decimal place in:
(a) a single shift of 4 h;
(b) 3 days each having four shifts;
(c) a working week of 144 h.

3.D20 The power in kilowatts used by the electric motors driving food production machinery in a small factory is shown below:

    4 mixers, each 1.83 kW
    2 spinners, each 0.64 kW
    3 presses, each 1.48 kW
    2 conveyors, each 2.13 kW

(a) Calculate the total power consumption in kilowatts correct to 1 decimal place.
(b) How much additional power is consumed by the mixers over that consumed by the conveyors? (Give the answer correct to 2 significant figures.)

Exercise 3.E Services to People I: Community Services

3.E1 A piece of carpet 2.85 m long is cut from a roll because of damage. The remaining carpet is then cut into two pieces of lengths 12.67 m and 13.44 m. What was the length of the original roll?

3.E2 Of the total number of students attending a college, 0.67 arrive by bus and 0.14 arrive by car or bicycle. The rest walk the whole way. If there are 2 800 students at the college, how many travel by (a) bus, (b) car or bicycle and (c) walk the whole way?

3.E3 (a) After buying some fishing tackle for £27.49 a fisherman was left with £12.63. How much money had he to start with?
(b) A shopper tenders two £20 notes and one £10 note for purchases of a tennis racket at £19.96, a box of tennis balls at £7.63 and a pair of tennis shoes at £15.87. How much change did he receive?

3.E4 (a) To cover the floor spaces of three rooms, 130 m² of carpet were required. The first room required 32.5 m², the second 48.75 m². How much carpet was required for the third room?
(b) The floor space in an office block is 330 m². 0.25 of the area is carpeted, 0.375 is covered in carpet tiles, and the rest lino tiled. What area is lino tiled? (Give the answer in square metres correct to 1 decimal place.)

3.E5 Education takes 0.56 of the total rates of a county, social services takes 0.12, police take 0.1, roads and transport take 0.1, whilst other services take 0.09. The rest is held in reserve for contingencies. How much is held in reserve from a total rate of £643 million?

3.E6 A small nursing home receives the following quarterly electricity bills in a year: £157.49, £76.32, £87.93 and £193.87.
(a) How much is paid for electricity in a full year?
(b) What is the difference between the highest and lowest quarterly bill?

3.E7 (a) A season ticket for a football ground costs £63.50. If the purchaser attends only 38 times, how much has it cost him per attendance?
(b) A mini-bus carries 16 members of a sports club on a touring holiday. Each member contributes £83.60 towards expenses. The total cost of the holiday is £1 130:
    (i) How much money is left over after paying all expenses?
    (ii) If the remaining money is distributed equally amongst the members, how much would each receive?

3.E8 (a) Carpet tiles are priced at £6.37 each, and 79 are required to cover the floor of a hotel lobby. How much does it cost to cover the floor?
(b) A youth club buys nine long-playing records at a total cost of £44.07. What was the cost per record?

3.E9 To win a 12-lap bicycle race, the last lap was covered in one-fifteenth of the total time. If the total time was 13.65 min, what was (a) the time for the last lap and (b) the time for the other 11 laps?

3.E10 The table shows the number of patients attending the casualty ward of a hospital over the first 6 months of a year.

Jan	836
Feb	723
Mar	719
Apr	527
May	418
Jun	404

(a) Rewrite the table showing the attendance figure for each month correct to 2 significant figures.
(b) Use the values in your table to estimate the total number of hours spent in attending to casualty patients, assuming that each patient received 0.78 h of treatment.

3.E11 The table shows the power consumption in kilowatts of electrical equipment in a hairdressing salon.

    6 dryers, each 0.94 kW
    8 blowers, each 0.42 kW
    3 heaters, each 1.55 kW
    2 water heaters, each 2.49 kW

(a) Find the total power consumption if all the equipment is used at once. (Give the answer correct to 1 decimal place.)
(b) How much additional power is consumed by the dryers over that consumed by the blowers?

3.E12 A flask of setting lotion contains 0.48 litre.
(a) How much would remain in a flask after using 0.226 litre?
(b) How much would remain in a flask if 0.8 of the contents had been used?
(c) How much setting lotion would be contained in 1 doz flasks?

3.E13 A medicine bottle contains 0.055 litres and one-third of the contents have been used. Calculate the remaining contents of the bottle correct to 3 decimal places.

3.E14 An ambulance answering an emergency call took 5.7 min to travel the distance of 3.28 miles between the ambulance depot and the patient's home. The time taken to embark the patient in the ambulance was 3.5 min and the journey of 6.37 miles between the patient's home and the hospital took 11.6 min.

(a) For what time was the ambulance being driven?
(b) How many miles did the ambulance cover?
(c) What was the total time taken between the ambulance leaving its depot and arriving at the hospital?

3.E15 The switchboard at a County Hall received 1 400 incoming calls in a week. 0.46 of the calls were to the Housing Department, 0.22 to Education, 0.18 to Social Services and the remainder to other departments. How many calls were to (a) the housing department, (b) education, (c) social services and (d) other departments?

3.E16 The batting average of a cricketer is found from the formula

$$\text{average runs} = \frac{\text{total runs}}{\text{number of finished innings}}$$

Use this formula to find the batting averages, correct to 1 decimal place, of the following players in a youth club team:

PLAYER	TOTAL RUNS	NUMBER OF FINISHED INNINGS
P. Duran	126	4
R. Starr	61	3
R. Stewart	29	5
M. Brown	84	5
I. Diditt	33	4
G. Evans	58	3
T. Last	19	4

3.E17 A public health inspector receives an allowance of 16.82 p per mile when travelling in his own car. Calculate the value of the allowance when he travels (a) 345 miles, (b) 93.5 miles and (c) 178 miles. (Give the answers to the nearest penny.)

3.E18 A heavy lorry consumes 28.5 gal of diesel fuel in making a journey of 327.8 miles. Given that

$$\text{miles per gallon} = \frac{\text{number of miles}}{\text{number of gallons}}$$

calculate the fuel consumption of the lorry in miles per gallon. (Give the answer correct to 1 decimal place.)

3.E19 A coach travels 147.6 miles to a seaside resort. On the return journey the driver makes a detour which adds 26.9 miles to the distance.

(a) What is the total mileage covered?
(b) Given that 1 mile = 1.61 km, calculate the total distance covered in kilometres.

3.E20 The table shows the distance in miles from Manchester to five coach destinations. Rewrite the table giving the distances in kilometres correct to 1 decimal place. (1 mile = 1.61 km.)

	Preston	Birmingham	London	Lincoln	Liverpool
Manchester	33	167	198	88	36

Exercise 3.F Services to People II: Food and Clothing

3.F1 A coat costing £76 is reduced in a sale to £39.50. How much was the reduction?

3.F2 To make a jumper the following materials are required:

12 balls of wool at 39.75 p per ball
2 cards of buttons at 32.5 p per card
1 zip at £1.37
1 pattern at £1.47
2 pairs of needles at 83.5 p per pair

What is the total cost of the materials?

3.F3 Two pairs of curtains are made from a piece of cloth 17.5 m long. The first pair are 4.25 m long and the second 3.125 m long.

(a) How much cloth is left after the curtains have been cut from the original piece?
(b) If the cloth was priced at £1.86 per metre, what would be the cost of the remaining piece?

3.F4 Find the total cost of:

5 tins of peas at $17\frac{1}{2}$ p per tin
3 jars of coffee at £1.92 each
2 jars of honey at 73 p each

3.F5 (a) Each day 32.25 kg is taken from a bag of dog biscuits containing 200 kg. What mass of biscuits remains after 6 days?
(b) A meal in a restaurant costs £16.89 for two people.

(i) What change would be given from £20?
(ii) What is the cost per person?

3.F6 (a) Apples are sold at £8.63 per box. What would be the price of 24 boxes?
(b) How many pots of tea can be made from a packet containing 0.25 kg, if each pot requires 0.003 kg?

3.F7 Hotel charges are £29.75 for bed and breakfast and an additional £7.30 for the evening meal. How much will it cost three persons to stay at the hotel for 14 days if they have (a) bed and breakfast only and (b) bed and breakfast and evening meal?

3.F8 Find the price of each item and total the bill:

4 kg of butter at £2.35 per kilogramme
8 kg of sugar at £0.30 per kilogramme
2.5 kg of flour at £0.23 per kilogramme

3.F9 (a) A roll of printed material 27 m long has a flaw in its pattern over a length of 2.38 m from one end. How much material will be left when the flawed portion is removed?
(b) A housewife buys three loaves of bread at $49\frac{1}{2}$ p each, a packet of sausages at $63\frac{1}{2}$ p, a dozen eggs at 6.25 p each, a container of margarine at £1.37. How much did she spend?
(c) What change would the housewife receive from a £5 note?

3.F10 (a) If six eggs cost 38.5 p, how much does one egg cost?
(b) Grapefruits cost 17.5 p each. How much does it cost for eight?
(c) A litre of cooking oil cost £1.52. How much would 10 litres cost?

3.F11 (a) A box containing 6 doz biscuits costs £2.62. How much does each biscuit cost? (Give the answer correct to the nearest penny.)
(b) Nine young people buy fish and chips at 78p per person. How much is the total bill?
(c) A woman buys 2.2 kg of sugar, 0.75 kg of tea, 1.1 kg of butter and 4.4 kg of flour. What is the total mass of her purchases? (Give the answer in kilogrammes correct to 1 decimal place.)

3.F12 Calculate the total mass in kilogrammes of the ingredients used in the following recipe for Cumberland Sausage:

2.2 kg lean pork
0.675 kg fat pork
0.01 kg powdered sage
0.025 kg white pepper
0.075 kg salt
0.120 kg breadcrumbs

3.F13 The table shows the weekly production of wholemeal loaves at a local bakery.

1st week	1 461
2nd week	1 384
3rd week	1 528
4th week	1 704
5th week	1 273
6th week	1 345

(a) Rewrite the table showing the number of loaves produced each week correct to 2 significant figures.
(b) If the cost of producing one loaf is 18.7 p, use the figures from your table to estimate the total cost of producing wholemeal loaves for the 6-week period.
(1 in = 2.54 cm.)

3.F14 In a boutique a skirt is marked as 34 inch hips. What is the size of the skirt to the nearest centimetre? (1 in = 2.54 cm.)

3.F15 A pre-packed portion of cheese is marked 225 g. Find:

(a) the mass of the cheese in kilogrammes (1 kg = 1 000 g);
(b) the total mass in kilogrammes of 4 doz similar portions.

3.F16 Calculate the total mass in kilogrammes of:

150 g butter
450 g sugar
1.2 kg flour
225 g cheese

(1 kg = 1 000 g.)

3.F17 It costs an airline £2.37 to produce one pre-packed meal. What is the total cost of providing pre-packed meals for an aircraft carrying 218 passengers?

3.F18 Rewrite the following shopping list giving the quantities in pounds weight (use 1 kg = 2.2 lb).

5   kg potatoes
2   kg carrots
1.5 kg swedes
2.4 kg sprouts
0.5 kg tomatoes

3.F19 An ice-cream churn contains 16 litres.

(a) How many whole portions of 0.6 litres can be served from the churn?
(b) Calculate the capacity of the churn in gallons. Use 1 gal = 4.55 litres, and give the answer correct to 2 decimal places.

3.F20 (a) How many lengths of 2.3 m can be cut from a roll of cloth 32.2 m long?
(b) The table shows the width in inches of different coloured ribbon in stock at a haberdashery store. Rewrite the table giving the ribbon widths to the nearest millimetre.

Red	1 in
Blue	0.75 in
Yellow	0.5 in
Green	1.25 in
White	1.5 in
Pink	0.25 in

(1 in = 25.4 mm.)

4 Conversion of Fractions, Decimals and Percentages

4.1 Fractions to Decimals

The fraction $\frac{4}{5}$ can be written as $4 \div 5$ and the answer to this division sum can be given as a decimal fraction

$$5 \overline{)4.0}^{0.8} \qquad \therefore \frac{4}{5} = 0.8$$

Thus to convert a vulgar fraction to a decimal fraction, we simply divide the numerator by the denominator.

Example 4.1 Convert to decimal fractions: (a) $\frac{3}{4}$, (b) $\frac{5}{8}$ and (c) $\frac{13}{20}$.

(a) $\frac{3}{4} = 3 \div 4$.

$$\begin{array}{r} 0.75 \\ 4\overline{)3.00} \\ \underline{2\ 8} \\ 20 \\ \underline{20} \\ - \end{array} \qquad \frac{3}{4} = 0.75 \quad (Ans.)$$

(b) $\frac{5}{8} = 5 \div 8$.

$$\begin{array}{r} 0.625 \\ 8\overline{)5.000} \\ \underline{4\ 8} \\ 20 \\ \underline{16} \\ 40 \\ \underline{40} \\ - \end{array} \qquad \frac{5}{8} = 0.625 \quad (Ans.)$$

(c) $\frac{13}{20} = 13 \div 20$

$$\begin{array}{r} 0.65 \\ 20\overline{)13.00} \\ \underline{12\ 0} \\ 1\ 00 \\ \underline{1\ 00} \\ - \end{array} \qquad \frac{13}{20} = 0.65 \quad (Ans.)$$

Example 4.2 Convert to decimals giving the answers correct to 3 decimal places: (a) $\frac{2}{3}$ and (b) $\frac{5}{13}$.

(a) $\frac{2}{3} = 2 \div 3$

To obtain 3 d.p. in the answer, the division sum must be worked out to 4 d.p. and then rounded-off.

$$\begin{array}{r} 0.6666 \\ 3\overline{)2.0000} \\ \underline{1\ 8} \\ 20 \\ \underline{18} \\ 20 \\ \underline{18} \\ 20 \\ \underline{18} \\ 2 \end{array}$$

$\frac{2}{3} = 0.667$ correct to 3 d.p. (*Ans.*)

(b) $\frac{5}{13} = 5 \div 13$

$$\begin{array}{r} 0.3846 \\ 13\overline{)5.0000} \\ \underline{3\ 9} \\ 1\ 10 \\ \underline{1\ 04} \\ 60 \\ \underline{52} \\ 80 \\ \underline{78} \\ 2 \end{array}$$

$\frac{5}{13} = 0.385$ correct to 3 d.p. (*Ans.*)

4.2 Decimals to Fractions

When converting a decimal fraction to a vulgar fraction we use the place value of the last figure of the decimal as the denominator of the new fraction. The whole of the decimal portion becomes the numerator.

Example 4.3 Convert 0.5168 to a vulgar fraction.

Place value	units	$\frac{1}{10}$	$\frac{1}{100}$	$\frac{1}{1000}$	$\frac{1}{10000}$
	0 .	5	1	6	8

Thus

$0.5 = \frac{5}{10}$
$0.51 = \frac{51}{100}$
$0.516 = \frac{516}{1000}$
$0.5168 = \frac{5168}{10000}$

The fraction can then be cancelled down to give the answer in its lowest terms

$$\frac{5168 \div 8}{10000 \div 8} = \frac{646 \div 2}{1250 \div 2} = \frac{323}{625}$$

$0.5168 = \frac{323}{625}$ (Ans.)

Example 4.4 Convert 3.424 to a mixed number.

The whole number 3 does not change

$0.424 = \frac{424}{1000} = \frac{53}{125}$

$3.424 = 3\frac{53}{125}$ (Ans.)

4.3 Percentage

If the fraction of faulty parts produced in a plastic extrusion process is $\frac{7}{100}$, this means there are 7 faulty parts in each 100 parts. Another term meaning one hundred is a century. Thus the fraction of faulty parts could be stated as 7 per century, which may be abbreviated to 7 per cent.

$\therefore \frac{7}{100} = 7$ per cent, usually written as 7%

All fractions having a denominator of 100 may be stated in this way.

$\frac{3}{100} = 3\%$

$\frac{21}{100} = 21\%$

$\frac{89}{100} = 89\%$

Fractions having other denominators can be replaced by equivalent fractions having a denominator of 100 and then stated as percentages.

$\frac{3}{4} = \frac{3 \times 25}{4 \times 25} = \frac{75}{100} = 75\%$

$\frac{9}{20} = \frac{9 \times 5}{20 \times 5} = \frac{45}{100} = 45\%$

4.4 Fractions to Percentages

Vulgar fractions can be converted to percentages by multiplying by 100.

Example 4.5 Convert $\frac{7}{20}$ to a percentage.

$\frac{7}{20} \times 100 = \frac{700}{20} = 35\%$ (Ans.)

Example 4.6 Convert $\frac{3}{8}$ to a percentage.

$\frac{3}{8} \times 100 = \frac{300}{8} = 37.5\%$ (Ans.)

4.5 Percentages to Fractions

Percentages can be converted to vulgar fractions by dividing by 100.

Example 4.7 Convert 72% to a vulgar fraction.

$\frac{72}{100} = \frac{18}{25}$ (Ans.)

Example 4.8 Convert 33.5% to a vulgar fraction.

$\frac{33.5}{100} = \frac{335}{1000} = \frac{67}{200}$ (Ans.)

Example 4.9 Convert 135% to a mixed number.

$\frac{135}{100} = \frac{27}{20} = 1\frac{7}{20}$ (Ans.)

4.6 Decimals to Percentages

Decimal fractions can be converted to percentages by multiplying by 100.

Example 4.10 Convert 0.37 to a percentage.

$0.37 \times 100 = 37\%$ (Ans.)

Example 4.11 Convert 0.639 to a percentage.

$0.639 \times 100 = 63.9\%$ (Ans.)

4.7 Percentages to Decimals

Percentages can be converted to decimal fractions by dividing by 100.

Conversion of Fractions, Decimals and Percentages

Example 4.12   Convert 57% to a decimal fraction.

$$\frac{57}{100} = 0.57 \quad (Ans.)$$

Example 4.13   Convert 28.6% to a decimal fraction.

$$\frac{28.6}{100} = \frac{286}{1\,000} = 0.286 \quad (Ans.)$$

4.8 Percentage of a Quantity

The percentage of a quantity can be found by multiplying the quantity by the fraction equivalent of the percentage.

Example 4.14   Find 16% of 800.

$$16\% = \frac{16}{100}$$

$$\frac{16}{100} \times 800 = 128 \quad (Ans.)$$

Example 4.15   Find 65% of 80.

$$65\% = \frac{65}{100}$$

$$\frac{65}{100} \times 80 = 52 \quad (Ans.)$$

Example 4.16   Find the value of (a) 23% of 200 m, (b) 87% of 500 kg, (c) $12\frac{1}{2}$% of 320 mm and (d) 30% of £4.50.

(a) $\frac{23}{100} \times 200 = 46\,\text{m} \quad (Ans.)$

(b) $\frac{87}{100} \times 500 = 435\,\text{kg} \quad (Ans.)$

(c) $\frac{12\frac{1}{2}}{100} = \frac{25}{200} = \frac{1}{8}$

$\frac{1}{8} \times 320 = 40\,\text{mm} \quad (Ans.)$

(d) £4.50 = 450p

$\frac{30}{100} \times 450 = 135\text{p}$

$\quad\quad\quad = £1.35 \quad (Ans.)$

Example 4.17   The petrol tank of a family car has a capacity of 50 litres. How much petrol is contained in the tank when it is 45% full?

$$\frac{45}{100} \times 50 = 22.5 \text{ litres} \quad (Ans.)$$

Example 4.18   The total labour force of a department store is 150 people. 40% of the workers are men, how many women are employed in the store?

Percentage of women = $100 - 40 = 60\%$

Number of women = $\frac{60}{100} \times 150 = 90 \quad (Ans.)$

4.9 Expressing a Quantity as a Percentage

One quantity may be found as a percentage of another by showing the relationship of the two quantities, first as a fraction, then converting to a percentage by multiplying by 100.

Example 4.19   Give 25 as a percentage of 125.

$$\frac{25}{125} \times 100 = 20\% \quad (Ans.)$$

Example 4.20   In a batch of 300 loaves, 45 were found to be underweight. What percentage of the batch was underweight?

$$\frac{45}{300} \times 100 = 15\% \quad (Ans.)$$

Example 4.21   Express (a) 15 as a percentage of 500, (b) 24 m as a percentage of 120 m and (c) 14 kg as a percentage of 50 kg.

(a) $\frac{15}{500} \times 100 = 3\% \quad (Ans.)$

(b) $\frac{24}{120} \times 100 = 20\% \quad (Ans.)$

(c) $\frac{14}{50} \times 100 = 28\% \quad (Ans.)$

Exercises 4.A Core

4.A1 Convert to decimal fractions:

(a) $\frac{1}{2}$ (b) $\frac{3}{10}$ (c) $\frac{11}{20}$ (d) $\frac{1}{4}$ (e) $\frac{4}{5}$ (f) $\frac{5}{8}$ (g) $\frac{3}{8}$ (h) $\frac{7}{20}$
(i) $\frac{19}{50}$ (j) $\frac{7}{40}$ (k) $\frac{9}{10}$ (l) $\frac{47}{100}$ (m) $\frac{731}{1000}$ (n) $\frac{21}{50}$ (o) $\frac{39}{100}$
(p) $\frac{17}{40}$

4.A2 Convert to decimal fractions giving the answer correct to 3 decimal places:

(a) $\frac{2}{3}$ (b) $\frac{5}{6}$ (c) $\frac{4}{7}$ (d) $\frac{5}{11}$ (e) $\frac{6}{13}$ (f) $\frac{8}{21}$ (g) $\frac{8}{9}$ (h) $\frac{5}{7}$

4.A3 Convert to decimal numbers:

(a) $1\frac{3}{4}$ (b) $2\frac{2}{5}$ (c) $7\frac{7}{10}$ (d) $8\frac{9}{20}$ (e) $14\frac{3}{5}$ (f) $12\frac{7}{40}$
(g) $6\frac{9}{50}$ (h) $2\frac{31}{100}$

4.A4 Convert to vulgar fractions, give the answer in the lowest terms:

(a) 0.3 (b) 0.25 (c) 0.88 (d) 0.050 (e) 0.125
(f) 0.875 (g) 0.95 (h) 0.64 (i) 0.328 (j) 0.375
(k) 0.488 (l) 0.796

4.A5 Convert to mixed numbers:

(a) 1.5 (b) 1.375 (c) 2.25 (d) 4.35 (e) 2.48
(f) 7.015 (g) 3.364 (h) 8.105

4.A6 Convert to percentages:

(a) 0.3 (b) 0.55 (c) 0.48 (d) 0.75 (e) 0.625
(f) 0.319 (g) 0.806 (h) 0.254

4.A7 Convert to percentages:

(a) $\frac{3}{4}$ (b) $\frac{5}{8}$ (c) $\frac{17}{20}$ (d) $\frac{7}{10}$ (e) $\frac{9}{40}$ (f) $\frac{13}{25}$
(g) $\frac{79}{100}$ (h) $\frac{11}{16}$

4.A8 Convert to fractions (give answer in the lowest terms):

(a) 70% (b) 24% (c) 17% (d) 96% (e) 55%
(f) 12.5% (g) 37.25% (h) 85%

4.A9 Convert to mixed numbers:

(a) 150% (b) 250% (c) 175% (d) 325%
(e) 140% (f) 125% (g) 455% (h) 635%

4.A10 Convert to percentages:

(a) 0.65 (b) 0.28 (c) 0.317 (d) 0.405 (e) 0.123
(f) 0.454 (g) 0.072 (h) 0.005

4.A11 Convert to decimals:

(a) 20% (b) 65% (c) 32% (d) 19.5% (e) 28.7%
(f) 86.4% (g) 93.9% (h) 4.75%

4.A12 Find:

(a) 5% of 200 (b) 11% of 300 (c) 17% of 600
(d) 28% of 1 000 (e) 33% of 500 (f) 50% of 84
(g) 70% of 360 (h) 44% of 900

4.A13 Find:

(a) 25% of 600 m (b) 18% of £400
(c) 87% of 300 cm (d) 60% of 25 g (e) $12\frac{1}{2}$% of 64p
(f) 30% of 180 kg (g) $22\frac{1}{2}$% of 800 g
(h) 35.5% of 120 mm

4.A14 In a factory's labour force of 800 workers there are 48 who hold first-aid certificates. What percentage of the labour force is qualified in first-aid?

4.A15 The capacity of a machine coolant tank is 40 litres. How much coolant is contained when the tank is 30% full?

4.A16 Which is the greatest in each of the following pairs:

(a) 30% of 600 or 10% of 1 750?
(b) 15% of 70 g or 80% of 12 g?
(c) $4\frac{1}{2}$% of £1 500 or 60% of £130?
(d) 11% of 90 kg or 35% of 30 kg?

4.A17 Express:

(a) 36 as a percentage of 60
(b) 120 as a percentage of 720
(c) 10 m as a percentage of 80 m
(d) 45 g as a percentage of 200 g
(e) 19 mm as a percentage of 400 mm
(f) 80p as a percentage of £3.20
(g) £24 as a percentage of £60
(h) 15p as a percentage of £3

4.A18 A batch of 500 machined components is checked for size; 30 are found to be undersize and 20 oversize. Determine the percentage of the batch that are (a) undersize, (b) oversize and (c) correct size.

4.A19 A casting having a mass of 40 kg is produced from an alloy of 85% copper and 15% tin. Determine the amounts of copper and tin in the casting.

4.A20 The employees or a rubber works receive a pay increase of 5%. Determine the new weekly rates of pay for each of the following workers:

(a) press operator   £100 per week
(b) roll operator   £120 per week
(c) plant fitter   £140 per week
(d) fork-lift driver £110 per week

Exercises 4.B Business, Administration and Commerce

4.B1 A small business firm agreed to purchase a photocopying machine costing £786 in the following manner: one-quarter of the purchase price as deposit and the remainder divided into 24 equal monthly payments.

(a) What was the amount of the deposit?
(b) What was the amount of each monthly payment?
(c) If each of the monthly payments were made on time, a rebate of 2% of the deposit was given. What was the amount of this rebate?

4.B2 An assistant manager of a building society branch is paid a salary of £7 636 p.a. He receives a pay increase of $\frac{3}{16}$ of his salary plus £27. What is his new annual salary?

4.B3 The total weekly income of a household is £186. One-sixth is spent on rent, one-fifth is spent on rates and one-quarter is spent on heating, lighting and cooking. If 10% of the total income is saved each week, how much is spent on food, clothing and pleasure? Find:

(a) the amount of rent;
(b) the amount of rates;
(c) the cost of heating, lighting and cooking;
(d) the amount saved each week;
(e) the amount remaining for food and other items.

4.B4 An office supplies firm has 2 000 A4 ring files in stock. One-quarter of these are green, one-fifth are blue and the remainder are black.

(a) How many files are black?
(b) What percentage of the files are green?
(c) Express the number of blue files as a decimal fraction of the total number of files.
(d) If 20% of the green files, 10% of the blue files and 50% of the black files are sold, how many files remain in stock?

4.B5 The list below shows the services used by 1 200 customers of a local branch of a bank.

Current accounts	1 080
Deposit accounts	636
Cheque cards	960
Cash cards	660
Personal loans	180
Mortgages	72

Use this information to complete the following table giving the number of customers using each service as a vulgar fraction, a decimal fraction and a percentage of the total number of customers.

SERVICE	FRACTION	DECIMAL	PERCENTAGE
Current accounts	$\frac{9}{10}$	0.9	90%
Deposit accounts			
Cheque cards			
Cash cards			
Personal loans			
Mortgages			

4.B6 (a) A typist works an 8-h day, she spends $\frac{5}{8}$ of the time on typing, $\frac{3}{16}$ on answering telephone calls, $\frac{3}{32}$ on taking shorthand and the remainder on photocopying and letter stamping. How much time is spent on each part of her work?
(b) Express each of her duties as a percentage of her working day.

4.B7 The following items are purchased for use in an office:

$2\frac{3}{4}$ reams of paper at £1.30 per ream
$2\frac{1}{2}$ boxes of pins at £0.32 per box
$5\frac{1}{2}$ litres of duplicating spirit at £2.15 per litre
15 doz pencils at £0.20 each
24 plastic wallets at £0.75 each

(a) Find the total amount spent.
(b) What percentage of the total cost is for paper?

4.B8 (a) How much interest will be earned in 1 year by a £1 000 building society bond giving an interest rate of $9\frac{1}{2}$% per year?
(b) Express the amount of interest as a fraction of the amount invested.

4.B9 A workman pays 30% income tax on his overtime earnings. If he is paid for overtime at the rate of £3 per hour, how much tax will he pay on his overtime payments in each of the following weeks:

Week 1   6 h overtime
Week 2   8 h overtime
Week 3 $7\frac{1}{2}$ h overtime
Week 4 $10\frac{1}{2}$ h overtime
Week 5   5 h overtime

4.B10 The table shows the breakdown of the total cost of supplies used in an accountant's office as decimal fractions.

Paper	0.107
Computer tapes	0.174
Files	0.087
Typewriter supplies	0.285
Duplicating supplies	0.309
Sundry items	remainder

Rewrite the list giving the cost of each item as a percentage of the total cost.

4.B11 The weekly expenses of a small business office are shown below:

Wages	£350
Rent of building	£85
Rates and water charges	£35
Heating and lighting	£45
Office supplies	£105

State each item as a percentage of the total expenses.

4.B12 (a) A building society pays 8% interest on a 1-year deposit of £1 025. What is the total amount that the depositor will have in his account at the end of 1 year?
(b) A bank account earns 11% interest on £360. How much money is in the account after the interest has been added?
(c) Income tax at the rate of 30% must be paid on the interest earned by the bank account. How much tax must be paid?

4.B13 Five people form a company and invest sums of £1 250, £2 750, £3 200, £3 500 and £4 125. The company returns a 12% bonus to investors, how much would each investor receive?

4.B14 A switchboard handles calls of the following approximate duration: 70 calls of 2 min, 42 calls of 3 min, 17 calls of 4 min and 26 calls of 5 min. What percentage of the total time is taken by calls of less than

4-min duration? (Give the answer correct to 2 decimal places.)

4.B15 A worker earns a basic wage of £100 per week for 52 weeks in a year. He earns overtime payments amounting to £800 in the same year. What percentage of his yearly income is earned by overtime? (Give the answer correct to 1 decimal place.)

4.B16 From the annual salary of £16 000 earned by a business man, 38% is paid in income tax. How much income tax does he pay?

4.B17 Articles sold on a market stall are: 63 at less than £1, 195 between £1 and £3 and 42 at over £3. What percentage of the total number of articles sold fall into each price range?

4.B18 A pensioner receives £40 per week state pension. He spends £19 on rent, £7.50 on heat and light and £8.50 on food. What percentage of the pension is left for other expenses?

4.B19 A law clerk receives a salary of £5 000 p.a. in his first year. At the end of the first year he receives an increase of 8%, and at the end of the second year an increase of 10%. What is his yearly salary at the beginning of (a) the second year and (b) the third year?

4.B20 Calculate the amount of deposit required to pay each of the following items on hire purchase:

		PRICE (£)	DEPOSIT REQUIRED (%)
(a)	Refrigerator	120	5
(b)	Electric cooker	310	15
(c)	Home computer	134	20
(d)	Sports car	6 150	30
(e)	Dining suite	464	12
(f)	Electric typewriter	370	15
(g)	Bicycle	84	12

Exercises 4.C Technical Services: Engineering and Construction

4.C1 The petrol tank of a car when full holds 10 gal of petrol, but a leak in the fuel-line pipe causes $\frac{3}{8}$ of the contents to leak away before the driver finds it.

(a) How much petrol was lost?
(b) What percentage of the full tank remains?

4.C2 (a) A mechanic travelling by car to a breakdown covers 10.37 miles in $\frac{3}{4}$ h. What is his average speed in miles per hour?

$$\text{Average speed} = \frac{\text{distance travelled}}{\text{time taken}}$$

(b) The petrol tank on a small car holds 37.24 litres, and the petrol gauge indicates a $\frac{1}{3}$-full tank.
(i) How much petrol is in the tank?
(ii) What percentage of the tank is filled?

4.C3 Sketch each of the components shown in Fig. 4.1 and give the dimensions in decimals correct to 3 decimal places. (All dimensions are in inches.)

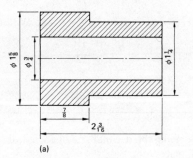
(a)

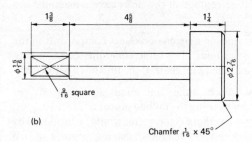
(b)

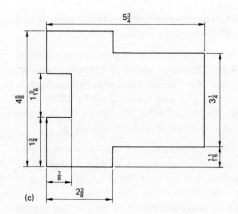
(c)

Fig. 4.1

4.C4 The gear ratio of a gearbox is $\frac{29}{7}$. Express the ratio in decimal form and calculate the speed of the output shaft when the input shaft is rotating at 1 200 r/min. (Give the answer correct to 1 d.p.) (Speed of output shaft = speed of input shaft × gear ratio.)

4.C5 A dimension on a drawing of a component reads 0.92 mm. A template used to produce the component is to be made $2\frac{3}{4}$ times full size. What will be the actual dimension on the template?

4.C6 The table shows the stock of small twist drills in fractional sizes held in a toolroom stores. (All dimensions are in inches.)

DRILL SIZE	NUMBER IN STOCK
$\frac{1}{16}$	14
$\frac{1}{8}$	31
$\frac{3}{16}$	15
$\frac{1}{4}$	41
$\frac{5}{16}$	12
$\frac{3}{8}$	35
$\frac{7}{16}$	19
$\frac{1}{2}$	33

(a) Rewrite the list giving the decimal size of each drill correct to 3 decimal places.
(b) What percentage of the drills have a diameter less than 0.250 in?

4.C7 A shaping machine produces components at the rate of 8 per hour. The same component can be produced on a milling machine at the rate of 12 per hour.

(a) How much time, in minutes, is saved per component by using the milling process?
(b) How many more components would be produced by the milling process than the shaping process in $3\frac{1}{4}$ h machining time?
(c) Express the time taken to mill one component as a percentage of the time taken to shape one component.

4.C8 (a) A mechanical hacksaw cuts through a piece of steel 50.4 mm thick at a speed of 7 mm per minute. How long does it take the saw to cut through the steel?
(b) If the cutting rate was speeded up by 10%, what would be the new speed? How long would the saw take to cut through a steel bar of 92.4 mm thickness?

4.C9 (a) Swarf is removed from an engineering works in wagon loads of 3.6 t per load. If $4\frac{3}{4}$ full loads were removed from a works, how much swarf was taken away?
(b) In a turning operation pins 17.6 cm long are turned down to a smaller diameter for $\frac{5}{8}$ of their length. How long is the turned-down section?

4.C10 4 m^3 of concrete is to be mixed in the following proportions:

Sand 45%
Cement 12%
Aggregate 43%

How much of each material is required? (Give the answer in cubic metres correct to 1 decimal place.)

4.C11 A builder estimates the cost of repairing a brick garage at £270.

(a) What will be his estimate if he adds 20% for profit?
(b) If V.A.T. at 15% is then to be added, what will be the final cost of the repair?

4.C12 A motor mechanic purchases the following car spares:

Bumper overider, list price £9.00
Rear light assembly, list price £6.75
Boot locking handle, list price £8.25

(a) What will be the total cost if he receives 10% trade discount?
(b) How much will he pay if V.A.T. at 15% is then added to the bill?

4.C13 A piece of plywood has an area of 4.75 m^2. When used to panel a wall the plywood is trimmed to size by removing 0.25 m^2. What percentage of the plywood was wasted?

4.C14 A home mechanic buys a used car at an auction for £715. He overhauls the car at a cost of £385. He then sells the car making a profit of 20% of his total expenses. What was the selling price of the car?

4.C15 A car engine requires 4.5 litres of oil in its sump to maintain correct running conditions. After a journey of 600 km the engine required topping-up with 0.9 litres of oil.

(a) What percentage of oil had been consumed on the journey?
(b) How many kilometres could be travelled before the engine had consumed 40% of the contents of its sump?

4.C16 A tension spring stretches a distance of 0.4 cm for every kilogramme of load.

(a) By how much will the spring stretch when loaded with 16.5 kg?
(b) If the spring has an unloaded length of 30 cm, express the stretch of the spring as a percentage of its original length.

4.C17 When checking an electrical circuit, an inaccurate voltmeter gave a reading of 53.7 V, whilst an accurate one gave a reading of 49.3 V. Express the error in the first voltmeter as a percentage of the voltage in the circuit.

4.C18 A finished component having a mass of 64.8 kg was machined from a casting having a mass of 90 kg. Calculate what percentage of the original metal had been removed by machining.

4.C19 (a) A casting having a mass of 150 kg is produced from an alloy of 85% copper, 13% tin and 2% lead. Calculate the amount of each metal in the casting.
(b) A batch of 800 machined components are checked for size. Forty-eight are found to be undersize and 56 oversize. Determine the percentage of the batch that were (i) undersize, (ii) oversize and (iii) correct size.

4.C20 (a) In 300 kg of brass there are 90 kg of zinc and the remainder is copper. What is the percentage of copper in the brass?
(b) A casting consists of 16% tin, 4% zinc and the rest is copper. If the casting has a mass of 120 kg, what will be the mass of the copper?
(c) A rod in a tensile test was extended to 255 mm from an original length of 250 mm. What is the percentage elongation?

Exercises 4.D General Manufacturing and Processing

4.D1 A gas escape pipe is 37.37 m in length and is to be erected in a vertical position with $\frac{3}{4}$ of its length visible above the ground. What length is exposed above ground?

4.D2 An off-shore oil rig is 107.5 km from shore. A helicopter pilot flying to the rig is asked to turn back to land having completed only $\frac{5}{8}$ of the distance. What distance has the helicopter flown when it returns to shore?

4.D3 (a) An oil storage tank holds 630 litres and is $\frac{1}{4}$ full. If it is topped up with 315 litres, what fraction of the tank would remain empty?
(b) A bottle of cough medicine holds 750 ml. If $\frac{5}{8}$ of the medicine is used, how much remains in the bottle?

4.D4 A company manufacturing audio tapes has the following cassettes in stock:

C120 1 800
C90   7 400
C60   8 100
C20   1 280

If the company wishes to increase its present stock levels by 15%, how many tapes of each size must they keep in stock?

4.D5 An electronics firm supplies microcomputers at a wholesale price of £180. The retailer adds his profit of 20% to this cost and then 15% V.A.T. is added. What is the retail price of the microcomputer?

4.D6 A manufacturer makes cardboard containers in three sizes. Size A makes up two-fifths of the total production, size B one-quarter and the remainder are size C.

(a) If the weekly production rate is 3 680 containers, how many containers of each size are made?
(b) What percentage of the total production are size C containers?

4.D7 A furniture manufacturer exports units of furniture to the following markets:

Common Market countries	2 400 units
USA/Canada	1 100 units
Far East	600 units
Middle East	900 units

What percentage of the total exports was supplied to each market?

4.D8 On a summer day a farmer works the 18 h of daylight available. He spends $\frac{1}{8}$ of the day on milking and attending to the dairy cows, $\frac{5}{8}$ of the day in field work, $\frac{3}{16}$ of the day in machinery and equipment maintenance, etc., and the remainder resting and eating, etc. How much time does he spend on each activity?

4.D9 (a) Find the total cost of the following gardening items:

$2\frac{1}{2}$ kg of lawn sand at £1.09 per kilogramme
$3\frac{1}{4}$ litres of weed killer at £2.17 per litre
$2\frac{5}{8}$ kg of peat at £0.73 per kilogramme
$13\frac{7}{16}$ m of fence at £3.63 per metre
$4\frac{1}{16}$ m of plastic netting at £1.32 per metre

(b) If V.A.T. at 15% is added to the cost, what will be the total amount to be paid?

4.D10 A wall-paper manufacturer is to increase the price of some lines by 12%. The present prices are shown below, rewrite the list giving the new prices correct to the nearest penny.

		PRICE PER ROLL (£)
(a)	Luxury Tartan	£5.80
(b)	Floral	£3.50
(c)	Kitchen King	£4.60
(d)	Gold Line	£8.20
(e)	Hi-Living	£5.30
(f)	Summer Scene	£4.40

4.D11 (a) To protect a chemical plant 4% of corrosion inhibitor is added to water used in the chemical process. How much corrosion inhibitor is required to treat 3 650 litres of water?
(b) Plastic parts produced in a compression moulding process have a reject rate of $2\frac{1}{2}$%. How many parts are likely to be rejected in a production batch of 2 800?

4.D12 A toy manufacturer offers a discount of 18% to large retail chains ordering more than 100 items of particular lines. What would be the total cost to a retailer placing the following order:

200 Supertrains, list price £4.50 each
150 Ruby Dolls, list price £8.40 each
150 Talking Bears, list price £10.20 each
200 Turbo Cars, list price £5.70 each?

4.D13 The side of a field is 52 m long and is fenced off with posts and wire. Large posts are situated every $\frac{1}{8}$ of the length and smaller posts every $\frac{1}{3}$ of the length between the large posts. Calculate the distance between (a) the large posts and (b) the small posts.

4.D14 (a) A flower bed in a garden is 5.3 m long, its width is $\frac{3}{8}$ of its length. What is the width of the flower bed in metres?
(b) A farmer has a herd of 438 cattle, of which $\frac{2}{3}$ are dairy cows. How many dairy cows does he have?
(c) A water trough holds 91.37 m^3 of water and is $\frac{7}{8}$ full. How much water is in the trough?

4.D15 (a) A job is rated on piecework at 0.47p per item. There is a bonus of 18% on completion of the work within the time allowed. What will be the earnings of an operator who completes 179 items within the time allowed.
(b) An automatic machine produces pins from brass bar. A fault in the mechanism causes the machine to produce 9.5% scrap. In a batch of 5 600 pins, how many would be scrap?

4.D16 (a) The material costs to decorate a room are £53. The decorator adds on 40% for labour and then adds 15% V.A.T. What is the total cost?
(b) A car tyre cost £17.50 wholesale and is sold by a garage for £24.50 plus V.A.T. What percentage profit does the garage make?

4.D17 (a) The moisture content of a sample of peat is 22%. Calculate the mass of moisture in 350 kg of peat.
(b) A sample of coal contains 73% carbon and 24% oxygen. Find the amount of each element in 240 kg of coal.

4.D18 The cost of raw materials to a small manufacturer increases by 8% each year. If the cost of raw materials in the first year was £39 500 what will be the cost in the second and third year?

4.D19 The employees of a furniture manufacturer receive a pay increase of 6%. Determine the new weekly rates of pay for each of the following workers:

(a) machinist   £110 per week
(b) assembly worker   £108 per week
(c) upholsterer   £120 per week
(d) delivery driver   £112 per week
(e) varnisher   £135 per week
(f) inspector   £140 per week

4.D20 (a) Mass-produced electrical resistors of nominal resistance 50 ohms are produced within a tolerance of ±10%. What are the maximum and minimum values of resistance that would be accepted?
(b) A gauge used for setting rolls at a rubber works has a thickness of $\frac{7}{64}$ in. Give this dimension as a decimal correct to 3 decimal places.

Exercises 4.E Services to People I: Community Services

4.E1 (a) When travelling to work by car I spend 35 min in heavy traffic, the rest of the journey is spent on traffic-free roads. If the whole journey takes 1 h 10 min, what percentage of the total time do I spend on traffic-free roads?
(b) When buying three items of sports gear for £19.96, £7.63 and £15.87, I had to pay 15% V.A.T., what was my total bill?

4.E2 (a) A 12-lap race is run in 13.63 min, with the last lap being run in 0.9 min. What percentage of the total race time was taken by the last lap?
(b) A men's hairdresser attends to an average of three customers each hour, each customer is dealt with in an average time of 18 min. For what percentage of his time is he waiting between customers?

4.E3 A holiday package offered by a travel agency cost £276.50 per person last year. The price for this year is to be increased by 9.2%.

(a) What will be the new cost?
(b) What will be the increase in price?

4.E4 (a) 83% of monies raised by young people from a sponsored walk were given to a charity, and the remainder used to cover expenses. If a total sum of £272 was raised, how much was donated to the charity?
(b) If a girl spends 42% of her savings of £55 on some sports wear, how much does she spend?

4.E5 A population census of a small town shows 41% are men, 36% are women and the rest children. The total population of this town is 76 000. Calculate the number of (a) men, (b) women and (c) children.

4.E6 A hotel charges £31.50 per night per person for bed and breakfast. A further charge of 12% is added for service. What would the bill be for (a) two persons staying 2 nights and (b) three persons for 9 nights?

4.E7 (a) A council estate has 1 155 houses, 42 of these are to be sold to tenants. What percentage of the total houses on the estate will still be rented?
(b) The total floor area of a house is 130 m^2. 0.25 of this area is carpeted and $\frac{3}{8}$ is carpet tiled. The rest is lino tiled. What percentage is (i) carpeted, (ii) carpet tiled and (iii) lino tiled?

4.E8 On an estate 20% of households have newspaper A delivered, 25% have newspaper B, 15% have newspaper C and 12% have newspaper D. If the estate has 135 houses on it, how many households do not have a newspaper delivered?

4.E9 (a) All goods purchased from a store are wrapped in either paper or plastic bags. If $\frac{7}{16}$ of a days sales were wrapped in paper bags, what percentage were wrapped in plastic bags?
(b) When travelling between two towns, 120 miles are travelled on a bus, 70 miles by train and the rest on foot. From a total journey of 210 miles what percentage is travelled (i) by bus, (ii) by train and (iii) on foot?

4.E10 A roll of carpet had a length of 2.875 m cut from its end. If the original length of the roll was 29 m, what percentage was cut off?

4.E11 An allotment of 2.82 ha was divided into plots for local gardeners as follows:

2 plots at $\frac{3}{8}$ of a hectare
3 plots at $\frac{5}{16}$ of a hectare
4 plots at $\frac{1}{8}$ of a hectare

How much of the original allotment was left for compost storage? Give the answer as a decimal fraction of a hectare correct to 2 decimal places.

4.E12 Seven youth-club members enter a sponsored charity walk of 48 km. After $\frac{3}{16}$ of the full distance two members drop out and after $\frac{11}{32}$ of the full distance a further three members drop out. At $\frac{3}{4}$ of the full distance another member retires and only one is able to complete the full course.

(a) What was the total distance walked by the seven entrants?
(b) If each of the youth-club members was sponsored at 12.5p per kilometre, what was the total amount raised for charity?

4.E13 (a) A bus fare from home to work costs £1.17 per day. The bus company is to put up the fare by $\frac{1}{8}$, what will be the new fare?
(b) A central-heating-oil storage tank holds 98.32 gal and its gauge reads $\frac{5}{8}$ full. How much oil is in the tank?

4.E14 (a) A refill bottle for a camping gas stove cost £4.35. On summer holidays a group of Scouts uses $7\frac{1}{4}$ bottles of gas. What was the total cost of the gas?
(b) If the cost was split up between 15 Scouts, how much did each have to pay?

4.E15 (a) A local football team scored 63 goals in one season. The skipper scored $\frac{1}{7}$ of them and the centre-forward scored $\frac{1}{9}$ of them. How many goals were scored by the rest of the team?
(b) On a touring holiday lasting 20 days, 8 days were sunny, 4 days had heavy rain and the rest had mixed weather. What percentage of the days were (i) sunny and (ii) mixed weather?

4.E16 (a) A hospital has 240 beds, of which 210 are occupied. What percentage of the beds are occupied?
(b) At a major ambulance depot 18 out of 24 ambulances are serviceable and the rest are off the road for maintenance or repair.

(i) What fraction of the total number of ambulances are serviceable?
(ii) What percentage of the total number of ambulances are not available for duty?

4.E17 Calculate (i) the increase and (ii) the new cost for each of the following:

(a) a quarterly telephone bill of £48 increased by 6%;
(b) petrol at £1.80 per gallon increased by 9%;
(c) a car tyre at £26.50 increased by 12%;
(d) annual rates of £370 increased by $5\frac{1}{2}$%;
(e) a carpet at £8.70 per square metre increased by 18%.

4.E18 Calculate the amount of V.A.T. at 15% on each of the following retail items:

(a) a hair dryer, £28.50;
(b) barber's scissors, £5.90;
(c) electric clippers, £32.30;
(d) 1 litre of hair conditioner, £2.90;
(e) an electric till, £417.

4.E19 Calculate the area of a playing field $44\frac{1}{2}$ m long and $36\frac{1}{4}$ m wide. Give the answer as a decimal number correct to 2 decimal places. (Area = length × width.)

4.E20 In a morning period of 4 h a health visitor spends $\frac{3}{5}$ of the time attending to patients and the remainder of the time travelling.

(a) How much time does she spend travelling?
(b) What percentage of the total time is spent with patients?
(c) State the time spent in travelling, as a decimal fraction of the total time.

Exercises 4.F Services to People II: Food and Clothing

4.F1 Find the amounts of:

(a) 15% of 250 litres of milk;
(b) 75% of 12 doz eggs;
(c) 20% of 500 loaves;
(d) 9% of 30 kg of sugar;
(e) 16% of 20 kg of flour.

4.F2 (a) From a piece of cloth 4.75 m long, a length of 0.5 m is cut off. What percentage of the original roll is left?
(b) A packet of sweets has a mass of 100 g. If the packet is 6% of the total mass, what is the mass of sweets?

4.F3 92% of a fruit drink is water. In a quantity of 7 litres of the drink, how much is water?

4.F4 A 25-litre container is 89% full.
(a) How many litres will fill it?
(b) What percentage of the container is empty?

4.F5 A café adds 12% service charge to a bill of £4.50. What will be the total bill?

4.F6 (a) An electric oven costs £435.50 plus V.A.T. at 15%, how much is the total cost of the oven?
(b) The ingredients for a cake have a total mass of 3.8 kg. If the amount of mixed fruit is 17% and the amount of flour is 42%, what is the mass in kilogrammes of the remaining ingredients?
(c) A cooker was bought for £435 and sold to make a profit of 9%. What was the selling price?

4.F7 (a) A pair of shoes was bought for £27.85. To make 18% profit, what price must they be sold for?
(b) If V.A.T. at 15% must be added to selling price, what is the total cost to the customer?

4.F8 Copy and complete the following table showing how the selling price of goods is calculated by a retailer. (Give amounts to the nearest penny.)

COST TO RETAILER (£)	RETAILER'S PERCENTAGE PROFIT	PRICE BEFORE V.A.T. (£)	V.A.T. (%)	SELLING PRICE TO CUSTOMER (£)
7.50	12	8.40	12	9.40
12.90	31		15	
16.32	27		8	
27.20	40		15	
36.90	33		12.5	
117.35	19		12.5	

4.F9 (a) A sewing machine has a recommended retail price of £392.85. If it is sold for £302.50, what percentage discount has been given?
(b) A set of pans having a list price of £64 is bought in a sale for £48. What is the saving as a percentage of the list price?

4.F10 A steel tape-measure expanded in length because it had been left in the hot sun. A piece of cloth was cut from a roll using the tape to measure the length of 3 m. When the cloth was measured later it was found to be 1.2 cm short. What was the percentage error in the length of the cloth? (1 m = 100 cm.)

4.F11 The costs in producing a dinner for 50 people were:

Food costs per cover	£1.50
Wages	£52
Other costs	£23

The dinner was priced at £4.80 per head; what was the percentage profit made by the hotel? (Percentage profit $= \dfrac{\text{profit}}{\text{cost}} \times 100$.)

4.F12 A canteen manageress is invoiced for the following purchases:

1 gross eggs at £0.63 per doz
$5\frac{1}{2}$ litres of milk at £0.22 per litre
4.2 kg of meat at £1.03 per 250 g
$3\frac{1}{2}$ kg of flour at £0.78 per kilogramme

The supplier gave a discount of $\frac{3}{20}$ on a trade account and a further discount of 4% for cash payment.
(a) What was the original cost of each invoiced item?
(b) How much discount did the trade account give?
(c) How much was the cash discount?
(d) What was the final price paid?

4.F13 Calculate the amount of V.A.T. at 15% to be paid on each of the following items in a boutique:
(a) a skirt, £4.60;
(b) a top, £6.80;
(c) jeans, £10.20;
(d) a trouser suit, £21.80;
(e) a blouse, £7.40.

4.F14 A chef's working day at a seaside restaurant varies in length according to the season. In the holiday season a working day is 16 h and in the quiet season 8 h. On a typical day he spends $\frac{5}{8}$ of his time on food preparation, $\frac{3}{16}$ on ordering food stocks and the remainder of his time attending to general kitchen work. How much time is spent on each task (a) on a holiday season day and (b) on a quiet season day?

4.F15 Find the cost to a restaurant manager of purchasing the following items from a butcher's shop.
(a) $\frac{1}{2}$ kg of liver at £1.72 per kilogramme;
(b) $1\frac{1}{4}$ kg of ham at £2.40 per kilogramme;
(c) $1\frac{5}{8}$ kg of sausage at £1.60 per kilogramme;
(d) $2\frac{7}{8}$ kg of pork at £3.20 per kilogramme;
(e) $5\frac{1}{4}$ kg of beef at £3.60 per kilogramme;
(f) the total cost of the five items.

4.F16 A large hotel has 400 guests. Three-eighths of the guests take breakfast only, $\frac{2}{5}$ take breakfast plus evening meal and the remainder take full board. What percentage of the guests take full board?

4.F17 A clothing stores chain orders 2 000 tee shirts from a manufacturer. The tee shirts are priced at £2.20 each less 10% discount for the bulk order.

(a) What is the total cost of the order?
(b) If the stores make a profit of 52p on each tee shirt and V.A.T. at 15% is then to be added, what is the selling price of each shirt?
(c) If $\frac{5}{8}$ of the shirts are printed, what percentage are unprinted?

4.F18 (a) A survey in a supermarket showed that only $\frac{1}{8}$ of the customers asked preferred Yellowdrip butter to Brand X. What percentage of the customers preferred Brand X?
(b) If 0.536 of the population of a town are male, what percentage are female?
(c) If $\frac{1}{4}$ of the males and $\frac{1}{2}$ of the females in the town regularly use the bus service, what percentage of the total population are regular bus users if the male and female populations are equal?

4.F19 A works canteen is to provide 165 meals, each having 135 g of lean meat. If $\frac{2}{5}$ of the meat delivered is bone and fat, what amount of meat in kilogrammes must be ordered? (1 kg = 1 000 g.)

4.F20 Find the total amount to be paid for each of the following bills. (Give the answer to the nearest penny.)

(a) One blouse, £8.40
One dressing gown, £11.99
Two shirts, £6.50 each
Plus V.A.T. at 15%
(b) One pair of shoes, £20.40
Two ties, £1.85 each
Six handkerchiefs, 60p each
One scarf, £2.30
Plus V.A.T. at 15%
(c) One pair of gloves, £5.70
One wallet, £7.55
One notepad, 67p
Plus V.A.T. at 10%
(d) One cardigan, £18.40
One pullover, £9.50
Two pairs socks, £1.54 each
One anorak, £17.49
Plus V.A.T. at 16%
(e) One trouser suit, £18.99
One dress, £9.50
Three pairs tights, 98p each
Two headscarves, £2.55 each
Plus V.A.T. at 12%

5 Estimates, Average, Ratio and Proportion

5.1 Estimates

So far in this book we have been concerned with obtaining accurate answers to problems. However, in many practical situations estimated values can be useful.

Estimates are often used to make a rough check of calculations involving decimals to ensure that the decimal point appears in the correct place in the answer.

Example 5.1 Find the value of 0.42×0.57.

To make a rough estimate of the answer we can take the two numbers correct to 1 decimal place giving

$$0.4 \times 0.6 = 0.24$$

The correct answer will have the same order of magnitude as our estimated answer, i.e. the correct answer is in the order of 0.24. The two numbers can now be multiplied together ignoring the decimal points, thus

$$42 \times 57 = 2394$$

The decimal point is then positioned in the same place as in the rough estimate, hence

$$0.42 \times 0.57 = 0.2394 \quad (Ans.)$$

Example 5.2 Find the value of 21.5×252.6.

Rough estimate $20 \times 250 = 5000$. Ignoring the decimal points

$$215 \times 2526 = 543090$$

Positioning the decimal point to make the answer of the same order as the estimate

$$21.5 \times 252.6 = 5430.9 \quad (Ans.)$$

Example 5.3 Find the value of $115.81 \div 31.3$.

Rough estimate $120 \div 30 = 4$.
$11581 \div 313 = 37$
$\therefore 115.81 \div 31.3 = 3.7 \quad (Ans.)$

Example 5.4 Find the value of $\dfrac{31.4 \times 18.9}{6.3}$.

Rough estimate $\dfrac{30 \times 20}{6} = 100$.

$$\frac{314 \times 189}{63} = 943$$

$$\therefore \frac{31.4 \times 18.9}{6.3} = 94.3 \quad (Ans.)$$

Example 5.5 Make rough estimates of the value of each of the following: (a) 296.7×0.028, (b) $7.89 \times 15.04 \times 21.7 \times 0.19$, (c) $31849 \div 7.6$ and (d) $\dfrac{4.83 \times 964}{0.086}$.

(a) 296.7×0.028.

Rough estimate $300 \times 0.03 = 9 \quad (Ans.)$

(b) $7.89 \times 15.04 \times 21.7 \times 0.19$.

Rough estimate $8 \times 15 \times 20 \times 0.2 = 480 \quad (Ans.)$

(c) $31849 \div 7.6$.

Rough estimate $32000 \div 8 = 4000 \quad (Ans.)$

(d) $\dfrac{4.83 \times 964}{0.086}$.

Rough estimate $\dfrac{5 \times 1000}{0.1} = 50000 \quad (Ans.)$

Example 5.6 Make a rough estimate of the value of each of the following:
(a) the area of a rectangle of length 18.95 cm and width 5.13 cm;
(b) the cost of 2053 articles at 7.3p each;
(c) the total area of 385 allotments, each having an area of 0.619 acres;
(d) the total output in 48 weeks of a factory working a 5-day week of $8\frac{1}{4}$ h per day and producing 319 components per hour.

(a) Area = length $\times$ breadth = 18.95×5.13.

Rough estimate $20 \times 5 = 100 \text{ cm}^2 \quad (Ans.)$

(b) Total cost = $2053 \times 7.3\text{p}$.

Rough estimate = $2000 \times 7 = 14000\text{p} = £140 \quad (Ans.)$

(c) Total area = 385×0.619.

Rough estimate = $400 \times 0.6 = 240$ acres $\quad (Ans.)$

60 Spotlight on Numeracy

(d) Total output = $48 \times 5 \times 8\frac{1}{4} \times 319$.

Rough estimate = $50 \times 5 \times 8 \times 300 =$
600 000 components   (Ans.)

5.2 Average

The **average** or **mean** of a set of values is given by

$$\text{average (or mean)} = \frac{\text{sum of the values}}{\text{number of values in the set}}$$

Example 5.7   Find the average of 20, 8, 34, 42, 16.

$$\text{Average} = \frac{20 + 8 + 34 + 42 + 16}{5}$$

$$= \frac{120}{5} = 24 \quad (Ans.)$$

Example 5.8   Find the mean of 34, 20, 4, 17, 49, 11, 12.

$$\text{Mean} = \frac{34 + 20 + 4 + 17 + 49 + 11 + 12}{7}$$

$$= \frac{147}{7} = 21 \quad (Ans.)$$

Example 5.9   The diameter of a machined shaft is measured at five positions along its length and the results are:

mm: 30.42   30.44   30.55   30.48   30.41.

Calculate the mean diameter of the shaft.

Mean diameter
$$= \frac{30.42 + 30.44 + 30.55 + 30.48 + 30.41}{5}$$

$$= \frac{152.30}{5} = 30.46 \, \text{mm} \quad (Ans.)$$

Example 5.10   The number of vehicles serviced by a garage in 4 working days was:

Monday   Tuesday   Wednesday   Thursday
9   11   13   11

Calculate the average number of vehicles serviced per day.

$$\text{Average} = \frac{9 + 11 + 13 + 11}{4}$$

$$= \frac{44}{4} = 11 \quad (Ans.)$$

Example 5.11   The resistance, in ohms, of a batch of 10 electrical resistors is:

15.2   15.1   15.0   15.3   15.1   15.2   15.1   15.4
15.3   15.0

Calculate the average resistance of the batch.

For the convenience listing the variation of each resistor from a nominal value of 15.0 ohms gives:
0.2   0.1   0   0.3   0.1   0.2   0.1   0.4   0.3   0

Average variation
$$= \frac{0.2 + 0.1 + 0 + 0.3 + 0.1 + 0.2 + 0.1 + 0.4 + 0.3 + 0}{10}$$

Average resistance = nominal value + average variation
$$= 15.0 + 0.17$$
$$= 15.17 \, \text{ohms} \quad (Ans.)$$

Example 5.12   The viewing time of each of six video films is:

1 h 45 min   1 h 41 min   2 h 8 min
2 h 2 min   1 h 57 min   2 h 15 min

Calculate the average viewing time.

For convenience convert the viewing times to minutes:

105 min   101 min   128 min
122 min   117 min   135 min

Average viewing time
$$= \frac{105 + 122 + 101 + 117 + 128 + 135}{6}$$

$$= \frac{708}{6}$$

$$= 118 \, \text{min} = 1 \, \text{h} \, 58 \, \text{min} \quad (Ans.)$$

5.3 Ratio

The relationship between two quantities having the same units may be expressed as a **ratio**.

For example, two metal rods, A and B, have the lengths:

A = 2 m,   B = 3 m

Thus the length of A is to the length of B as 2 is to 3, which can be written as the ratio 2:3.

Example 5.13   Find the ratio between the diameters of two pulleys, A and B, where A is 30 cm in diameter and B is 24 cm in diameter.

Estimates, Average, Ratio and Proportion   61

Diameter of A is to diameter of B
as 30 is to 24
30:24
cancelling down 5:4 (Ans.)

Example 5.14 Express as a ratio the masses of two crates, A and B, where A = 150 kg and B = 250 kg.

Mass A is to mass B
as 150 is to 250
150:250
3:5 (Ans.)

If the ratio of the lengths of two rods, A and B, is 2:3, then for every 2 units of length of A, rod B has 3 units. Hence,

length of B = $\dfrac{\text{length of A}}{2} \times 3$

B = $A \times \dfrac{3}{2}$

length of A = $\dfrac{\text{length of B}}{3} \times 2$

A = $B \times \dfrac{2}{3}$

Example 5.15 The length of two rods, A and B, is in the ratio 5:7. If the length of A is 360 mm what is the length of B?

B = $A \times \dfrac{7}{5}$

= $360 \times \dfrac{7}{5}$

= 504 mm (Ans.)

Example 5.16 Two numbers are in the ratio 8:13. If the second number is 390 what is the first number?

First number = $390 \times \dfrac{8}{13}$

= 240 (Ans.)

Example 5.17 The ratio between the time taken to make a journey by train and the time taken to make the same journey by coach is 7:12. If the journey takes 1 h 45 min by train, how long does it take by coach?

1 h 45 min = 105 min

Coach time = train time $\times \dfrac{12}{7}$

= $105 \times \dfrac{12}{7}$

= 180 min = 3 h (Ans.)

Example 5.18 Two metals, A and B, combine in the ratio 2:9 to form an alloy. Calculate the amount of metal B required to combine with 18 kg of metal A to form the alloy.

Mass of B = mass of $A \times \dfrac{9}{2}$

= $18 \times \dfrac{9}{2}$

= 81 kg (Ans.)

Example 5.19 An alloy is made up of three metals, A, B and C, in the ratio 2:3:5. Calculate the amount of each metal in an alloy casting having a mass of 30 kg.

The alloy consists of

2 parts A + 3 parts B + 5 parts C = 10 parts

Thus

A is 2 parts in 10 = $\dfrac{2}{10}$

B is 3 parts in 10 = $\dfrac{3}{10}$

C is 5 parts in 10 = $\dfrac{5}{10}$

Mass of A = $30 \times \dfrac{2}{10}$ = 6 kg

Mass of B = $30 \times \dfrac{3}{10}$ = 9 kg

Mass of C = $30 \times \dfrac{5}{10}$ = 15 kg (Ans.)

Check 6 + 9 + 15 = 30 kg.

Example 5.20 A line of length 1 200 mm is divided into three sections A, B and C, in the ratio 4:5:6. Calculate the length of each section.

4 + 5 + 6 = 15 parts

Length of A = $1\,200 \times \dfrac{4}{15}$ = 320 mm

Length of B = $1\,200 \times \dfrac{5}{15}$ = 400 mm

Length of C = $1\,200 \times \dfrac{6}{15}$ = 480 mm (Ans.)

Check 320 + 400 + 480 = 1 200 mm.

5.4 Proportion

When two quantities increase or decrease at the same rate they are said to be in **direct proportion** to each other.

For example, if the total cost of 4 articles is 20p then the total cost of 8 articles would be 40p. The number of articles and the total cost are in direct proportion and have increased at the same rate. When the number of articles was doubled the total cost doubled.

Example 5.21 If the cost of 6 sausage rolls is 90p, find the cost of 25 sausage rolls.

6 sausage rolls cost 90p

$\therefore$ 1 sausage roll costs $\frac{90}{6} = 15$p

25 sausage rolls cost $15 \times 25 = 375$p
$= £3.75$ (*Ans*)

Alternative method:

25 sausage rolls cost $90 \times \frac{25}{6} = 375$p

$= £3.75$ (*Ans.*)

Example 5.22 If 480 set screws are used in the assembly of 20 gearboxes, how many set screws will be required for 45 gearboxes?

20 gearboxes require 480 set screws

1 gearbox requires $\frac{480}{20} = 24$ set screws

45 gearboxes require $24 \times 45 = 1\,080$ set screws

(*Ans.*)

Alternative method:

45 gearboxes require $480 \times \frac{45}{20} = 1\,080$ set screws

(*Ans.*)

Example 5.23 A car uses 4 gal of petrol to travel a distance of 120 miles. How many gallons will be used to travel a distance of 400 miles?

400 mile journey requires $4 \times \frac{400}{120} = 13.33$ gal

(*Ans.*)

When one quantity increases as the other decreases, the quantities are said to be in **inverse proportion** to each other.

Example 5.24 If 4 men take 16 h to prepare a football pitch, how long will 8 men take?

The answer to this particular problem is obvious. If twice the number of men are used then the job will be done in half the time, i.e. 8 h. As the number of men used increases, the number of hours required decreases. Two methods can be used:

1
Work content in man-hours
= number of men × number of hours
= 4 × 16
= 64 man-hours
$\therefore$ number of hours required

$= \frac{\text{work content in man hours}}{\text{number of men}}$

$= \frac{64}{8} = 8$ h (*Ans.*)

2 Alternative method:

number of hours required $= 16 \times \frac{4}{8}$

$= 8$ h (*Ans.*)

Example 5.25 If 2 agricultural workers can weed a field in 16 h, how many workers are required to do the same job in 8 h?

Number of workers required $= 2 \times \frac{16}{8} = 4$ workers

(*Ans.*)

Example 5.26 If 10 vans can deliver a consignment of newspapers in 5 h, how many vans would be required to complete the deliveries in 2 h?

Number of vans required $= 10 \times \frac{5}{2} = 25$ vans

(*Ans.*)

Exercise 5.A Core

5.A1 Make rough estimates of the value of each of the following:
(a) 2.98×18.75 (b) $304.2 \div 49.6$
(c) $31.45 \times 6.78 \times 11.09$ (d) $38.39 \div 7.86$
(e) 0.095×2.862 (f) $79.4 \times 3.109 \times 0.076$
(g) 13.38×6.894 (h) $17\,398 \times 0.987$

5.A2 Make rough estimates of the value of each of the following:

(a) $\dfrac{0.878 \times 2.143 \times 49.6 \times 19.45}{24.77}$

(b) $\dfrac{0.125 \times 9.64 \times 108.3}{0.984 \times 3.069}$

(c) $\dfrac{43.1 \times 19.6 \times 0.08}{0.029}$

(d) $\dfrac{78.71 \times 10.06 \times 5.32}{23.45}$

5.A3 Give approximate answers to each of the following:

(a) $\frac{2}{9}$ of 173.6 g (b) $\frac{3}{4}$ of £202.73 (c) $\frac{6}{19}$ of 4 000 m
(d) $\frac{8}{21}$ of 2 000 kg (e) $\frac{9}{11}$ of £420 (f) $\frac{7}{8}$ of £55.70

5.A4 State approximate values for each of the following:

(a) the diameter of a 10p coin in centimetres;
(b) the mass of a 1 lb bag of sugar in grammes;
(c) the average height of man in metres;
(d) the length of a family car in metres;
(e) the height of a house in metres;
(f) the length and breadth of this page in centimetres;
(g) the mass of a car in kilogrammes;
(h) the number of litres in a gallon;
(i) the thickness of this book in millimetres;
(j) the mass of this book in grammes.

5.A5 Given that 1 in = 25.4 mm, state the approximate sizes in millimetres of:

(a) an envelope 6 in by 9 in;
(b) a rectangle 15 in by 23 in;
(c) a circle of 19 in diameter;
(d) a plank 2 in × 4 in × 2 ft 6 in;
(e) a window 3 ft 7 in by 5 ft 8 in;
(f) a metal bar 3 in in diameter by 2 ft long;
(g) the total length of 19 pieces of cloth, each 19 in long;
(h) the length of a set screw $1\frac{3}{4}$ in long;
(i) the diameter of a 6-in saucepan;
(j) the length of a windscreen wiper $8\frac{1}{2}$ in long.

5.A6 Given that 1 kg = 2.2 lb, state the approximate mass in kilogrammes of:

(a) 2 lb sugar;
(b) 112 lb potatoes;
(c) 50 lb grass seed;
(d) 1 ton (2 240 lb) of coal;
(e) a 3-lb hammer head;
(f) thirty 2-lb bags of flour.

5.A7 Find the average of:

(a) 4, 10, 17, 8, 21;
(b) 11, 9, 6, 12, 7;
(c) 90, 117, 118, 91, 120, 112;
(d) 22, 74, 10, 41, 63;
(e) 38, 97, 216, 129, 95;
(f) 0.38, 0.77, 0.65, 0.08, 0.14, 0.44;
(g) 15.1, 15.2, 14.8, 14.9, 15.2, 15.2, 15.1, 15.2, 15.4, 15.3;
(h) 10.08, 10.04, 10.09, 10.14, 10.13, 10.00, 10.22, 10.18;
(i) 50.27, 50.11, 50.12, 50.10, 50.18, 50.14, 50.20.

5.A8 Give the mean of:

15.61, 13.85, 14.27, 17.02, 20.11, 16.43

correct to 2 decimal places.

5.A9 The table shows the number of customers in a department store counted at six times during a working day.

10.00	11.00	12.00	13.00	14.00	15.00
621	738	405	393	201	372

Calculate the average number of customers using the store at any time.

5.A10 The test values of a sample of seven electrical resistors are given below in ohms.

50.4 51.1 51.5 51.4 49.6 49.0 50.0

Calculate the average resistance of the sample correct to 1 decimal place.

5.A11 Find the average of:

(a) £5.40, £7.85, £13.20, £6.90, £5.75;
(b) 13 m, 76 m, 29 m, 66 m;
(c) 108 kg, 206 kg, 331 kg, 78 kg, 32 kg;
(d) 652 mm, 397 mm, 127 mm.

5.A12 Express each of the following pairs of quantities as a ratio in the lowest terms:

(a) 10 30;
(b) 18 72;
(c) 125 625;
(d) 30 kg 16 kg;
(e) 35p £1.50;
(f) 9 mm 162 mm;
(g) 1.5 m 3.5 m;
(h) $\frac{1}{2}$ g 4 g.

5.A13 (a) Express as a ratio the number of men and women employed at a small factory

48 men 72 women

(b) If the number of men employed is increased by 12, how many more women should be employed to maintain the same ratio.

5.A14 (a) The length of two metal bars, A and B, is in the ratio 7:9. If the length of bar A is 420 mm, what is the length of bar B?
(b) If the length of bar B is reduced by 180 mm what will be the new ratio of lengths?

5.A15 The ratio between the time taken to brush paint an office fitting and the time taken to spray paint it is 5:2.

If the time taken for brush painting is 1 h 15 min, how long would be taken by spray painting?

5.A16 Two metals, A and B, combine in the ratio 2:13 to form an alloy. Calculate:
(a) the amount of metal B required to combine with 8 kg of metal A;
(b) the amount of metal A required to combine with 39 kg of metal B.

5.A17 (a) A compound lubricating oil is a mixture of three oils, A, B and C, in the ratio 2:5:8. Calculate the amount of each oil required to produce 75 litres of the compound oil.
(b) Divide a line of length 880 mm into three sections in the ratio 5:8:7.

5.A18 Find the cost of:
(a) 8 cakes if 6 cakes cost 42p;
(b) 15 oranges if 4 oranges cost 60p;
(c) 300 wood screws if 8 wood screws cost 12p;
(d) 2 doz pens if 3 pens cost 21p;
(e) 3 cans of paint if 8 cans cost £12.16;
(f) 9 gal of petrol if 4 gal cost £7.32;
(g) 1 gross of pencils if 1 doz pencils cost £2.75;
(h) 225 files if 45 files cost £26.55.

5.A19 (a) If 6 men can do a job in 22 h, how long will it take 11 men to do the same job?
(b) A car uses 30 litres of petrol to travel a distance of 240 km. How many litres will be used on a journey of 560 km?

5.A20 A stationery firm produces three grades of duplicating paper: A, B and C. The cost of the different grades is in the ratio 2:3:5. Use this information to complete the following table.

	COST PER REAM (£)		
Paper	A	B	C
White	1.50		
Blue	1.64		
Green	1.32		
Luxury finish	2.10		
Document grey			3.99

Exercise 5.B Business, Administrating and Commerce

5.B1 A typing pool works a 5-day week for 49 weeks in a year. In one day 97 documents are typed. Estimate the number of documents typed in a full year.

5.B2 The table shows the number of working hours lost through sickness each month at a factory.

Jan	Feb	Mar	Apr	May	Jun
1039	897	764	705	729	608

Jul	Aug	Sep	Oct	Nov	Dec
483	419	563	744	892	907

Estimate the total number of working hours lost in the full year.

5.B3 Estimate the cost of:
(a) 1513 pencils at $9\frac{1}{2}$p each;
(b) 294 reams of typing paper at £2.07 per ream;
(c) 189 files at £1.95 each;
(d) 63 large envelopes at $15\frac{1}{2}$p each;
(e) 4839 small envelopes at 38p per dozen;
(f) 1944 postage stamps at $12\frac{1}{2}$p each;
(g) $9\frac{1}{2}$ gross of elastic bands at $3\frac{1}{2}$p per dozen;
(h) 2864 erasers at £2.16 per box of 50.

5.B4 If a photocopying process costs $5\frac{1}{2}$p per copy, estimate the cost of:
(a) 719 copies   (b) 2031 copies   (c) 876 copies
(d) 194 copies   (e) 3121 copies

5.B5 The table shows the number of telephone calls received at a switchboard in a 6-hour period.

9.00–10.00	10.00–11.00	11.00–12.00
44	57	51

12.00–13.00	13.00–14.00	14.00–15.00
24	37	45

Calculate the average number of calls received per hour.

5.B6 The costs of three different makes of word processors are:

£1014.58   £964.13   £1428.72

What is the average cost?

5.B7 The ages of eight typists in a general office are:
21 years 5 months   32 years 7 months
18 years 6 months   20 years 11 months
42 years 3 months   19 years 10 months
24 years 4 months   35 years 6 months

What is the average age?

5.B8 The contents of seven boxes of paper clips are found to be:

248   253   232   261   247   255   249

What is the average number of paper clips in a box?

5.B9 Telephone calls received by a two-line switchboard were registered as:

9.00–10.00	10.00–11.00	11.00–12.00	12.00–13.00
57	36	56	27

13.00–14.00	14.00–15.00	15.00–16.00
33	45	42

Estimates, Average, Ratio and Proportion

The average length of a call was 2 minutes.

(a) What was the average number of calls each hour?
(b) What time was spent over the 7-hour period?

5.B10 A clerical officer on flexi-time worked the following times in one week:

Monday	6 h 30 min
Tuesday	7 h 45 min
Wednesday	5 h 20 min
Thursday	9 h 10 min
Friday	8 h 40 min

(a) How long was his average working day?
(b) On which days did he work less than his average time?

5.B11 A bank account showed the following balances at the end of each month over a 6-month period:

Jan	Feb	Mar	Apr	May	Jun	Jul
£112.56	£89.25	£14.83	£72.20	£156.92	£104.32	£89.78

What was the average amount in balance at the end of a month?

5.B12 (a) A secretary takes shorthand for $\frac{1}{2}$ h and 3 600 words are dictated during that time. What is the average dictation speed in words per minute?
(b) An estate agent's clerk spends 2 h 45 min answering 55 telephone calls during a working day. What was the average length of a call?
(c) A duplicating machine produces 500 copies in 2 min 5 seconds. What is the average time taken per copy?

5.B13 Income tax is deducted at the rate of 30p for every pound earned over £37 per week. A worker receives the following wages over a period of time:

week 1	£140	week 2	£156	week 3	£161
week 4	£136	week 5	£98	week 6	£131
week 7	£159	week 8	£101		

What was the average weekly tax paid?

5.B14 A collection was taken in an office for the manager's leaving present. Six people gave £1.50, eight gave £1.00 each, four gave 75p each and two gave 50p each. What was the amount of the average contribution?

5.B15 (a) The ratio of incoming calls to outgoing calls at a bank switchboard in a 5-day working week was 4:11. If the number of incoming calls was 128, how many outgoing calls were made?
(b) What was the average number of calls of both types per day?

5.B16 (a) A local bank has 900 customers holding current accounts and 150 customers holding deposit accounts. What is the ratio of current account to deposit accounts?

(b) If 24 more customers opened deposit accounts how many more current accounts must be opened to maintain the same ratio?

5.B17 A retired government officer invests a lump sum of £6 000 in unit trusts, government stock and saving certificates in the ratio 3:2:5. How much does he invest in each?

5.B18 (a) A contract for a £240 000 building project was awarded jointly to three firms, A, B and C, in the ratio 3:4:5. What was the value of the contract awarded to each firm?
(b) A manufacturing company spends £300 000 on building alterations, £150 000 on new machinery and £50 000 on research. Express these costs as a ratio.

5.B19 (a) An office worker's day of $7\frac{1}{2}$ h is spent on typing and duplicating in the ratio 4:5. How much time was spent on each?
(b) An office equipment supplier sells manual and electric typewriters in the ratio 2:7. If he sells 21 electric typewriters in a month, how many manual typewriters does he sell?

5.B20 (a) Armchairs in an office are made from wood and plastic in the ratio 7:3. If the mass of one chair is 50 kg what is the mass of each material.
(b) A van moving office furniture from one office site to another is loaded with 4.5 t of furniture. The total mass of the loaded van is 22.5 t. What is the ratio between the masses of the unloaded van and the furniture?
(c) The ratio of skilled typists to junior staff is 9:2. If the office has 33 staff in it, how many staff are juniors?

Exercise 5.C Technical Services: Engineering and Construction

5.C1 The table shows the number of set screws of different sizes used in the assembly of one conveyor mechanism. Estimate the total number of screws of all sizes required to assemble 96 conveyor mechanisms.

DIAMETER (MM)	NO. OF SCREWS
5	83
8	47
10	68
12	29

5.C2 Estimate the area in square metres of each of the floor plans shown in Fig. 5.1. (Area = length × breadth.)

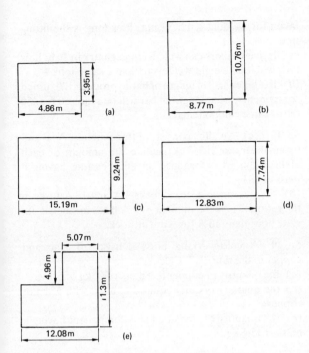

Fig. 5.1

5.C3 Estimate the cost of tiling for each of the floor plans shown in Fig. 5.1 at a cost of £6.15 per square metre.

5.C4 The time taken by different machining operations on a single component is:

Turning	14.8 min
Shaping	27.2 min
Milling	19.7 min
Grinding	12.2 min
Drilling	2.8 min

Estimate the total machining time for 103 components.

5.C5 Estimate the total labour cost of:
(a) 8 fitters working $4\tfrac{3}{4}$ h each at £2.41 per hour;
(b) 3 motor mechanics working $18\tfrac{1}{2}$ h each at £2.17 per hour;
(c) 18 process workers working $21\tfrac{3}{4}$ h each at £1.93 per hour.

5.C6 Estimate the length of bar required to produce each of the following batches of turned components:
(a) 28 components, each 37 mm long;
(b) 17 components, each 43 mm long;
(c) 48 components, each 17 mm long;
(d) 82 components, each 9.5 mm long;
(e) 21 components, each 57.5 mm long;
(f) 63 components, each 83 mm long;
(g) 9 components, each 146 mm long;
(h) 108 components, each 7.5 mm long.

5.C7 The diameter of a turned bar is measured at five positions along its length and the results are:

65.53 65.50 65.48 65.46 65.51

Calculate the mean diameter of the bar.

5.C8 (a) In a load test on a large electric motor the following speeds were recorded:

650 r/min 750 r/min 900 r/min 820 r/min

What was the average speed of the motor?
(b) An engineering inspector measured the length of five components as:

21.12 mm 23.32 mm 22.22 mm 21.33 mm
22.21 mm

What was the average length?

5.C9 (a) Four pieces of timber are cut to the lengths of:

86 mm 49 mm 110 mm 142 mm

What is the average length of the pieces?
(b) Four hundred felt nails are required to hold down roofing felt on a garage roof. A joiner hammers in all the nails in 47 min. What would be the average time taken per nail in seconds?

5.C10 (a) A tank containing 150 gal of water is emptied (through an outlet) in 55 min. What is the average flow of water in gallons per minute?
(b) A production worker produces 342 components in an 8-hour day. What is the average time taken per component in minutes?
(c) Copper pipes were cut off by a plumber's mate in lengths of:

116 mm 127 mm 133 mm 147 mm 262 mm
312 mm

What was the average length of pipe?

5.C11 The table shows the number of bags of cement sold at a builder's yard over a period of 10 weeks. Calculate the average daily sales for each of the ten weeks. (Give the answers to 1 decimal place.)

WEEK	MON	TUE	WED	THUR	FRI	SAT
(a)	26	27	30	163	421	210
(b)	28	35	97	175	270	113
(c)	34	33	107	210	327	92
(d)	42	34	167	193	302	115
(e)	103	33	142	93	260	187
(f)	67	61	102	180	292	281
(g)	71	107	63	190	315	210
(h)	78	93	111	210	256	191
(i)	82	163	171	82	302	126
(j)	90	96	66	166	217	132

Estimates, Average, Ratio and Proportion

5.C12 The production of machined components of all types on 4 working days of a small factory is:

	Mon	Tue	Wed	Thur
Number of components	396	402	104	528

Calculate the average number of machined components produced per day.

5.C13 The test values of the resistance in ohms of a sample of seven electrical resistors are:

51.2 51.7 52.2 52.2 50.2 49.7 50.7

Calculate the average resistance of the sample correct to 1 decimal place.

5.C14 (a) The results of a tool test on eight cutting tools are:

Life in minutes before breakdown
34 29 27 21 23 28 32

Calculate the average tool life.
(b) The times taken by a motor mechanic to carry out a routine service on each of five vehicles are:

3 h 15 min   2 h 45 min   2 h 50 min   4 h 10 min
3 h 30 min

Determine the average time taken to service a vehicle.

5.C15 (a) Find the ratio of the masses of two castings, A and B, where A has a mass of 81 kg and B a mass of 45 kg.
(b) Express as a ratio in its simplest form (i) 80 mm and 7.2 mm, (ii) 700 g and 400 g and (iii) 4 m, 10 m and 14 m.

5.C16 The production of a component requires the following machining times:

Turning   45 min
Milling   20 min
Drilling 15 min

Express as a ratio:
(a) turning time to drilling time;
(b) drilling time to milling time;
(c) milling time to turning time.

5.C17 A coolant mixture is made up of soluble oil and water in the ratio 1:40. Determine:
(a) the amount of soluble oil that must be added to 60 litres of water;
(b) the amount of water required to produce 164 litres of coolant.

5.C18 (a) Concrete for a machine foundation is mixed from 2 parts of cement, 5 parts of sand and 7 parts of stone.
 (i) Express the mixture as a ratio.
 (ii) Find the quantities of each material required to mix 280 kg of cement.
(b) The ratio between the time taken welding and fitting in the assembly of a conveyor is 3:2. If the time taken for welding is 1 h 30 min, how long is the fitting time?
(c) If five welders can weld a crane gantry in 16 h, how long will it take eight welders to do the same job?
(d) If it takes an 8 m length of bar to produce 120 turned components, what length of bar will be required for 450 of the same components?

5.C19 (a) An alloy consists of three metals, A, B and C, in the ratio 2:3:7. Calculate the amount of each metal required to produce an alloy casting having a mass of 48 kg.
(b) An electrician's store stocks 3 A and 13 A fuses in the ratio 2:11. If the number of 3 A fuses in stock is 48, state how many 13 A fuses are in stock.

5.C20 A cold-working brass consists of zinc and copper in the ratio 3:7. Calculate:
(a) the quantity of copper in a mass of 60 kg of brass;
(b) the quantity of zinc required to alloy with 35 kg of copper;
(c) the quantity of zinc in a brass component having a mass of 15.5 kg.

5.C21 A casting alloy is made from copper, tin and zinc in the ratio 45:5:1. What is the mass of each metal used to make a casting having a mass of 153 kg?

5.C22 The ratio between the masses of an iron casting and its pattern is 58:5. How much iron would need to be poured into the casting mould if the pattern had a mass of 4.5 kg?

5.C23 If 20 m of brass wire has a mass of 0.6 kg, what is the mass of the following lengths of the same wire:
(a) 50 m, (b) 70 m and (c) 90 m?

5.C24 The ratio between the revolutions of a propellor shaft and the back axle of a car is 3.5:1. How many revolutions of the propellor shaft will be needed to drive the back axle through 40 revolutions?

5.C25 (a) An engineering assembly is manufactured from ferrous metal, non-ferrous metal and plastic in the ratio of masses 12:5:1. Determine the quantity of each type of material if the total mass of the assembly is 144 kg.
(b) If an additional steel component having a mass of 4 kg is added to the assembly, what would be the new ratio of masses?

Exercise 5.D   General Manufacturing and Processing

5.D1 The table shows the number of employees in different types of occupations at the six factories of a large manufacturer.

FACTORY LOCATION	NUMBER OF EMPLOYEES			
	Manual	Technical	Office	Sales and Management
London	1 306	189	43	29
Birmingham	894	207	38	67
Manchester	1 017	311	52	41
Bristol	732	108	24	85
Glasgow	1 121	309	29	58
Cardiff	963	174	48	69

Give a rough estimate of:
(a) the total number of workers at each factory;
(b) the total number of workers in each type of occupation;
(c) the total number of workers employed by the manufacturer.

5.D2 Use the following average weekly salaries to estimate the weekly wage bill at each of the factories in Exercise 5.D1.

Manual	£109.52
Technical	£149.75
Office	£98.84
Sales and management	£194.68

5.D3 (a) A plastic injection machine produces golf tees at the rate of 32 per minute. Estimate the total output of nine of these machines in a 5-day working week of $8\frac{1}{4}$ h per day.
(b) The total cost of producing 711 plastic articles is £59.72, estimate the cost of producing 50 articles.

5.D4 Estimate the total output of each of the following processes:

	NUMBER OF DAYS	HOURS PER DAY	OUTPUT PER HOUR
(a)	3	$7\frac{1}{4}$	86
(b)	$5\frac{1}{2}$	$8\frac{1}{4}$	217
(c)	13	$7\frac{3}{4}$	97
(d)	28	$6\frac{1}{2}$	183
(e)	51	$9\frac{3}{4}$	378

5.D5 Estimate the area in square metres of each of the fields shown in Fig. 5.2. (Area = length × breadth.)

5.D6 Use the table given in Exercise 5.D1 to calculate to the nearest whole number the average number of workers per factory in each of the following occupations: (a) Manual, (b) Technical, (c) Office and (d) Management.

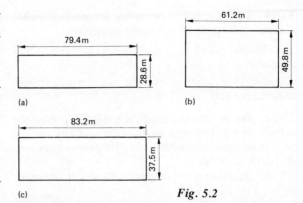

Fig. 5.2

5.D7 The table shows the daily output of a brass foundry in a working week.

	NUMBER OF CASTINGS
Monday	105
Tuesday	87
Wednesday	98
Thursday	112
Friday	73

Calculate the average number of castings produced per day.

5.D8 Measurements are taken at hourly intervals of the thickness in millimetres of rubber sheeting being produced at a rubber works. Calculate the mean thickness of the sheet.

Time	8.30	9.30	10.30	11.30	12.30
Thickness	4.63	4.85	4.82	4.51	4.49

Time	13.30	14.30	15.30	16.30	17.30
Thickness	4.67	4.73	4.78	4.80	4.69

5.D9 The number of working days lost due to sickness at a large factory was recorded for the first 6 months of a year and was:

Jan	Feb	Mar	Apr	May	Jun
738	621	405	372	393	201

Calculate the average number of absences due to sickness per month.

5.D10 Measurements of the temperatures of a processing oven were recorded at half-hourly intervals over a period of 4 h.

Time	13.10	13.40	14.10	14.40	15.10
°C	217	231	220	215	208

Time	15.40	16.10	16.40	17.10
°C	229	230	228	220

Calculate the mean temperature of the oven over that period.

Estimates, Average, Ratio and Proportion

5.D11 The time taken per week for routine maintenance of five items of a chemical plant are:

2 h 10 min   2 h 50 min   3 h 30 min   2 h 40 min
3 h 10 min

Calculate the average weekly maintenance time required per item.

5.D12 Two powder chemicals, A and B, are combined in the ratio 8:13 to form a plastic material. Calculate:
(a) the amount of A and B required to produce 84 kg of the plastic material;
(b) the amount of A required to combine with 130 kg of B;
(c) the amount of B required to combine with 160 kg of A.

5.D13 Men and women are employed at an oil refinery in the ratio 9:4.
(a) Calculate the number of men and women if the total workforce is 260.
(b) If 36 additional men are engaged, how many more women must be employed to maintain the same ratio?

5.D14 A model is made of a new fishing trawler with dimensions in the ratio 2:25.
(a) Determine:
 (i) the length of the model if the length of the trawler is 50 m;
 (ii) the beam of the trawler if the beam of the model is 0.8 m.
(b) Express the length and beam of the trawler as a ratio.

5.D15 The displacement of three fishing trawlers is given as:

Lilac Lady	360 t
Northwater	480 t
Salty Sal	300 t

Express the displacements as a ratio in the lowest terms.

5.D16 A replacement part is required for coal-cutting equipment. The time taken for dismantling, producing the part and re-assembly is in the ratio 3:4:2. Find the time taken by each operation if the whole job takes 6 h 27 min.

5.D17 (a) A farmer has 84 sheep and 60 cattle. What is the ratio of sheep to cattle?
(b) If the farmer bought 15 more cattle, how many more sheep must he buy to maintain the same ratio?

5.D18 (a) If 4 men can shear a flock of sheep in 3 days, how long will it take 6 men?
(b) Six farm workers together earn £432 in 4 days, how much can 15 workers earn in the same time?

5.D19 (a) The gardeners and packers at a grower's nursery are paid in the ratio 7:5. If a gardener's weekly wage is £104.51, how much does a packer earn?

(b) The earnings of a market gardener from the sale of fruit, plants and vegetables are in the ratio 3:1:7. If his total earnings for the year are £8 261, how much does he earn from each type of produce?

5.D20 (a) A manufactured product is made from wood, metal and plastic in the ratio 7:2:3. Determine the amount of each material if the total mass of the product is 36 kg.
(b) A fertiliser consists of three chemicals in the ratio 2:15:8. Determine the amount of each chemical required to manufacture 1 tonne of fertiliser. (1 tonne = 1 000 kg.)

5.D21 Two pipes deliver fluids to a mixing chamber in the ratio 15:2, the larger pipe supplies 3 000 litres per hour, what is the rate of delivery from the smaller pipe?

5.D22 (a) Two men on piecework at a production plant work at different rates. The first man does as much work in 4 h as the second man does in 5 h. If, between them, they earn £360 for a week's work, how much should each man receive?
(b) Chemicals are mixed in the following masses: 2.5 kg of A, 1.25 kg of B, 0.25 kg of C. What is the ratio of the chemicals in the mixture?

5.D23 (a) Three partners buy shares in a fishing boat for £1 500, £2 500 and £3 500. The profit at the end of a seasons fishing is £9 000. If the profit is divided between the men in the same ratio as their shares, how much would each man receive?
(b) If repairs to the boat during the fishing season have cost £600, how much should each partner have paid?

5.D24 A production run on tins of beans produces 2 880 tins in 2 h 40 min. If the time of the production run had been reduced by 30 min, what would have been the quantity produced?

5.D25 (a) A farmer employs 9 men to help with his harvest and they take 8 days to do the job. If he had employed 3 more men, how long would the job have taken?
(b) The ratio of coal to mining waste arriving at a pit head is 5:3. If 2 400 t are mined during a shift, how many tonnes of good coal are mined?

Exercise 5.E Services to People I: Community Services

5.E1 The table shows the number of people eligible to vote in four local election wards.

Southbend	1 074
Eastlee	986
Weston	2 130
Northby	1 778

If approximately $\frac{5}{8}$ of those eligible actually voted, estimate the number of votes cast in each ward.

5.E2 In a metropolitan borough, supplementary benefits averaging £28.17 were paid to 1 893 people. Estimate the total amount of money paid.

5.E3 In one day a central post office sold 3 844 stamps at 16p and 1 747 stamps at $12\frac{1}{2}$p. Estimate the total amount of money taken.

5.E4 The table shows the number of people using five branch libraries in one week.

	MEN	WOMEN	CHILDREN
York Street	856	1 261	394
Parkside	1 345	2 059	278
Elm Square	1 984	2 144	481
Avon Road	621	963	192
Devon Drive	783	1 886	209

Make rough estimates of:
(a) the total number of people using each branch;
(b) the numbers of men, women and children who borrowed books in that week.

5.E5 Use the table given in Exercise 5.E1 to determine the average number of eligible voters per ward.

5.E6 Use the table given in Exercise 5.E4 to calculate to the nearest whole number the average number of readers in the categories (a) men, (b) women and (c) children.

5.E7 (a) A school purchased 300 books at a total cost of £1 005. What was the average cost of a book?
(b) Annual rates paid by five householders were:

£384.60 £298.75 £401.35 £550.60 £412.40

What was the average rate of the five houses?

5.E8 The table shows the mileage covered by four family cars before major repairs became necessary.

Roadmaster de luxe	24 734 miles
Hasty Super	17 892 miles
Dragon 180	19 304 miles
Superior Coupe	38 650 miles

Calculate the average mileage covered before a major repair was required.

5.E9 The table shows the number of out-patients treated each month at a hospital.

Jan	Feb	Mar	Apr	May	Jun
1 384	1 202	1 149	1 285	962	1 254

Calculate the average number of out-patients treated per month.

5.E10 Distribution of leaflets to advertise a product is carried out in 7 h 20 min by the local Scout group, which has 17 members. How long would it have taken if the job had been done by a group of 25?

5.E11 An automatic telephone exchange handles 322 calls per minute. How many calls will it handle in (a) 1 h, (b) 31 min and (c) 2 h 41 min.

5.E12 (a) The ratio of the lengths of the sides of two square fields is 3:4. What is the ratio of their areas?
(b) The area of a playing field is $26 \, m^2$. The spectator area surrounding it is $54 \, m^2$. What is the ratio of playing area to spectator area?

5.E13 (a) It costs £1.35 to travel a distance of 18 miles by train. What will it cost for a journey of 64.6 miles?
(b) The rates bill for a house with a rateable value of £216 is £372.43. What will be the rate bill for a house whose rateable value is £279?

5.E14 The occupancy of hospital beds in a ward with 42 beds is:

Mon	Tue	Wed	Thur	Fri	Sat	Sun
38	37	42	40	39	36	34

(a) What was the ratio of used beds to available beds on each day of the week?
(b) What was the ratio of used to unused beds in a full week?

5.E15 (a) The plans of a rectangular playing area are drawn to a size of 14 cm by 22 cm. The scale of the plan is 1:100. What are the true dimensions of the playing area?
(b) The scale on a road map is 5 miles to 1 in. To estimate the distance between Preston and Warrington a rule is used and measures a distance of 5.3 in. What is the approximate distance in miles between Preston and Warrington?

5.E16 Two towns are shown on a map 3.3 cm apart. What is the actual distance between the towns if the scale of the map is (a) 1:25 000, (b) 1:35 000 and (c) 1:45 000.

5.E17 A housing estate has bungalows and houses on it in the ratio 5:3. If the total number of properties on the estate is 576, how many houses and bungalows are there?

5.E18 A swimming group comprises 45 young people. The ratio of girls to boys is 5:4. How many of each sex are there in the group?

5.E19 After building a new by-pass to avoid a town centre, the usual journey of a regular user of the route found that his journey was reduced by 35 min. His journey now took him 3 h 30 min. What is the ratio of his old journey to his new one?

Estimates, Average, Ratio and Proportion

5.E20 (a) The rate of exchange of one pound sterling to Australian dollars is 1:1.6. How many Australian dollars do I receive for £76?

(b) A school has a staff to student ratio of 2:37. If the total population of the school is 1 521, how many staff are there?

Exercise 5.F Services to People II: Food and Clothing

5.F1 For V.A.T. at 15p per pound, make a rough estimate of the total amount of tax to be paid on the following sales:
(a) 18 blouses at £11.65 each;
(b) 21 pairs of jeans at £9.34 each;
(c) 13 coats at £27.80 each;
(d) 32 dresses at £18.94 each.

5.F2 Make a rough estimate of the total mass of the following deliveries:
(a) 33 bags of potatoes, each 28 kg;
(b) 14 bags of sprouts, each $2\frac{1}{2}$ kg;
(c) 21 bags of flour, each $1\frac{1}{2}$ kg;
(d) 83 bags of sugar, each $1\frac{1}{4}$ kg.

5.F3 Make a rough estimate of the total cost of:
(a) 39 packets of tea at 29p each;
(b) 17 large loaves at 48p each;
(c) 33 boxes of cornflakes at 41p each;
(d) 27 tins of salmon at £1.19 each;
(e) 11 tins of corned beef at 98p each.

5.F4 Make a rough estimate of the cost of each of the following items:
(a) $5\frac{1}{2}$ lb leg of lamb at £1.12 per pound;
(b) $3\frac{3}{4}$ lb lamb chops at 89p per pound;
(c) $1\frac{3}{4}$ lb braising steak at £1.38 per pound;
(d) $4\frac{1}{4}$ lb pork sausages at 68p per pound;
(e) 18 pairs of kippers at $64\frac{1}{2}$p per pair.

5.F5 A bakery produced 1 639 white loaves and 865 brown loaves in one day. If three-quarters of the loaves were sold, make a rough estimate of the number of loaves remaining.

5.F6 (a) If 160 g of stewing steak is required to produce one portion of beef goulash, estimate the amount of steak required to produce 87 portions.

(b) A list of ingredients for making a rhubarb tart is shown below. Estimate the amount of each ingredient required to produce 3 doz tarts.

$1\frac{1}{4}$ lb rhubarb
$7\frac{1}{2}$ oz plain flour
$4\frac{1}{2}$ oz self-raising flour
$3\frac{3}{4}$ oz margarine
$2\frac{1}{4}$ oz lard

5.F7 (a) A box containing 48 oranges has a mass of 23 kg. If the box alone has a mass of 7 kg, what is the average mass per orange?

(b) A waiter pours wine into 7 glasses from a one-litre bottle. He leaves 0.22 litre in the bottle. What is the average quantity of wine in each glass?

5.F8 (a) The price in pence of ten food items on a supermarket shelf is: 140, 52, 122, 246, 163, 445, 508, 515, 1222, 499. What is the average cost per item?

(b) A salesgirl makes out the following sales invoices: £18.60, £19.48, £18.78, £17.54, £18.37, £15.55. What is the average sum per invoice?

5.F9 (a) The total cost of a restaurant meal for 7 persons was £87.73. What was the average cost per person?

(b) A snack bar sells sandwiches of beef, ham and chicken. The takings after a day are recorded as:

46 beef	£17.48
57 ham	£18.81
75 chicken	£26.25

If all the sandwiches are sold during a 3-h period:
(i) How many sandwiches were sold per hour?
(ii) What was the average takings per hour?
(iii) What is the charge for (1) a beef sandwich, (2) a ham sandwich and (3) a chicken sandwich.

5.F10 (a) The number of bedrooms occupied each day of a week at a hotel having 45 bedrooms was:

Mon 20 Tue 40 Wed 35 Thur 30 Fri 27 Sat 42 Sun 18

What was the ratio of occupied bedrooms to unoccupied bedrooms on each day of the week?
(b) What was the ratio over the week?

5.F11 (a) A standard price of 22p is charged for a school meal costing 50p, the remainder of the cost is subsidised by local authority rates. What is the ratio of the subsidy to the standard price?

(b) When cutting dresses from a roll of cotton material the ratio of used material to waste is 3:1. If a dress length is 3.36 m, what will be the amount of cloth wasted?

5.F12 (a) The freezers at a hotel contain enough frozen food to provide 30 guests with 8 meals each. If 60 guests arrive unexpectedly how many meals can be served from the freezer to each guest?

(b) A snack bar buys crisps in boxes of 48 for £4.20. They are sold at 12p per packet. What is the profit on the sale of a box of crisps?

5.F13 A salesman in restaurant crockery receives commission in proportion to his sales. A recent sale of £2 250 gave him a commission of £150. He made similar sales of (a) £945, (b) £1 800 and (c) £2 775. What would be his commission on each sale?

5.F14 A caterer is required to serve 15 kg of meat. From experience he knows that the ratio of waste to lean meat is $2\frac{1}{2}:1$. How much meat does he order?

5.F15 (a) If 5 kg of apples cost £3.70, what is the cost of 13 kg?
(b) A team of 15 waitresses have been employed to serve a banquet for 160 people. If an extra 50 people are expected at the banquet, how many more waitresses need to be added to the team?

5.F16 A caterer ordered 45 kg of meat. After trimming, the meat produced one hundred and two 130-g portions. What was the ratio of waste to lean meat?

5.F17 A recipe for fish pie quotes the following amounts to serve four people:

　200 g fish
　　1 egg
　250 ml bechamel
　　50 g mushrooms
　200 g potatoes
　　　parsley, salt, pepper to taste

What amounts of each ingredient would be required to serve six people?

5.F18 Chicken soup is made from the following ingredients:

　　1 litre chicken stock
　　50 g flour
　　100 g onion, leek, celery
　　250 ml milk
　　50 g butter or margarine
　　25 g chicken (diced)
　　　salt/pepper to taste

These quantities will make enough soup for 4 portions. What will be the quantities for (a) 9 portions, (b) 15 portions and (c) 32 portions?

5.F19 (a) Men and women guests at a retirement dinner are in the ratio 7:5. If there are 60 guests at the dinner, determine the number of men and women.
(b) Six waiters can lay the tables in a banquet hall in 2 h 40 min. How long will it take four waiters?

5.F20 (a) When making bread, 20 g of fresh yeast is used with 700 g of plain flour. Express as a ratio the amount of yeast to the amount of flour.
(b) How much yeast would be required when using 1 050 g of flour?

6 Length, Area and Perimeter

6.1 Units of Length

The basic unit of length in the SI system of units is the metre. The preferred multiple of the metre is the kilometre which equals one thousand metres, and the preferred sub-multiple is the millimetre which equals one-thousandth of a metre.

Preferred units

$$1 \text{ metre} = 1 \text{ m}$$
$$1\,000 \text{ metres} = 1 \text{ kilometre} = 1 \text{ km}$$
$$\frac{1}{1\,000} \text{metre} = 1 \text{ millimetre} = 1 \text{ mm}$$

In certain occupations the centimetre, which is equal to one-hundredth of a metre, is still commonly used.

$$1 \text{ centimetre} = 1 \text{ cm} = \frac{1}{100} \text{m or } 10 \text{ mm}$$

Example 6.1 Convert: (a) 1.95 km to metres, (b) 2.64 m to millimetres, (c) 47 cm to millimetres, (d) 432 mm to metres and (e) 850 mm to centimetres.

(a) $1 \text{ km} = 1\,000 \text{ m}$

$1.95 \times 1\,000 = 1\,950 \text{ m}$ (*Ans.*)

(b) $1 \text{ m} = 1\,000 \text{ mm}$

$2.64 \times 1\,000 = 2\,640 \text{ mm}$ (*Ans.*)

(c) $1 \text{ cm} = 10 \text{ mm}$

$47 \times 10 = 470 \text{ mm}$ (*Ans.*)

(d) $1\,000 \text{ mm} = 1 \text{ m}$

$\frac{432}{1\,000} = 0.432 \text{ m}$ (*Ans.*)

(e) $10 \text{ mm} = 1 \text{ cm}$

$\frac{850}{10} = 85 \text{ cm}$ (*Ans.*)

6.2 Length Conversion Factors

Many older items of machinery and equipment used in industry was manufactured in the Imperial or 'inch' system of length units. When replacement parts are required for such equipment, it is often necessary to convert dimensions between the Imperial and the SI system in order to produce 'inch' dimensioned parts on metric machine tools. Length conversion factors are used to convert dimensions between the two systems.

Imperial units:

$$12 \text{ in} = 1 \text{ ft}$$
$$3 \text{ ft} = 1 \text{ yd}$$
$$1\,760 \text{ yd} = 1 \text{ mile}$$

Conversion factors:

$$1 \text{ in} = 25.4 \text{ mm}$$
$$1 \text{ ft} = 0.304\,8 \text{ m}$$
$$1 \text{ yd} = 0.914\,4 \text{ m}$$
$$1 \text{ mile} = 1.609\,3 \text{ km}$$

Example 6.2 Convert: (a) 3.9 in to millimetres, (b) 8 miles to kilometres, (c) 114.3 mm to inches, (d) 36 ft to metres, (e) 1 ft 6 in to centimetres and (f) 3 miles to metres.

(a) $1 \text{ in} = 25.4 \text{ mm}$

$3.9 \times 25.4 = 99.06 \text{ mm}$ (*Ans.*)

(b) $1 \text{ mile} = 1.609\,3 \text{ km}$

$8 \times 1.609\,3 = 12.874\,4 \text{ km}$ (*Ans.*)

(c) $25.4 \text{ mm} = 1 \text{ in}$

$\frac{114.3}{25.4} = 4.5 \text{ in}$ (*Ans.*)

(d) $1 \text{ ft} = 0.304\,8 \text{ m}$

$36 \times 0.304\,8 = 10.972\,8 \text{ m}$ (*Ans.*)

(e) $1 \text{ ft } 6 \text{ in} = 18 \text{ in}$

$1 \text{ in} = 25.4 \text{ mm}$
$18 \times 25.4 = 457.2 \text{ mm}$ (*Ans.*)

(f) $3 \text{ miles} = 3 \times 1\,760 = 5\,280 \text{ yd}$

$1 \text{ yd} = 0.914\,4 \text{ m}$
$5\,280 \times 0.914\,4 = 4\,828 \text{ m}$ (*Ans.*)

Alternative method:

1 mile = 1.609 3 km
3 × 1.609 3 = 4.828 km
1 km = 1 000 m
4.828 × 1 000 = 4 828 m   (*Ans.*)

Example 6.3 Convert to millimetres: (*a*) 4.175 in, (*b*) $7\frac{5}{8}$ in and (*c*) $\frac{3}{8}$ in. Give the answers correct to 2 decimal places.

(*a*)   1 in = 25.4 mm

   4.175 × 25.4 = 106.045 mm
              = 106.5 mm correct to 2 d.p.   (*Ans.*)

(*b*)   7 in = 7 × 25.4 = 177.8 mm

   $\frac{5}{8}$ in = $\frac{5}{8}$ × 25.4 = 15.875 mm
   $7\frac{1}{2}$ in = 193.675 mm
              = 193.68 mm correct to 2 d.p.   (*Ans.*)

(*c*) $\frac{3}{8}$ × 25.4 = 9.525
              = 9.53 mm correct to 2 d.p.   (*Ans.*)

Example 6.4 The dimensions of the component shown in Fig. 6.1 are given in inches. Make a sketch of the component showing the dimensions in millimetres correct to the nearest 0.01 mm. (1 in = 25.4 mm.)

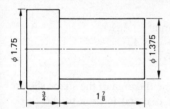

Fig. 6.1

Conversion of diameter:

1.75 × 25.4 = 44.45 mm
1.375 × 25.4 = 34.93 mm

Conversion of lengths:

$\frac{3}{4}$ × 25.4 = 19.05 mm
$\frac{15}{8}$ × 25.4 = 47.63 mm

The component with metric dimensions is shown in Fig. 6.2.

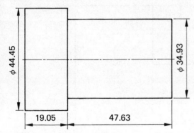

Fig. 6.2

In many instances simple conversion factors can be used to give estimated sizes.

Example 6.5 Figure 6.3 shows the dimensions of a rectangular field in yards. Estimate the size of the field in metres and find its approximate area in square metres. (Take 1 yd = 0.9 m.)

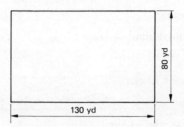

Fig. 6.3

Length = 130 yd × 0.9 = 117 m
Breadth = 80 yd × 0.9 = 72 m
Area = length × breadth
     = 117 × 72
     = 8 424 m²

6.3 Units of Area

The area of a shape is given by the number of square units that it contains. A square metre is defined as the area contained by a square of 1 m side. Conversion factors for area can be derived from the units of length by considering two squares of the same area. In the same manner the following conversion factors may be derived:

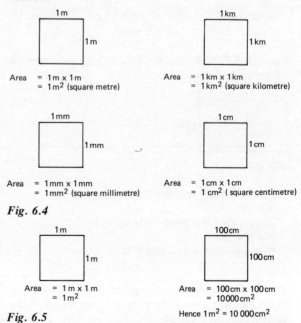

Fig. 6.4

Fig. 6.5

$1\,\text{km}^2 = 1\,000\,000\,\text{m}^2$
$1\,\text{m}^2 = 1\,000\,000\,\text{mm}^2$
$1\,\text{cm}^2 = 100\,\text{mm}^2$
$1\,\text{in}^2 = 645.16\,\text{mm}^2$

Example 6.6 Convert to square millimetres: (*a*) $0.07\,\text{m}^2$, (*b*) $43\,\text{cm}^2$ and (*c*) $3\,\text{in}^2$.

(*a*) $1\,\text{m}^2 = 1\,000\,000\,\text{mm}^2$

$0.07 \times 1\,000\,000 = 70\,000\,\text{mm}^2$ (*Ans.*)

(*b*) $1\,\text{cm}^2 = 100\,\text{mm}^2$

$43 \times 100 = 4\,300\,\text{mm}^2$ (*Ans.*)

(*c*) $1\,\text{in}^2 = 645.16\,\text{mm}^2$

$3 \times 645.16 = 1\,935.48\,\text{mm}^2$ (*Ans.*)

6.4 Rectangle and Square

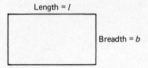

Fig. 6.6

The area of the rectangle shown in Fig. 6.6 is given by

area of rectangle = length × breadth
$A = lb$

The perimeter of the rectangle is the sum of its four sides:

perimeter of rectangle = $l + b + l + b$
$P = 2l + 2b$

Example 6.7 Calculate the area and the perimeter of the rectangle shown in Fig. 6.7.

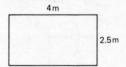

Fig. 6.7

Length = 4 m, breadth = 2.5 m
$A = lb$
Area = $4 \times 2.5 = 10\,\text{m}^2$ (*Ans.*)
$P = 2l + 2b$
Perimeter = $(2 \times 4) + (2 \times 2.5)$
= $8 + 5 = 13\,\text{m}$ (*Ans.*)

The square is a rectangle in which all four sides are of equal length, its area and perimeter are found in the same way as the rectangle.

Example 6.8 Calculate the area and the perimeter of the square shown in Fig. 6.8.

Fig. 6.8

Length = 3 m, breadth = 3 m
Area = $3 \times 3 = 9\,\text{m}^2$ (*Ans.*)
Perimeter = $(2 \times 3) + (2 \times 3) = 12\,\text{m}$ (*Ans.*)

Example 6.9 Calculate the area and the perimeter of the rectangle shown in Fig. 6.9.

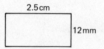

Fig. 6.9

Length = 2.5 cm = 25 mm
Breadth = 12 mm
Area = $25 \times 12 = 300\,\text{mm}^2$ (*Ans.*)
Perimeter = $(2 \times 25) + (2 \times 12)$
= $74\,\text{mm}$ (*Ans.*)

Example 6.10 Determine the area of the shape shown in Fig. 6.10. The shape is equal in area to the three rectangles, A, B and C, shown by the dotted lines.

Area of shape = area of A + area of B + area of C
Area of A = $10 \times 40 =\ \ 400\,\text{mm}^2$
Area of B = $20 \times 15 =\ \ 300\,\text{mm}^2$
Area of C = $20 \times 30 =\ \ 600\,\text{mm}^2$

Area of shape $\qquad\ =1\,300\,\text{mm}^2$ (*Ans.*)

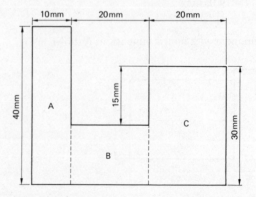

Fig. 6.10

6.5 Parallelogram

The area of a parallelogram is equal to the area of a rectangle having the same dimensions of length and breadth. The rectangle and the parallelogram shown in Fig. 6.11 are equal in area.

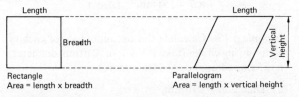

Fig. 6.11

Example 6.11 Determine the area of the parallelogram shown in Fig. 6.12.

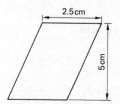

Fig. 6.12

Length = 2.5 cm, vertical height = 5 cm
Area = length × vertical height
= 2.5 × 5
= 12.5 cm² (*Ans.*)

6.6 Triangle

Figure 6.13 shows a parallelogram *ABCD* which is divided into two equal triangles by the dotted line *AC*. The area of each triangle is equal to one-half of the area of the parallelogram.

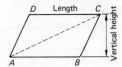

Fig. 6.13

Area of triangle = ½ area of parallelogram
= ½ × length × vertical height

This formula applies to all triangles and is usually given as

Area of triangle = ½ × length of base × vertical height

Example 6.12 Calculate the area of each of the triangle shown in Fig. 6.14.

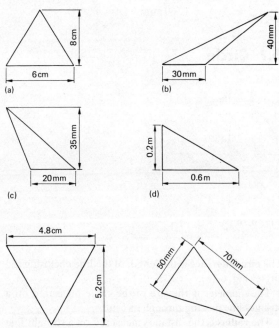

Fig. 6.14

Area of a triangle = ½ × length of base × vertical height

(a) Area = ½ × 6 × 8 = 24 cm² (*Ans.*)
(b) Area = ½ × 30 × 40 = 600 mm² (*Ans.*)
(c) Area = ½ × 20 × 35 = 350 mm² (*Ans.*)
(d) Area = ½ × 0.6 × 0.2 = 0.06 m² (*Ans.*)
(e) Area = ½ × 4.8 × 5.2 = 12.48 cm² (*Ans.*)
(f) Area = ½ × 50 × 70 = 1 750 mm² (*Ans.*)

Example 6.13 Figure 6.15 shows a rectangular metal plate which has a triangular hole punched out. Calculate the remaining area of the surface of the plate.

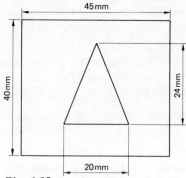

Fig. 6.15

Length, Area and Perimeter

Remaining area = area of rectangle − area of triangle
= (45 × 40) − ($\frac{1}{2}$ × 20 × 24)
= 1 800 − 240
= 1 560 mm² (*Ans.*)

6.7 Circle

Fig. 6.16

The **circumference** is the length of the line enclosing the circle (Fig. 6.16).

The **diameter** is the size of the circle measured on a straight line passing through its centre.

The **radius** is the distance measured on a straight line from the centre of the circle to its circumference, the radius is one-half of the diameter.

Let area of circle = A, circumference = C, diameter = d, radius = r, then
$d = 2r$

The ratio of the circumference of any size circle to its diameter is $3\frac{1}{7}:1$. Thus, for every circle

circumference = $3\frac{1}{7}$ × diameter
$C = 3\frac{1}{7}d$

The value of the constant $3\frac{1}{7}$ is given the Greek symbol π (pi).

$\pi = 3\frac{1}{7}$

or $\pi = \frac{22}{7}$ as an improper fraction or $\pi = 3.142$ as a decimal correct to 3 d.p.

Circumference = π × diameter
$C = \pi d$

or

$C = 2\pi r$

The area of a circle is given by

Area = π × radius²
$A = \pi r^2$

Example 6.14 Calculate the circumference and the area of a circle of diameter 14 mm (take $\pi = \frac{22}{7}$).

Diameter = 14 mm, radius = 7 mm
Circumference = π × diameter
= πd
= $\frac{22}{7}$ × 14 = 44 mm (*Ans.*)

Area = π × radius²
= πr^2
= $\frac{22}{7}$ × 7 × 7 = 154 mm² (*Ans.*)

Example 6.15 Calculate the circumference of each of the following circles (take $\pi = \frac{22}{7}$): (*a*) 210 mm diameter, (*b*) 35 cm radius and (*c*) 84 mm radius.

(*a*) Diameter = 210 mm.
$C = \pi d$
= $\frac{22}{7}$ × 210 = 660 mm (*Ans.*)

(*b*) Radius = 35 cm, diameter = 70 cm.
$C = \pi d$
= $\frac{22}{7}$ × 70 = 220 cm (*Ans.*)

(*c*) Radius = 84 mm, diameter = 168 mm
$C = \pi d$
= $\frac{22}{7}$ × 168 = 528 mm (*Ans.*)

Example 6.16 Calculate the area of each of the following circles. (take $\pi = 3.142$): (*a*) 20 cm diameter, (*b*) 40 mm radius and (*c*) 8 cm diameter.

(*a*) Diameter = 20 cm, radius = 10 cm.
$A = \pi r^2$
= 3.142 × 10 × 10 = 314.2 cm² (*Ans.*)

(*b*) Radius = 40 mm.
$A = \pi r^2$
= 3.142 × 40 × 40 = 5 027.2 mm² (*Ans.*)

(*c*) Diameter = 6 cm, radius = 3 cm.
$A = \pi r^2$
= 3.142 × 3 × 3 = 28.278 cm² (*Ans.*)

Example 6.17 Calculate the cross-sectional area of the metal tube shown in Fig. 6.17 (take $\pi = \frac{22}{7}$).

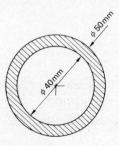

Fig. 6.17

The cross-sectional area

= area of outer circle − area of inner circle
= $(\pi \times 25^2) - (\pi \times 20^2)$
= $(\frac{22}{7} \times 25 \times 25) - (\frac{22}{7} \times 20 \times 20)$
= $1964 - 1257$
= 707 mm^2 (*Ans.*)

Exercises 6.A Core

6.A1 Convert to millimetres:

(*a*) 1.3 cm (*b*) 2.5 m (*c*) 0.8 m (*d*) 4.65 m
(*e*) 1.238 m (*f*) 4.7 cm (*g*) 23.6 cm (*h*) 0.07 m

6.A2 Convert to metres:

(*a*) 210 cm (*b*) 650 mm (*c*) 75 cm (*d*) 125 mm
(*e*) 48.6 cm (*f*) 1035 mm (*g*) 149 cm
(*h*) 173.2 cm

6.A3 Convert:

(*a*) 2.7 km to m (*b*) 3844 m to km
(*c*) 1965 mm to m (*d*) 675 cm to m
(*e*) 385 mm to cm (*f*) 0.95 m to mm
(*g*) 2675 mm to m (*h*) 1.047 m to cm

6.A4 Convert to millimetres:

(*a*) 4 in (*b*) 12 in (*c*) 3 ft (*d*) 1.5 in (*e*) 7.2 in
(*f*) 0.9 in (*g*) 1 yd (*h*) 15.6 in

6.A5 Convert to inches:

(*a*) 101.6 mm (*b*) 279.4 mm (*c*) 91.44 mm
(*d*) 355.6 mm (*e*) 7.62 mm (*f*) 6.35 cm
(*g*) 125.73 mm (*h*) 38.1 mm

6.A6 Convert:

(*a*) 6 ft to metres (*b*) 20 yd to metres
(*c*) 4 miles to kilometres (*d*) 15 in to centimetres
(*e*) 0.05 in to millimetres (*f*) $4\frac{3}{4}$ in to millimetres
(*g*) $2\frac{1}{2}$ ft to centimetres (*h*) 55 yd to metres

6.A7 The dimensions of the machined block shown in Fig. 6.18 are given in inches. Sketch the block and show its dimensions in millimetres correct to the nearest 0.01 mm.

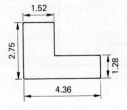

Fig. 6.18

6.A8 The dimensions of the rectangular car park shown in Fig. 6.19. are given in yards. Sketch the car park and show its dimensions in metres correct to the nearest 0.1 m.

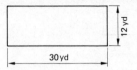

Fig. 6.19

6.A9 Calculate the area of each of the rectangles shown in Fig. 6.20.

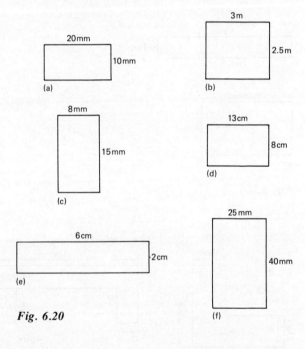

Fig. 6.20

6.A10 Calculate the area of each of the rectangles shown in Fig. 6.21.

6.A11 Complete the table showing areas of rectangles.

LENGTH		WIDTH		AREA
4	cm	1.8	cm	7.2 cm²
10	mm	15	mm	
9	cm	4.3	cm	
3.5	cm	1.5	m	
48	mm	19.5	mm	
135	mm	85	mm	

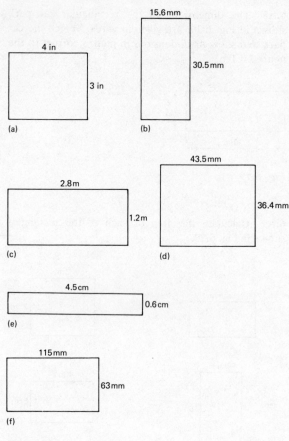

Fig. 6.21

6.A12 Determine the area of each of the following:
(*a*) a rectangle, 13.5 cm long and 5.8 cm wide;
(*b*) a square, 8 cm side;
(*c*) a rectangle, 4 cm long and 12.5 mm wide;
(*d*) a square, 13 mm side;
(*e*) a rectangle, 1.3 m long and 80 cm wide;
(*f*) a rectangle, 145 mm long and 3.5 cm wide;
(*g*) a square, 1.25 m side.

6.A13 Determine the area of each of the shapes shown in Fig. 6.22.

6.A14 Calculate the area of each of the triangles shown in Fig. 6.23.

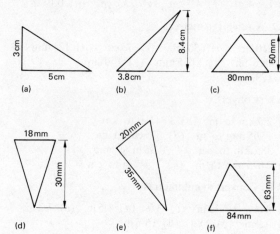

Fig. 6.23

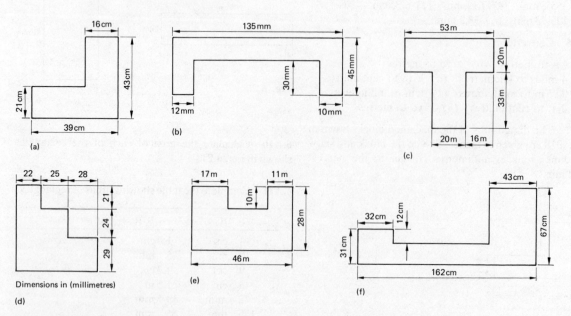

Fig. 6.22

80 Spotlight on Numeracy

6.A15 Determine the area of the shapes shown in Fig. 6.24.

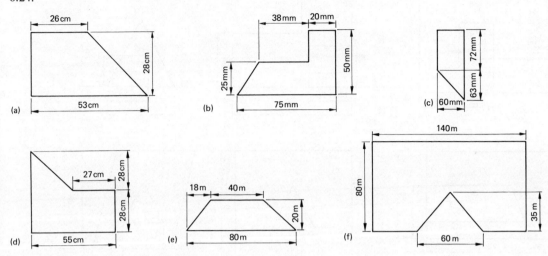

Fig. 6.24

6.A16 Determine the area of each of the parallelograms shown in Fig. 6.25.

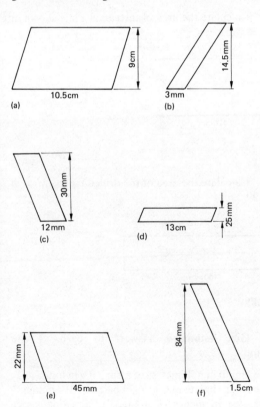

Fig. 6.25

6.A17 Each of the shapes in Fig. 6.26 shows a plate of metal which has a hole punched out. Determine the remaining area of each plate. (All dimensions are in millimetres.) Give answer in square centimetres.

6.A18 Calculate the area of the following triangles:

(a) base = 13 mm, vertical height = 8 mm;
(b) base = 4.4 cm, vertical height = 2.3 cm;
(c) base = 105 mm, vertical height = 82 mm;
(d) base = 0.3 m, vertical height = 0.12 m;
(e) base = 19 cm, vertical height = 10.4 cm;
(f) base = 0.8 mm, vertical height = 0.35 mm;
(g) base = 464 mm, vertical height = 210 mm;
(h) base = 38 mm, vertical height = 16 mm.

6.A19 Calculate the circumference of each of the following circles (take $\pi = \frac{22}{7}$):

(a) 42 mm radius   (b) 98 mm diameter
(c) 147 mm diameter   (d) 14 cm radius
(e) 3.5 m diameter   (f) 10.5 mm diameter
(g) 84 cm radius   (h) 157.5 mm diameter

6.A20 Calculate the circumference of each of the following circles correct to 2 decimal places (take $\pi = 3.142$):

(a) 14.6 mm radius   (b) 25.4 mm diameter
(c) 1.09 m radius   (d) 3.19 cm diameter
(e) 49.3 mm diameter

Length, Area and Perimeter 81

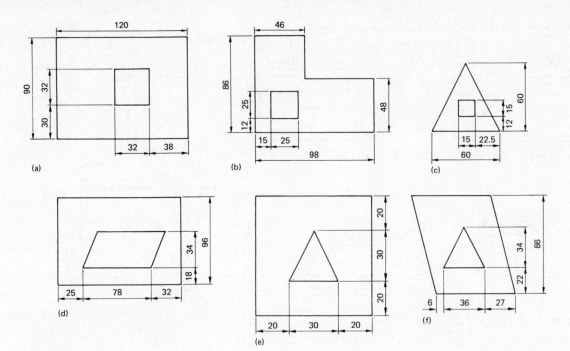

Fig. 6.26

6.A21 Calculate the area of such of the following circles (take $\pi = \frac{22}{7}$):

(a) 14 mm diameter (b) 21 mm radius
(c) 28 cm diameter (d) 147 mm radius
(e) 154 mm diameter

6.A22 Calculate the area of each of the following circles (take $\pi = 3.142$) and give the answer correct to 4 significant figures:

(a) 464 mm diameter (b) 1.46 m diameter
(c) 519 mm diameter (d) 47.25 mm radius
(e) 83 cm diameter

6.A23 Calculate the cross-sectional area of each of the tubes shown in Fig. 6.27.

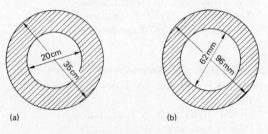

Fig. 6.27

6.A24 Calculate the area of the carver plate shown in Fig. 6.28 (take $\pi = \frac{22}{7}$).

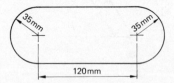

Fig. 6.28

6.A25 Calculate the area of the drilled plate shown in Fig. 6.29.

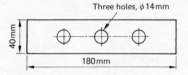

Fig. 6.29

6.A26 Give **estimated** answers to each of the following:

(a) the length in millimetres of a bar 3 ft 6 in long;
(b) the length in feet of a straight pathway 104 m long;
(c) the size in inches of a piece of drawing paper 800 mm by 45 mm;

82 Spotlight on Numeracy

(d) the area in square metres of a rectangular playing field 150 yd by 80 yd;
(e) the area in square miles of a circular region of 2 km radius;
(f) the cross-sectional area in square millimetres of a square bar of $1\frac{1}{2}$ in side.

Exercises 6.B Business, Administration and Commerce

6.B1 State the metric unit of length to be used when measuring each of the following (km, m, cm, mm):

(a) the thickness of a sheet of paper;
(b) the length of a typewriter ribbon;
(c) the thickness of a sheet of card;
(d) the length of a corridor;
(e) the size of type on a typewriter;
(f) the height of a filing cabinet;
(g) the distance from the office to the canteen;
(h) the distance between two branch offices in neighbouring towns;
(i) the width of a note pad;
(j) the length of a pencil;
(k) the thickness of a rubber;
(l) the size of an envelope.

6.B2 (a) State the length of each of the following items in millimetres:

(i) a book, length 24 cm;
(ii) a rule, length 30 cm;
(iii) a desk, length 1.5 m;
(iv) a stapler, length 17.5 cm;
(v) a file, length 32 cm.

(b) State the length of each of the following items in centimetres:

(i) a drawing board, length 0.9 m;
(ii) a pencil, length 165 mm;
(iii) a rule, length 150 mm;
(iv) a bookcase, length 1.8 m;
(v) a notice board, length 0.8 m.

(c) State the length of the following items in metres:

(i) a desk drawer, length 60 cm;
(ii) a photocopier, length 72 cm;
(iii) a library shelf, length 180 cm;
(iv) a counter top, length 230 cm;
(v) a radiator, length 1 500 mm.

6.B3 A page of notepaper is 30 cm long and 21 cm wide. Give its dimensions in (a) millimetres and (b) inches (1 in = 25.4 mm).

6.B4 Estimate the length of the following items in inches (take 1 in = 25 mm).

(a) a desk diary, length 300 mm;
(b) a computer lead, length 250 cm;
(c) a wall chart, length 1 200 mm;
(d) a rule, length 30 cm;
(e) an office table, length 1.6 m;
(f) an envelope, length 40 cm.

6.B5 Estimate suitable metric dimensions for each of the following:

(a) the length, width and thickness of a telephone directory;
(b) the height of a three-storey office building;
(c) the main dimensions of a public telephone kiosk;
(d) the length and diameter of a pencil;
(e) the length and width of a £5 note;
(f) the height, length and width of a typist's desk;
(g) the distance between the holes on punched file paper;
(h) the dimensions of a rectangular eraser.

6.B6 (a) A 5.2-m strip is cut off a 12-m roll of Sellotape. How much tape is left on the roll (i) in metres and (ii) in yards.
(b) A roll of paper is 30 ft long and is to be cut into strips of 200-mm length. How many strips can be obtained from the roll?

6.B7 (a) Two hundred sheets of cardboard make a pile 42.3 cm thick. What is the thickness of each sheet in (i) millimetres and (ii) inches.
(b) A book has 276 pages and is 0.78 in thick. How thick is 1 page in millimetres?

6.B8 A box of 100 treasury tags contains, 20 at 25 mm long, 24 at 50 mm long, 28 at 70 mm long, 17 at 100 mm long and 11 at 152 mm long. Estimate the length of string in yards required to make them.

6.B9 Envelopes made in Imperial sizes are (a) 9×4 in, (b) $10\frac{5}{8} \times 8\frac{1}{2}$ in, (c) 12×10 in and (d) 15×10 in. What are the equivalent metric sizes?

6.B10 (a) A machine stapler used in an office uses an average of 10 000 staples per month. Each staple measures 22 mm over its total length. How many yards of wire would be required to manufacture sufficient staples for 3 months supplies?
(b) A duplicating machine produces 78 copies per minute. If the copies are 270 mm long. How many miles of paper would be fed into the machine in a total of 3 h duplicating?

6.B11 (a) A typewriter ribbon is 32 m long and each letter typed used 2 mm. How much ribbon is unused after typing 3 675 words of four letters?
(b) Wooden pencils when new are 18 cm long. One

sharpening removes an average 4 mm of length, but a point breakage removes 1.3 cm of length. How long would a pencil be after two breakages and seven sharpenings?

6.B12 Figure 6.30 shows the sizes of six advertising leaflets. Calculate the area of each leaflet in (i) square millimetres and (ii) square centimetres.

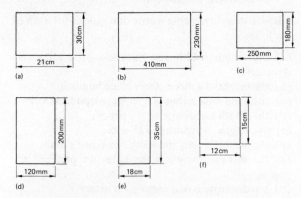

Fig. 6.30

6.B13 Figure 6.31 shows the floor plan of a suite of offices.

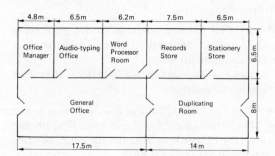

Fig. 6.31

(a) Calculate the floor area of each room in square metres.
(b) What is the total area occupied by the suite of offices in square yards?

6.B14 An electric typewriter ribbon is 125 m long and 70 mm wide. Calculate:

(a) the perimeter of the ribbon in metres;
(b) the area of the ribbon in square metres.

6.B15 (a) A paper punch makes holes 7 mm in diameter. Calculate the circumference and the area of each punched hole.
(b) Two holes are punched into each sheet of paper to enable it to be filed in a ring binder. What area of paper would be punched from the ream? (1 ream = 500 sheets.)

6.B16 A postage stamp measuring 16 mm by 23 mm is used on an envelope which measures 230 × 100 mm. Calculate:

(a) (i) the area of the stamp;
 (ii) the perimeter of the stamp;
(b) (i) the area of the envelope;
 (ii) the perimeter of the envelope;
(c) the area of the envelope not covered by the stamp.

6.B17 (a) Indicator stickers for progress of orders are available in several shapes and colours for quick identification. Calculate the areas of each of the following shapes:

 (i) a square, 2.4 cm side;
 (ii) a rectangle, 3.5 cm by 2.2 cm;
 (iii) a circle, 2.1 cm diameter.

(b) A square sticker has a perimeter of 973 cm. What is the length of one side?

6.B18 (a) A4 duplicating paper is 210 mm × 297 mm. What is the perimeter length?
(b) What is the area covered by a piece of A4 paper?
(c) A foolscap folder (13″ × 10″) is used to hold A4 paper. What will be the area of folder visible around the A4 paper? (Give the answer in square centimetres.)

6.B19 Figure 6.32 shows the plan view of an office.

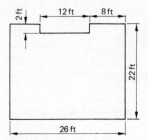

Fig. 6.32

(a) Calculate (i) the perimeter and (ii) the area on the floor.
(b) The office contains the following furniture: hat stand with a 2 ft diameter base; desk, 4 ft × 3 ft; filing cabinet, 2 ft 6 in × 18 in; circular table, 4 ft diameter. What area of floor space is left uncovered by furniture?

6.B20 (a) The clock on an office wall has a minute hand 12 cm long. What distance does the tip of the hand travel in 30 min?
(b) How far does the tip of the hand travel in 50 min?
(c) What area of the clock face has been passed over by the hand in 70 min?

6.B21 Figure 6.33 shows a number of plots of building land with dimensions given in metres. Calculate the area of each plot in square metres.

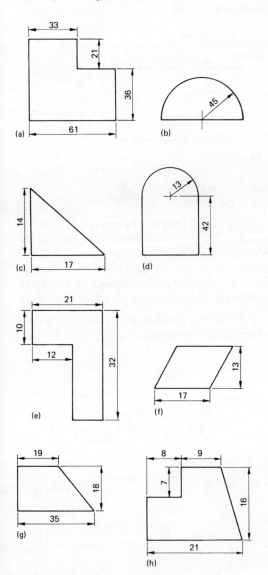

Fig. 6.33

Exercises 6.C Technical Services: Engineering and Construction

6.C1 State the metric units of length which would be used to measure each of the following (km, m, cm, mm):

(*a*) the length of a car;
(*b*) the length of a windscreen wiper blade;
(*c*) the spark-plug gap;
(*d*) the diameter of a turned shaft;
(*e*) the length of steel bar in a bar stores;
(*f*) the width of a lathe bed;
(*g*) the length of a garden path;
(*h*) the length of a brick;
(*i*) the thickness of a hacksaw blade;
(*j*) the thickness of a wooden plank;
(*k*) the diameter of a copper wire;
(*l*) the length of a piece of cable.

6.C2 State the SI units of mass which would be used for each of the following (kg, g):

(*a*) a bag of cement;
(*b*) a bar of resin core solder;
(*c*) a bag of nails;
(*d*) a packet of plastic plugs;
(*e*) a tin of putty;
(*f*) a bar of steel.

6.C3 Estimate a suitable metric dimension for the length of each of the following items:

(*a*) a car-jack handle;
(*b*) a heavy screwdriver;
(*c*) a split pin;
(*d*) a box spanner;
(*e*) a steel tape;
(*f*) a towing chain;
(*g*) a wood chisel;
(*h*) a power-tool cable;
(*i*) a taper shank sleeve;
(*j*) a bench-vice handle;
(*k*) a spade;
(*l*) a jack plane.

6.C4 The turned component shown in Fig. 6.34 is dimensioned in inches. Sketch the component and show the dimensions in millimetres correct to 0.01 mm.

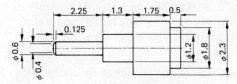

Fig. 6.34

6.C5 The milled component shown in Fig. 6.35 is dimensioned in millimetres. Sketch the component and show the dimensions in inches correct to 0.001 in.

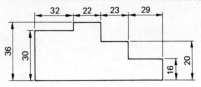

Fig. 6.35

6.C6 Use the approximation 1 in = 25 mm to estimate:

(a) the area in square millimetres of a metal plate $4\frac{1}{2}$ in long and $3\frac{1}{4}$ in wide;
(b) the length in millimetres of a steel bar 2 ft long;
(c) the diameter in millimetres of a bore of 8 in diameter;
(d) the length in millimetres of a shaft $11\frac{1}{2}$ in long;
(e) the area in square millimetres of a small surface plate 6 in by 4 in.

6.C7 Use the approximation 1 yd = 0.9 m to estimate the length in metres of each of the following:

(a) a brick wall, 9 ft long;
(b) a ladder, 30 ft long;
(c) fencing, 18 yd long;
(d) a garden path, $12\frac{1}{2}$ yd long;
(e) a plank, 7 ft long.

6.C8 Calculate the width in millimetres of the tenon shown in Fig. 6.36 if it is to fit the slot with 0.02 mm clearance.

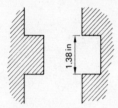

Fig. 6.36

6.C9 Calculate the diameter of a shaft in millimetres which would run in a bearing of 2.750 in diameter and give a clearance of 0.05 mm.

6.C10 A lathe tool holder is manufactured to use $\frac{3}{8}$ in square HSS tool bits. The nearest size available in metric tool bits is 10 mm. Calculate the increase in cross-sectional area in square millimetres of the tool bit when using the metric size.

6.C11 (a) A stepped shaft is turned on a lathe to the dimensions shown in Fig. 6.37.

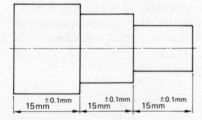

Fig. 6.37

(i) What is the maximum length of the shaft?
(ii) What is the minimum length of the shaft?
(b) A slot is milled in a rectangular block to the dimensions shown in Fig. 6.38.

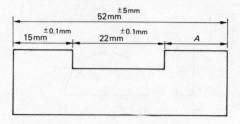

Fig. 6.38

(i) What is the maximum size of A?
(ii) What is the minimum size of A?

6.C12 (a) If one welding rod 18 in long is required to fill a weld run 2 in long, what length of welding rod in yards is required for a run 76 cm long?
(b) Estimate the number of times that a truck wheel of 28 in diameter would rotate in a distance of 2 km (take $\pi = \frac{22}{7}$ and use the approximation 1 ft = 0.3 m).

6.C13 Calculate the area in square millimetres of each of the sheet-metal patterns shown in Fig. 6.39.

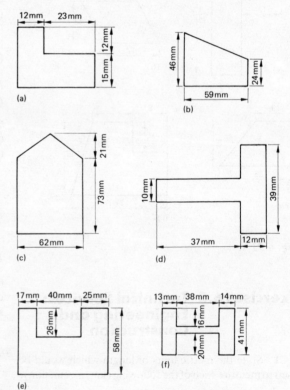

Fig. 6.39

6.C14 Bricks 25 cm long and 9 cm high are used to build a wall. Mortar 1 cm thick is laid on each side of the bricks. How many bricks are required for a wall 26 m long and 1 m high?

6.C15 Ceramic tiles 15 cm square are adhered to a wall with a gap of 2 mm surrounding each tile. How many tiles would be required to cover a wall 3 m high by 4 m long?

6.C16 Figure 6.40 shows a development pattern to be made from sheet metal.

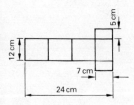

Fig. 6.40

(a) What is the length of the perimeter?
(b) What area of sheet metal is required?

6.C17 A rectangular room is 7.5 m long, 4.5 m wide and 3.5 m high. The room has a double door at one end, 3 m high by 2 m wide, and a window 2 m wide by 1.5 m high. A decorator uses wall-paper 0.5 m wide to cover all the remaining wall space. Estimate the length of paper that would be required.

6.C18 Calculate the cost of carpet at £7.60 per square metre required to cover the floor area shown in Fig. 6.41.

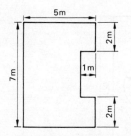

Fig. 6.41

6.C19 Calculate the areas of each of the following in square centimetres unless otherwise stated:

(a) a metal plate, 415 mm by 85 mm;
(b) a square surface plate, 250 mm wide;
(c) the floor area of a rectangular workshop, 13 m by 8.5 m (answer in square metres);
(d) a bench top, 2 m by 650 mm (answer in square metres);
(e) a scribing-block base, 65 mm by 50 mm;
(f) the cross-sectional area of a timber plank, 25 cm by 3.5 cm;
(g) the cross-sectional area of a steel pipe, 100 mm external diameter and 80 mm internal diameter;
(h) a rectangular machine base, 2.3 m by 1.6 m (answer in square metres).

6.C20 Each of the shapes shown in Fig. 6.42 shows a plate of metal which has a hole punched out. Determine the remaining area of each plate. (All dimensions are in centimetres.)

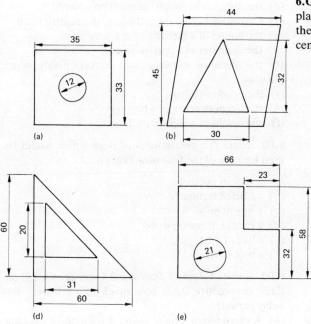
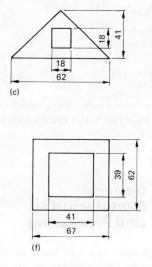

Fig. 6.42

6.C21 Calculate the area of the cover plate shown in Fig. 6.43. (Dimensions are in millimetres.)

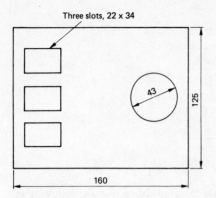

Fig. 6.43

6.C22 Calculate the area of the end face of the vee block shown in Fig. 6.44. (Dimensions are in millimetres.)

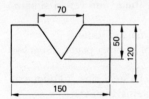

Fig. 6.44

6.C23 Calculate the area of the drilled plate shown in Fig. 6.45.

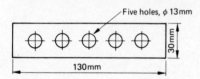

Fig. 6.45

6.C24 Calculate the cross-sectional areas of each of the extruded sections shown in Fig. 6.46. (Dimensions are in millimetres.)

Exercises 6.D General Manufacturing and Processing

6.D1 State the metric units of length which would be used to measure each of the following (km, m, cm, mm):

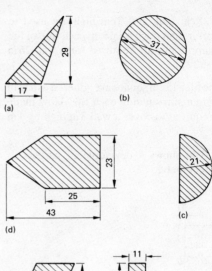

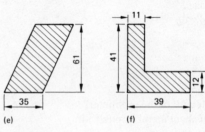

Fig. 6.46

(*a*) the length of a meadow;
(*b*) the depth of a mine shaft;
(*c*) the length of an average-size fish;
(*d*) the diameter of a medicinal capsule;
(*e*) the distance between the teeth on a comb;
(*f*) the distance from the pithead to the nearest town;
(*g*) the length of a furrow;
(*h*) the diameter of a barrow wheel;
(*i*) the distance between two lettuce plants when planting;
(*j*) the height of a barn;
(*k*) the circumference of a hen egg;
(*l*) the height of a factory roof.

6.D2 State the metric units of mass which would be used for each of the following (kg, g):

(*a*) a sack of potatoes;
(*b*) a load of fertiliser;
(*c*) a box of pills;
(*d*) a packet of cotton wool;
(*e*) a bag of coal;
(*f*) a bale of straw.

6.D3 (*a*) If 1.7 cm, 3.2 cm and 2.1 mm of rain fell on three consecutive days, how much rain fell over the 3-day period?
(*b*) A piece 68 cm long is sawn off a length of gas pipe 4.7 m long. What is the length of the remaining pipe?

6.D4 A delivery van makes the following journey:

Depot to factory A   17.6 miles
Factory A to factory B   13.9 miles
Factory B to factory C   7.57 km
Factory C to depot   18.8 miles

What was the total length of the journey in (*a*) miles and (*b*) kilometres (take 1 mile = 1.6 km).

6.D5 The diameter of steel rolls used at a rubber works are shown below in Imperial sizes. Give the size of each roll in millimetres.

(*a*) 36 in   (*b*) 24 in   (*c*) 18 in   (*d*) 15 in

6.D6 A cable drum holds 2.4 km of cable. What is the length of cable in (*a*) miles and (*b*) yards.

6.D7 Figure 6.47 shows the dimensions of a warehouse in yards.

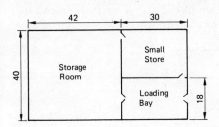

Fig. 6.47

(*a*) Sketch the warehouse and give the dimensions in metres.
(*b*) Find the area in square metres of (i) the storage room, (ii) the loading bay and (iii) the small stores.

6.D8 A manufacturer produces plastic plant tabs in the sizes shown in Fig. 6.48. Calculate the area of plastic sheet in square centimetres required to produce 100 of each size (1 cm^2 = 100 mm^2).

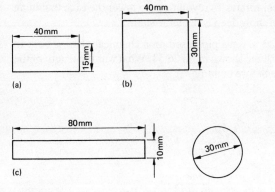

Fig. 6.48

6.D9 Calculate the perimeter of each of the allotment plots shown in Fig. 6.49.

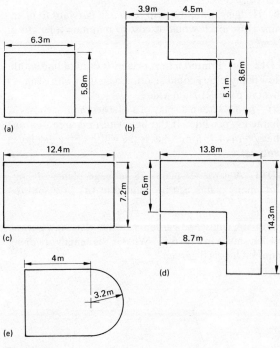

Fig. 6.49

6.D10 Figure 6.50 shows the dimensions of fields in yards. Estimate the area of each field in square metres (take 1 yd = 0.9 m).

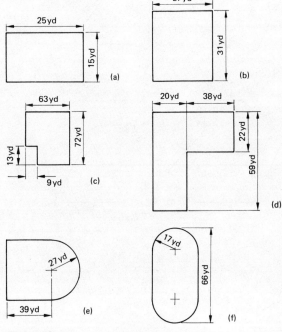

Fig. 6.50

Length, Area and Perimeter 89

6.D11 (a) A farmer sprays 2.5 m³ of liquid fertiliser on an area of 8 000 m². How much would he spray on an area of 10 000 m²?
(b) Hedging around a field costs 53p per yard to maintain. How much would it cost to maintain a length of 2 km?

6.D12 (a) Printed letters occupy 0.04 in of line width. How many letters could be printed on the width of an A4 sheet of paper 210 mm wide?
(b) The TV screen display of a microcomputer shows 32 characters per line. If the line width occupied by one character is 4.75 mm, what is the width of the display in centimetres?

6.D13 A gardener plants 18 cabbage plants to 1 m². How many plants can be planted in (a) 5 yd² and (b) 3 ft²?

6.D14 A landscape gardener draws plan views of the garden shown in Fig. 6.51. What is the length of edging required for each garden?

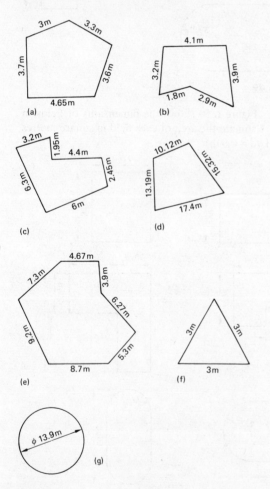

Fig. 6.51

6.D15 The garden shown in Fig. 6.52 is to be edged leaving a gap of 1.3 m to allow access for machinery. The edging stones are 0.7 m long and cost £2.35 each.
(a) How many edging stones will be required?
(b) What will be the total cost of the edging stones?

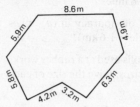

Fig. 6.52

6.D16 A wooden fence stake is 4 in square in section and protrudes above the ground by 3 ft 6 in. What surface area of the stake, in square feet, is protruding from the ground?

6.D17 Stakes are placed at 6-ft intervals around the fence line shown in Fig. 6.53.

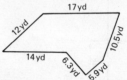

Fig. 6.53

(a) How many stakes are around the perimeter?
(b) If the stakes cost £5.73 each, what is the total cost of the stakes?

6.D18 (a) One hundred circular steel plates of diameter 1.4 m are each sprayed on both sides with two coats of paint. How many litres of paint will be used if 1 litre is sufficient to cover an area of 12 m²?
(b) Twenty steel tubes are galvanised on the outer surface. The tubes are 65 mm in diameter and are 22 m long. What is the total galvanized area in square metres?

6.D19 A cable wheel of 14 m diameter is used at the top of a mine shaft to raise and lower the cage. If the wheel rotates 78 revolutions to lower the cage to mining level, how deep is the mine shaft?

6.D20 A gas pipe used on a chemical plant is bent to the shape shown in Fig. 6.54. What was the length of the pipe before bending?

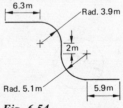

Fig. 6.54

6.D21 A slurry container is 35 m in diameter and 15.3 m high. Calculate:

(a) the area of the base;
(b) the surface area of the cylindrical portion;
(c) the circumference of the container.

6.D22 The fish tank shown in Fig. 6.55 has glass sides, a steel base and an aluminium lid. The glass is held in position by an angular steel frame. The tank is 1.6 m long, 1.12 m wide and 0.65 m deep.

(a) What length of angular frame is required to make the framework?
(b) What is the surface area of the steel base?
(c) What is the surface area of the glass?
(d) What is the surface area of the aluminium lid?

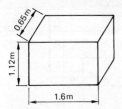

Fig. 6.55

Exercises 6.E Services to People I: Community Services

6.E1 State the metric units of length which would be used to measure each of the following (km, m, cm, mm):

(a) the length of a road;
(b) the length of carpet;
(c) the length of a cricket pitch;
(d) the distance between tee and green on a golf course;
(e) the length of the London marathon;
(f) the diameter of a golf ball;
(g) the length of a bus;
(h) the length of a bus ticket;
(i) the thickness of a tram ticket;
(j) the height of high jump;
(k) the length of an obstacle course;
(l) the distance between goal-posts.

6.E2 State the SI units of mass which would be used for each of the following (kg, g):

(a) a cricket ball;
(b) a volley ball;
(c) a boxer's punch bag;
(d) a road roller;
(e) a pair of boxing gloves;
(f) a pair of training shoes.

6.E3 Estimate a suitable metric dimension for the length of each of the following items:

(a) a golf club;
(b) a squash court;
(c) a bandage;
(d) a comb;
(e) a bicycle saddle;
(f) a delivery van;
(g) a stretcher;
(h) a walking stick;
(i) a cricket bat;
(j) a 10-min walk;
(k) a 4-h marathon run;
(l) a hospital bed.

6.E4 Figure 6.56 shows the floor plan of the hairdressing department in a large store, the dimensions are given in feet. Sketch the plan and give the dimensions in metres correct to 0.1 m.

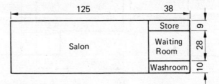

Fig. 6.56

6.E5 Convert the following dimensions to millimetres correct to 0.1 mm:

(a) a pair of scissors, length $6\frac{1}{4}$ in;
(b) a snooker cue, length 58 in;
(c) a bandage, width $1\frac{1}{4}$ in;
(d) a test tube, length $5\frac{1}{2}$ in, diameter $\frac{7}{8}$ in;
(e) a steering wheel, diameter $16\frac{1}{2}$ in.

6.E6 Use the approximation 1 in = 25 mm to estimate:

(a) the area in square millimetres of a brake pad $3\frac{1}{2}$ in long and $1\frac{1}{4}$ in wide;
(b) the length in millimetres of a gas pipe 3 ft long;
(c) the length in metres of a handrail 92 in long;
(d) the area in square millimetres of a medical dressing $4\frac{3}{4}$ in long and $2\frac{1}{2}$ in wide.

6.E7 A flight of stairs has 16 steps and is to be covered with stair carpet. Each step uses a length of 63 cm of carpet. Calculate the total length of carpet required in (a) metres and (b) yards.

6.E8 (a) A runner makes three laps of a 500-m running track. What distance did he run in yards?
(b) A mini-bus makes the following journey:

  Garage to railway station, 2 miles
  Railway station to airport, $4\frac{1}{2}$ miles
  Airport to garage, 6 miles

What was the total length of the journey in kilometres correct to 0.1 km?

Length, Area and Perimeter 91

6.E9 Calculate the perimeter and the area of each of the playing fields shown in Fig. 6.57. (All dimensions are in metres.)

(c) On the telegraph pole wires are attached at distances of 85 cm, 0.67 m and 325 mm from the top. How far from the ground is the lowest wire?

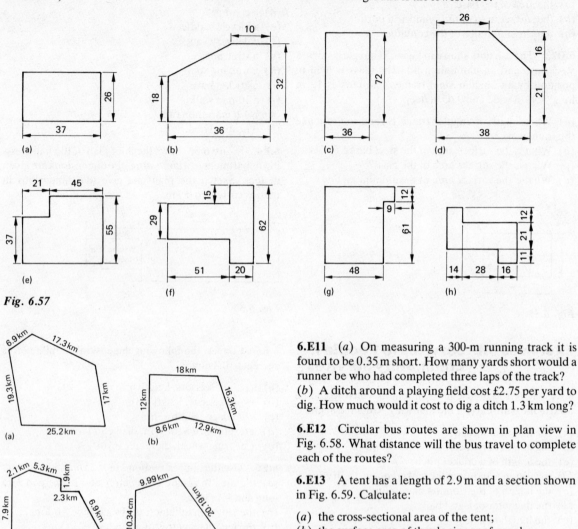

Fig. 6.57

Fig. 6.58

Fig. 6.59

6.E10 (a) An athlete drops out of a 5 000-m race after running 2.63 km. How far had he still to go?
(b) A telegraph pole 10 m long is sunk into the floor for 176 cm. What length is visible above ground?

6.E11 (a) On measuring a 300-m running track it is found to be 0.35 m short. How many yards short would a runner be who had completed three laps of the track?
(b) A ditch around a playing field cost £2.75 per yard to dig. How much would it cost to dig a ditch 1.3 km long?

6.E12 Circular bus routes are shown in plan view in Fig. 6.58. What distance will the bus travel to complete each of the routes?

6.E13 A tent has a length of 2.9 m and a section shown in Fig. 6.59. Calculate:

(a) the cross-sectional area of the tent;
(b) the surface area of the sloping roof panels;
(c) the perimeter of the base.

6.E14 The race track shown in Fig. 6.60 has two parallel sides with semi-circular ends. The sides are 200 m long and the radius of the semi-circular ends is 21 m. What is the length of one lap of the track?

92   **Spotlight on Numeracy**

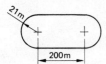

Fig. 6.60

6.E15 The living room of a house has a fireplace with a semi-circular hearth as shown by the plan view in Fig. 6.61. Calculate:

(*a*) the area of carpet in square feet required for the room;
(*b*) the perimeter of the carpet in feet.

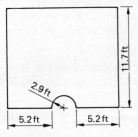

Fig. 6.61

6.E16 Calculate the area of each of the carpet pieces shown in Fig. 6.62.

6.E17 Determine the cross-sectional area of each of the pipes shown in Fig. 6.63.

(a)

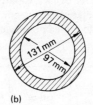
(b)

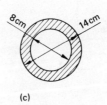
(c)

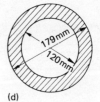
(d)

Fig. 6.63

6.E18 A hairdresser's gown, used to protect clients, is shaped as shown in Fig. 6.64. Calculate:

(*a*) the area of the gown in square metres;
(*b*) the cost of the gown at £1.32 per square metre.

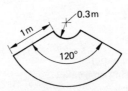

Fig. 6.64

6.E19 (*a*) The hospital ward shown in Fig. 6.65 is 35 m long and 24 m wide. At each end of the ward double doors 12 m wide open outwards. Each bed with a locker requires 2.5 m of wall space. How many beds can be located in the ward?
(*b*) If each bed and locker require 2.38 m² of floor. How much free floor space is there?

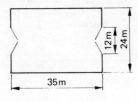

Fig. 6.65

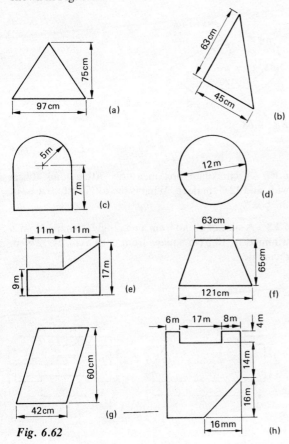

Fig. 6.62

Length, Area and Perimeter 93

6.E20 Figure 6.66 shows the floor plan of a service garage. Determine:

(a) the floor area of each room;
(b) the perimeter of each room;
(c) the total floor area of the garage.

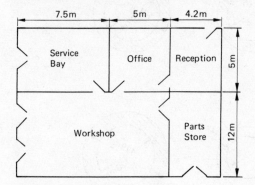

Fig. 6.66

Exercises 6.F Services to People II: Food and Clothing

6.F1 State the metric units of length which would be used to measure each of the following (km, m, cm, mm):

(a) the length of a roll of cloth;
(b) the width of a shoe;
(c) the length of cotton on a reel;
(d) the diameter of a needle;
(e) the diameter of a button;
(f) the width of an oven;
(g) the depth of a saucepan;
(h) the thickness of greaseproof paper;
(i) the thickness of a layer of icing on a cake;
(j) the length of a shop counter;
(k) the diameter of a clothes line;
(l) the length of a skirt.

6.F2 State the SI units of mass which would be used for each of the following (kg, g):

(a) a tin of peas;
(b) a sack of flour;
(c) a bag of sugar;
(d) an egg;
(e) a ball of wool;
(f) a box of apples.

6.F3 The cutting-out patterns shown in Fig. 6.67 are dimensioned in inches. Sketch each pattern and give its dimensions in centimetres. (1 in = 25.4 cm.)

6.F4 A tablecloth is 1.75 m long and 0.83 m wide and has a hem sewn around its edge with a stitch every 3 mm.

(a) What is the length of its perimeter?
(b) What is the area of the tablecloth?
(c) How many stitches are there around the edge?
(d) If the tablecloth was made for a table so that an overhang of 15 cm was on each side, what would be the size of the table?

6.F5 A bobbin for cotton is 1.8 cm in diameter. How long is a piece of cotton that is wrapped around its circumference 250 times?

6.F6 A cape is made to the pattern shown in Fig. 6.68. Calculate:

(a) the area of the cape;
(b) the cost of material at £1.36 per square metre.

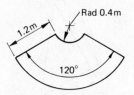

Fig. 6.68

6.F7 A chocolate box measures 30 cm long, 16 cm wide and 12.5 cm deep. What is the total surface area of the box?

6.F8 A slice of bread from a rectangular-shaped loaf is trimmed to 10.5 cm square from a 10.5 cm by 12.8 cm rectangle.

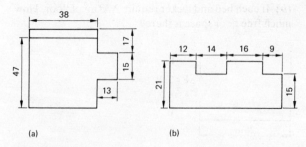

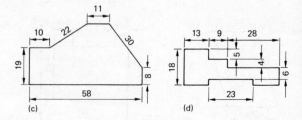

Fig. 6.67

(*a*) What surface area of bread is wasted?
(*b*) What surface area is left for making sandwiches?

6.F9 A training restaurant has 25 rectangular tables each 106 cm long and 92 cm wide. A distance of 60 cm around each side of the table is required for seating, and a further space of 60 cm is required at each side. What would be the minimum floor space required?

6.F10 The banquet room shown in Fig. 6.69 is to be layed out for 135 guests. Twelve guests will be sat on the top table and the rest on the three legs. The required spacings of places are 60 cm per person on the side tables and 90 cm per person on the top table. What length of tables are required?

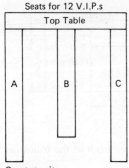

Fig. 6.69

6.F11 (*a*) Fifteen people are seated around a circular dining table at distances of 65 cm apart.

(i) What is the perimeter of the table?
(ii) What is the diameter of the table?
(iii) What is the surface area of the table?

(*b*) If a centre circle of 186 cm diameter was taken for the food dishes, what would be the area left for each person?

6.F12 The plan view of a kitchen is shown in Fig. 6.70.

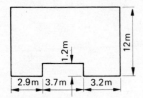

Fig. 6.70

(*a*) Calculate the area of the kitchen floor.
(*b*) The floor is covered with lino tiles 25 cm square.
 (i) What is the surface area of one tile?
 (ii) How many lino tiles would be needed to cover the floor?

(*c*) There is a skirting board around the kitchen walls with the exception of the door which is 75 cm wide. What is the length of the skirting board?
(*d*) The kitchen contains equipment occupying the following floor space: gas oven, 60 cm square; electric oven, 75 cm square; sink unit, 54 cm by 180 cm; working bench, 1.1 m by 1.2 m.
 (i) How much floor space is occupied by the equipment?
 (ii) How much floor space is left free?

6.F13 (*a*) An aluminium chip pan has a diameter of 30 cm. What will be the area of fat showing when the lid is removed?
(*b*) The depth of the pan is 18 cm, what is the curved surface area of the pan?

6.F14 The base of a roasting tin is rectangular in shape and measures 47 cm by 32 cm. Calculate:

(*a*) the perimeter of the base;
(*b*) the area of the base.

6.F15 (*a*) Cloth is sold at £1.60 per yard. What would be the cost of (i) 3.6 m and (ii) 5.5 m?
(*b*) If the cloth is 1.2 m wide, what is the area of a length of 6.5 m of the cloth?

6.F16 Calculate the area of each of the chocolate moulds shown in Fig. 6.71. (All dimensions are in millimetres.)

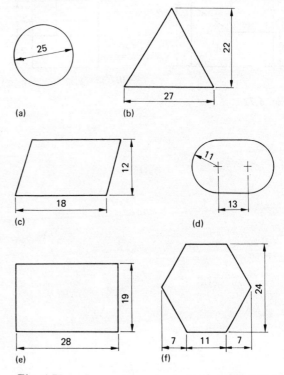

Fig. 6.71

Length, Area and Perimeter 95

6.F17 Find the area of each of the kitchen floor plans shown in Fig. 6.72. (All dimensions are in metres.)

6.F18 Figure 6.73 shows the floor plan of a boutique. Calculate the perimeter and the floor area of:

(a) the display area;
(b) the sales office;
(c) the store;
(d) changing room 1;
(e) changing room 2;
(f) the washroom.

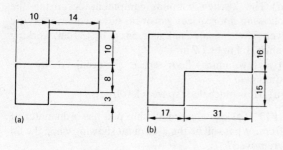

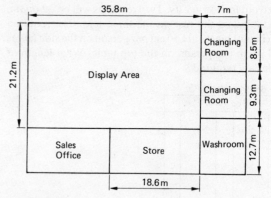

Fig. 6.73

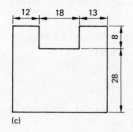
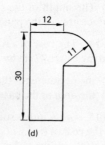

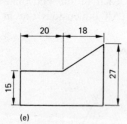

Fig. 6.72

6.F19 Find the area of cloth on each of the following rolls:

(a) 1.3 m wide by 20.5 m long;
(b) 0.8 m wide by 12 m long;
(c) 1.5 m wide by 31.5 m long;
(d) 1.25 m wide by 10 m long.

6.F20 (a) How many full yards of cloth can be cut from a roll 45 m long?
(b) A piece of suiting is 1.5 m wide by 4.8 m long. Calculate its area in (i) square metres and (ii) square feet.

7 Volume, Mass and Density

7.1 Units of Volume

The volume of a solid is given by the number of **cubic** units that it contains. A cubic metre is defined as the volume contained by a cube of 1 m side (Fig. 7.1). In the same manner other units of volume may be derived from the units of length (Fig. 7.2). The two cubes shown in Fig. 7.3 are exactly the same size, but their length dimensions are given in different units.

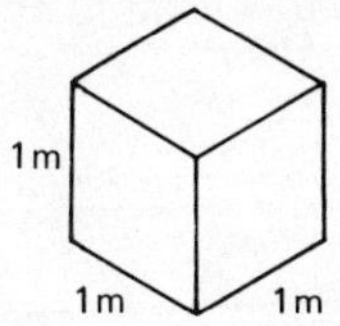

[Fig. 7.1 — cube with 1 m sides, Volume = 1 cubic metre = 1 m³]

Fig. 7.1

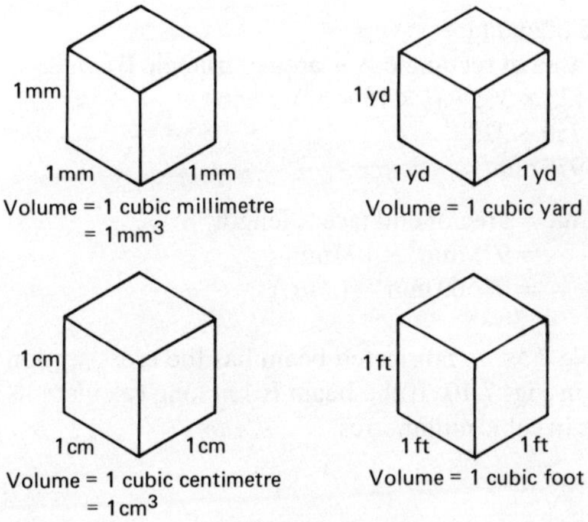

[Fig. 7.2 — cubes showing 1 mm³, 1 cubic yard, 1 cm³, 1 cubic foot]

Fig. 7.2

Fig. 7.3

Volume = 1 m × 1 m × 1 m
      = 1 m³

Volume = 100 cm × 100 cm × 100 cm
      = 1 000 000 cm³

As the volume of the two cubes are the same, then

1 m³ = 1 000 000 cm³

The same method may be used to convert between other units of volume.

Example 7.1 Calculate the number of cubic millimetres in 1 in³.

Using 1 in = 2.54 cm, we can show two cubes of exactly the same size with different units of length (Fig. 7.4).

 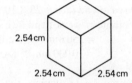

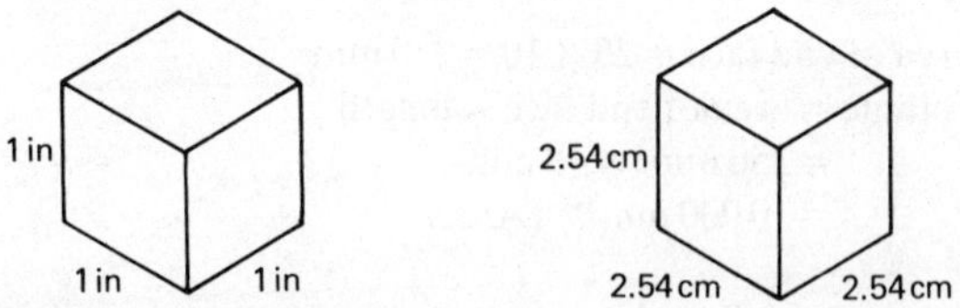

Fig. 7.4

Volume = 1 in × 1 in × 1 in
      = 1 in³

Volume = 2.54 cm × 2.54 cm × 2.54 cm
      = 16.387 cm³

∴ 1 in³ = 16.387 cm³   (*Ans.*)

7.2 Volume of a Prism

A **prism** is a solid which has the same shape and area of cross-section throughout its length. Figure 7.5 shows a rectangular prism, the end face of the prism is shaded. The cross-section at any cutting plane such as *AA* or *BB* will have the same shape and area as the end face. The volume of the prism is given by

volume = area of end face × length

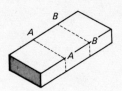

Fig. 7.5

Example 7.2 Calculate the volume of the rectangular prism shown in Fig. 7.6.

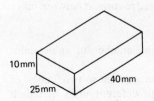

Fig. 7.6

Area of end face = 25 × 10 = 250 mm²
Volume = area of end face × length
= 250 mm² × 40 mm
= 10 000 mm³ (*Ans.*)

Example 7.3 Calculate the volume of the triangular prism shown in Fig. 7.7.

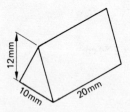

Fig. 7.7

Area of end face = area of triangle
= ½ × 10 × 12
= 60 mm²

Volume = area of end face × length
= 60 mm² × 20 mm
= 1 200 mm³ (*Ans.*)

Example 7.4 Calculate the volume of the cast-iron strip shown in Fig. 7.8.

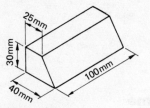

Fig. 7.8

Sketch the end face (Fig. 7.9).

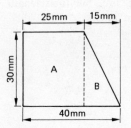

Fig. 7.9

Area of end face
= area of rectangle A + area of triangle B
= (25 × 30) + (½ × 15 × 30)
= 750 + 225
= 975 mm²

Volume = area of end face × length
= 975 mm² × 100 mm
= 97 500 mm³ (*Ans.*)

Example 7.5 A fabricated beam has the cross-section shown in Fig. 7.10. If the beam is 1 m long calculate its volume in cubic millimetres.

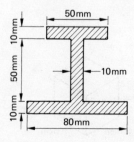

Fig. 7.10

98 Spotlight on Numeracy

The cross-section can be divided into three rectangles (Fig. 7.11).

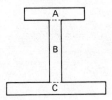

Fig. 7.11

Area of cross-section
= area of rectangle A + area of rectangle B + area of rectangle C
= (50 × 10) + (10 × 50) + (80 × 10)
= 500 + 500 + 800
= 1 800 mm²

Area of end face = area of cross-section
Length = 1 m = 1 000 mm
Volume = area of end face × length
= 1 800 mm² × 1 000 mm
= 1 800 000 mm³ (Ans.)

7.3 Volume of a Cylinder

A **cylinder** is a prism where a constant cross-section is a circle.

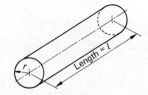

Fig. 7.12

Volume of cylinder = area of end face × length
Area of end face = πr^2
Volume = $\pi r^2 l$

Example 7.6 Calculate the volume of a solid cylinder 28 mm in diameter and 100 mm long.

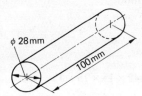

Fig. 7.13

Radius of cylinder = $\dfrac{\text{diameter}}{2} = \dfrac{28}{2} = 14$ mm

Area of end face = $\pi r^2 = \dfrac{22}{7} \times 14 \times 14 = 616$ mm²

Volume = area of end face × length
= 616 × 100
= 61 600 mm³ (Ans.)

Example 7.7 Calculate the volume of the shaft shown in Fig. 7.14.

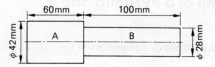

Fig. 7.14

Volume of shaft = volume of cylinder A + volume of cylinder B

Radius of cylinder A = $\dfrac{42}{2} = 21$ mm

Area of end face of cylinder A =

$\pi r^2 = \dfrac{22}{2} \times 21 \times 21 = 1 386$ mm²

Volume of cylinder A = 1 386 × 60 = 83 160 mm³

Radius of cylinder B = $\dfrac{28}{2} = 14$ mm

Area of end face of cylinder B =

$\pi r^2 = \dfrac{22}{7} \times 14 \times 14 = 616$ mm²

Volume of cylinder B = 616 × 100 = 61 600 mm³
Volume of shaft = 83 160 + 61 600
= 144 760 mm³ (Ans.)

Example 7.8 Calculate the volume of the cast-iron bush shown in Fig. 7.15. Take $\pi = 3.142$ and give the answer correct to 3 significant figures.

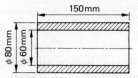

Fig. 7.15

Radius of outer circle = $\dfrac{80}{2} = 40$ mm

Radius of inner circle = $\dfrac{60}{2} = 30$ mm

Area of end face
= area of outer circle − area of inner circle
= (3.142 × 40 × 40) − (3.142 × 30 × 30)
= 5 027.2 − 2 827.8
= 2 199.4 mm²

Volume of bush = 2 199.4 × 150
= 329 910 mm³
= 330 000 mm³ to 3 s.f.   (*Ans.*)

7.4 Volume of a Pyramid and a Cone

The volume of a **pyramid** or a **cone** is given by

volume = $\frac{1}{3}$ × area of base × vertical height

Example 7.9 Calculate the volume of the square pyramid shown in Fig. 7.16.

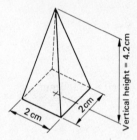

Fig. 7.16

Area of base = 2 cm × 2 cm = 4 cm²

Volume of pyramid
= $\frac{1}{3}$ × area of base × vertical height

= $\frac{1}{3}$ × 4 cm² × 4.2 cm

= 5.6 cm³   (*Ans.*)

Example 7.10 A pyramid has a rectangular base 25 mm × 18 mm and a vertical height of 35 mm, calculate its volume.

Area of base = 25 × 18 = 450 mm²

Volume = $\frac{1}{3}$ × 450 × 35

= 5 250 mm³   (*Ans.*)

Example 7.11 Calculate the volume of the cone shown in Fig. 7.17.

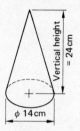

Fig. 7.17

Radius of base = $\frac{14}{2}$ = 7 cm

Area of base = $\pi r^2 = \frac{22}{7} \times 7 \times 7 = 154$ cm²

Volume of cone = $\frac{1}{3}$ × area of base × vertical height

= $\frac{1}{3}$ × 154 × 24

= 1 232 cm³   (*Ans.*)

7.5 Volume of a Sphere

Fig. 7.18

Volume of sphere = $\frac{4}{3}\pi r^3$

where *r* = radius of sphere.

Example 7.12 Determine the volume of a sphere 30 mm in diameter (take π = 3.142).

Radius of sphere = $\frac{30}{2}$ = 15 mm

Volume of sphere = $\frac{4}{3}\pi r^3$

= $\frac{4}{3}$ × 3.142 × 15 × 15 × 15

= 14 139 mm³   (*Ans.*)

Example 7.13 Calculate the volume of metal required to cast 150 solid brass balls 40 mm in diameter. (Give the answer correct to 3 significant figures.)

Radius of ball $= \dfrac{40}{2} = 20$ mm

Volume of ball $= \dfrac{4}{3}\pi r^3$

$ = \dfrac{4}{3} \times 3.142 \times 20 \times 20 \times 20$

$ = 33\,514.6$ mm^3

Volume of metal for 150 balls
$= 33\,514.6 \times 150$
$= 5\,027\,199$ mm^3
$= 5\,030\,000$ mm^3 to 3 s.f.   (*Ans.*)

7.6 Capacity of Containers

When a container is used to hold liquids, the total volume which it can contain is referred to as its **capacity** and may be given in **litres**. Smaller capacities may be measured in centilitres or millilitres.

1 000 litres = 1 cubic metre
1 000 litres = 1 m^3
    1 litre = 1 000 cubic centimetres
    1 litre = 1 000 cm^3
    1 litre = 100 centilitres
    1 litre = 100 cl
    1 litre = 1 000 millilitres
    1 litre = 1 000 ml

Example 7.14 Determine the capacity in litres of the oil storage tank shown in Fig. 7.19. (Neglect the wall thickness of the tank.)

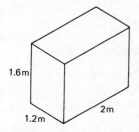

Fig. 7.19

Area of end face $= 1.2 \times 1.6 = 1.92$ m^2
Volume of tank $=$ area of end face $\times$ length
$ = 1.92 \times 2$
$ = 3.84$ m^3
1 m^3 = 1 000 litres
Capacity of tank $= 3.84 \times 1\,000$
$ = 3\,840$ litres   (*Ans*)

Example 7.15 Determine the capacity of an oil drum of internal dimensions of 0.6 m diameter and 0.8 m height. Take $\pi = 3.142$ and give the answer to the nearest whole litre.

Radius of drum $= \dfrac{0.6}{2} = 0.3$ m

Area of end face $= \pi r^2$
$ = 3.142 \times 0.3 \times 0.3$
$ = 0.282\,8$ m^2
Volume of drum $= 0.282\,8 \times 0.8$
$ = 0.226\,2$ m^3
Capacity of drum $\times 0.226\,2 \times 1\,000$
$ = 226.2$ litres
Drum could contain 226 litres   (*Ans.*)

7.7 Mass

The **mass** of a body is the quantity of material that it contains. The unit of mass in the S.I. system is the **kilogramme**. For larger masses the metric **tonne** may be used, and for small masses the **gramme**.

1 tonne = 1 000 kilogrammes
    1 t = 1 000 kg

1 kilogramme = 1 000 grammes
    1 kg = 1 000 g

In the British or Imperial system the unit of mass is the **pound** (lb).

2 240 pounds = 1 British ton
    2 240 lb = 1 ton

112 pounds = 1 hundredweight
    112 lb = 1 cwt

1 pound = 16 ounces
    1 lb = 16 oz

In everyday life the distinction between **mass** and **weight** is often ignored. By common usage people speak of a weight of so many kilogrammes, when they really mean a mass. At this stage we can consider both terms to refer to a quantity of material; e.g. a mass of 5 kg of potatoes or a weight of 5 kg of potatoes, both meaning the same quantity. (However, students who may later follow courses in engineering or science are reminded that mass is the correct term for quantity of material, and weight is an effect of mass and gravitational attraction.)

Example 7.16 Calculate the total mass of the contents of a delivery van containing:

  14 iron castings each having a mass of 31.5 kg
    8 steel bars each having a mass of 14.7 kg
  300 cutting tools each having a mass of 500 g

Iron castings, $14 \times 31.5 = 441$ kg
Steel bars, $8 \times 14.7 = 117.6$ kg
Cutting tools, $300 \times \dfrac{500}{1\,000} = 150$ kg
$\phantom{Cutting tools, 300 \times \dfrac{500}{1\,000} =}$ 708.6 kg (*Ans.*)

Example 7.17 Calculate the total weight of the following ingredients for making four Christmas puddings. (Give the answer in kilogrammes.)

225 g butter
275 g breadcrumbs
100 g plain flour
225 g demerara sugar
100 g candied orange peel
50 g candied lemon peel
100 g glacé cherries
½ kg raisins

Total weight
$= 225 + 275 + 100 + 225 + 100 + 50 + 100 + 500$ g
$= 1\,575$ g
$= 1.575$ kg (*Ans.*)

7.8 Density

Articles which have the same shape and size can have different masses (or weights) depending on whether the material from which they are made is a light material such as wood, or a heavy material such as iron. This property of the material is referred to as its **density** and is stated as the mass of unit volume of the material.

TABLE 7.1 Densities of Materials

MATERIAL	DENSITY (g/cm^3)
Aluminium	2.6
Brass	8.4
Brick	1.8
Cast iron	7.3
Concrete	2.1
Copper	8.9
Earth	1.6
Lead	11.0
Sand	1.5
Steel	7.8
Wood	0.75

Table 7.1 gives the average densities of some common metals and construction materials in grammes per cubic centimetre. The mass of an article can be calculated by multiplying its volume by the density of the material from which it is made:

mass = volume × density

Care must be taken to use the correct units, e.g. when the density is in grammes per cubic centimetre (g/cm^3) then the volume must be in cubic centimetres (cm^3).

Example 7.18 Calculate the mass in kilogrammes of a rectangular block of cast iron 40 mm wide, 30 mm thick and 100 mm long.

Converting dimensions to centimetres:

width = 4 cm, thickness = 3 cm, length = 10 cm

Volume of block $= 4 \times 3 \times 10 = 120\,cm^3$
Mass of block $=$ volume × density
$= 120\,cm^3 \times 7.3\,g/cm^3$ (density is
 taken from Table 7.1)
$= 876$ g
$= 0.876$ kg (*Ans.*)

Example 7.19 Calculate the mass in kilogrammes of a wooden plank 250 mm wide, 100 mm thick and 3 m long.

Converting dimensions to centimetres:

width = 25 cm, thickness = 10 cm, length = 300 cm

Volume of plank $= 25 \times 10 \times 300 = 75\,000\,cm^3$
Mass of plank $=$ volume × density
$= 75\,000\,cm^3 \times 0.75\,g/cm^3$ (density is
 taken from Table 7.1)
$= 56\,250$ g
$= 56.25$ kg (*Ans.*)

Exercises 7.A Core

7.A1 Determine the number of:

(*a*) cubic millimetres in a cubic centimetre;
(*b*) cubic inches in a cubic foot;
(*c*) cubic millimetres in a cubic metre;
(*d*) cubic feet in a cubic yard;
(*e*) cubic millimetres in a cubic inch;
(*f*) cubic centimetres in a cubic foot;
(*g*) cubic feet in a cubic metre.

7.A2 Convert:

(*a*) $3.5\,m^3$ to cm^3 (*b*) $1.6\,m^3$ to ft^3
(*c*) $1\,200\,mm^3$ to cm^3 (*d*) $1.2\,m^3$ to mm^3
(*e*) $2.8\,cm^3$ to mm^3 (*f*) $1.65\,m^3$ to litres
(*g*) 0.75 litres to ml   (*h*) 454 ml to cl
(*i*) 0.653 litres to cm^3

7.A3 Calculate the volume of each of the following rectangular prisms:

	WIDTH	HEIGHT	LENGTH
(a)	15 mm	8 mm	30 mm
(b)	1.2 m	0.6 m	2 m
(c)	6.4 cm	3 cm	12 cm
(d)	30 mm	20 mm	85 mm
(e)	0.35 m	0.1 m	0.9 m

7.A4 Calculate the volume of each of the rectangular blocks shown in Fig. 7.20. (Dimensions in millimetres.)

7.A7 Calculate the volume of the component shown in Fig. 7.22. (All dimensions in millimetres.)

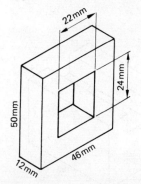

Fig. 7.22

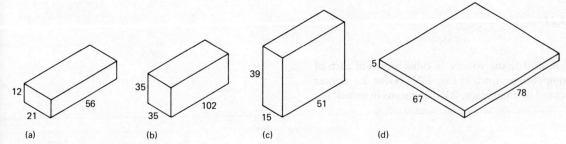

Fig. 7.20

7.A5 Calculate the volume of:

(a) a packing case, $0.7 \text{ m} \times 0.6 \text{ m} \times 1.1 \text{ m}$
(b) a steel bar, 20 mm square by 0.5 m long
(c) a wooden plank, $15 \text{ cm} \times 8 \text{ cm} \times 2 \text{ m}$
(d) a small office, 3 m wide, 4.5 m long and 2.8 m high

7.A6 Determine the volume of each of the solids shown in Fig. 7.21. (All dimensions in centimetres.)

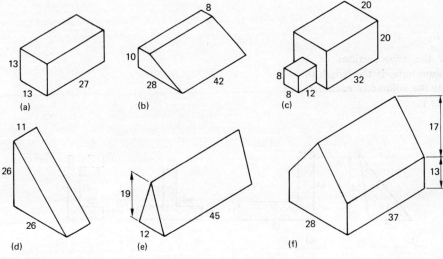

Fig. 7.21

Volume, Mass and Density 103

7.A8 Calculate the volume in cubic millimetres of each of the components shown in Fig. 7.23. (All dimensions in millimetres.)

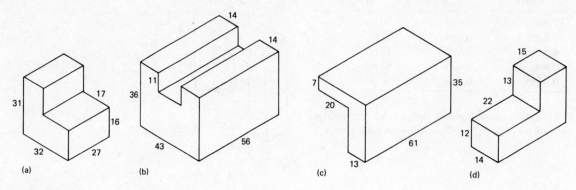

Fig. 7.23

7.A9 Calculate the volume in cubic inches of each of the components shown in Fig. 7.24. (Give the answer correct to 1 decimal place. All dimensions in inches.)

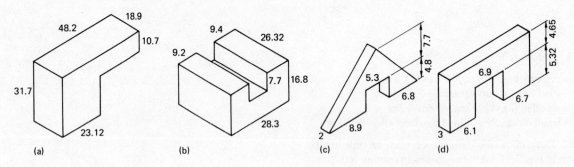

Fig. 7.24

7.A10 Figure 7.25 shows the cross-sections of a number of extruded aluminium bars. If the length of each bar is 300 mm, calculate the volume of each bar. (All dimensions in millimetres.)

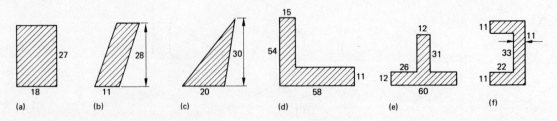

Fig. 7.25

7.A11 A beam is 1.5 m long and has the cross-section shown in Fig. 7.26. Calculate the volume of the beam in cubic millimetres. (All dimensions in millimetres.)

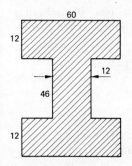

Fig. 7.26

7.A12 Determine the volume of each of the following solid cylinders. Give the answer correct to the nearest whole cubic millimetre.

	DIAMETER	LENGTH
(a)	14 mm	40 mm
(b)	20 mm	85 mm
(c)	40 mm	100 mm
(d)	25 mm	0.5 m
(e)	30 mm	20 cm

7.A13 Calculate the volume of each of the shafts shown in Fig. 7.27. Give the answer correct to the nearest whole cubic millimetre. (Dimensions in millimetres.)

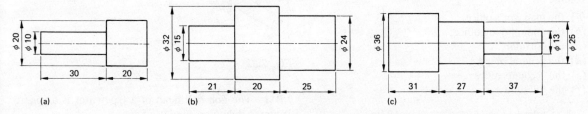

Fig. 7.27

7.A14 Calculate the volume of the hollow cylinder shown in Fig. 7.28.

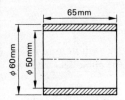

Fig. 7.28

7.A15 Determine the volume of a cone having a base of 40 mm diameter and a vertical height of 65 mm. (Take $\pi = 3.142$ and give the answer correct to 3 significant figures.)

7.A16 Calculate the volume of a pyramid having the base shown in Fig. 7.29 and a vertical height of 72 mm, to 1 decimal place.

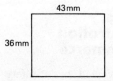

Fig. 7.29

7.A17 Find the volume of:

(a) a sphere, 7 cm radius;
(b) a pyramid, base 20 cm square, vertical height 45 cm;
(c) a cone base, 30 mm diameter, vertical height 60 mm;
(d) a sphere, 42 mm diameter.

7.A18 Determine the capacity of an oil drum of internal dimensions 0.8 m diameter and 1.2 m high. (Take $\pi = 3.142$ and give the answer to the nearest whole litre.)

7.A19 How many litres of water could be contained in the rectangular tank shown in Fig. 7.30? (Neglect the thickness of the metal.)

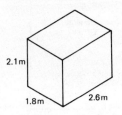

Fig. 7.30

Volume, Mass and Density

7.A20 Calculate the mass of each of the following articles (densities are given in Table 7.1 in Section 7.8):

(a) a brick, 30 cm × 15 cm × 10 cm;
(b) 1 m³ of concrete;
(c) a plank of wood, 16 cm × 8 cm × 0.5 m;
(d) a steel shaft, 20 mm diameter and 250 mm long;
(e) 15 m³ of sand;
(f) a lead block, 6.5 cm × 4 cm × 10 cm.

Exercises 7.B Business, Administration and Commerce

7.B1 Calculate the volume occupied by a safety deposit box 20 cm wide, 15 cm deep and 35 cm long.

7.B2 Determine the volume occupied by 16 box files, each 40 cm × 16 cm × 24 cm.

7.B3 Calculate the volume in cubic millimetres of each of the following rectangular parcels:

	WIDTH	HEIGHT	LENGTH
(a)	150 mm	70 mm	200 mm
(b)	30 mm	15 mm	45 mm
(c)	40 mm	20 mm	10 cm
(d)	65 mm	43 mm	250 mm

7.B4 Determine the available volume of storage space in a safe of internal dimensions 45 cm × 55 cm × 0.3 m.

7.B5 Figure 7.31 shows the floor plan of a suite of offices. The height of each room is 4.2 m. Calculate the volume of:

(a) the manager's office;
(b) the audio-typing office;
(c) the corridor;
(d) the general office;
(e) the accounts office;
(f) the full suite.

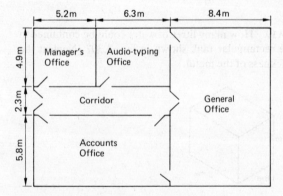

Fig. 7.31

7.B6 A stationery storeroom is 8 ft wide, 12 ft long and 10 ft high. Calculate its volume in (a) cubic feet and (b) cubic metres.

7.B7 A package containing one ream of duplicating paper has the following dimensions, 22 cm × 30 cm × 5 cm.

(a) Calculate the volume of the package.
(b) How much storage space in cubic feet would be occupied by 500 packages?

7.B8 Determine the volume of:

(a) a map case, 10 cm diameter and 0.8 m long;
(b) a cash box, 200 mm × 110 mm × 300 mm;
(c) a storage cupboard, 0.9 m × 1.2 m × 1.9 m.

7.B9 Complete the following table to show the volume in cubic metres occupied by each of the following packing cases:

CASE	WIDTH	LENGTH	HEIGHT	VOLUME (m³)
(a)	1.2 m	3.4 m	0.7 m	
(b)	0.5 m	2.6 m	0.3 m	
(c)	1.5 m	4.3 m	1.5 m	
(d)	50 cm	1.8 m	40 cm	
(e)	0.4 m	2.3 m	75 cm	

7.B10 Find the volume of the triangular ruler shown in Fig. 7.32.

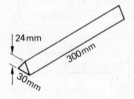

Fig. 7.32

7.B11 The golf-ball head of a typewriter is a sphere 38 mm in diameter. Find its volume.

7.B12 Calculate the volume of office space, in cubic metres, taken by each of the following items:

(a) a desk, 2 m × 1.6 m × 0.7 m;
(b) a safe, 0.6 m × 0.5 m × 0.4 m;
(c) a filing cabinet, 0.4 m × 0.65 m × 1.3 m;
(d) a photocopier, 0.9 m × 0.7 m × 0.45 m.

7.B13 Give the space occupied by each item of office equipment in Exercise 7.B12 in (a) cubic feet and (b) cubic yards.

7.B14 The central-heating system of a small shop has a rectangular oil storage tank with the dimensions shown in Fig. 7.33. Determine the capacity of the tank in litres, give the answer to the nearest whole litre.

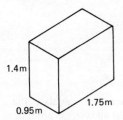

Fig. 7.33

7.B15 Calculate the capacity in litres of each of the following glass jars:

(a) 11 cm diameter, 25 cm high;
(b) 20 cm diameter, 32 cm high;
(c) 26 cm diameter, 40 cm high.

7.B16 A shop counter top is made from wood and has the dimensions $1.2\,m \times 3.5\,m \times 10\,cm$. Calculate the mass of the counter top. (Obtain the density from Table 7.1 in Section 7.8.)

7.B17 How many litres of water can be contained in a shop sink $450\,mm \times 350\,mm \times 220\,mm$?

7.B18 A rectangular can has the dimensions $310\,mm \times 220\,mm \times 150\,mm$ and is three-quarters full of duplicating fluid. How many litres of duplicating fluid are contained in the can? (Neglect the thickness of the metal.)

7.B19 Find the volume of:

(a) a typewriter-ribbon holder, 6 cm diameter, 2.5 cm wide;
(b) a desk diary, $160\,mm \times 190\,mm \times 3\,mm$;
(c) a pencil, 8 mm diameter, 180 mm long;
(d) a 10p coin, 2.8 cm diameter, 2 mm thick;
(e) a rubber eraser, $5\,cm \times 12\,mm \times 18\,mm$.

7.B20 Find the capacity in millilitres of:

(a) an ink bottle, 45 mm diameter, 60 mm high;
(b) a paper cup, 8 cm diameter, 10.5 cm high;
(c) a fountain pen sac, 8 mm diameter, 80 mm long.

Exercises 7.C Technical Services: Engineering and Construction

7.C1 Determine the volume of each of the metal blocks shown in Fig. 7.34. (Dimensions in millimetres.)

7.C2 Figure 7.35 shows a steel block before and after a shaping operation. Calculate the volume of:

(a) the original block;
(b) the machined block;
(c) the metal removed by machining.

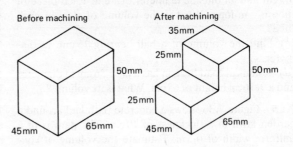

Fig. 7.35

7.C3 State the unit of volume or capacity which would be used to measure each of the following (litres, ml, cm^3):

(a) a tin of engine oil;
(b) a tin of paint;
(c) a tank of petrol;
(d) a tube of colour additive;
(e) a sachet of car shampoo;
(f) the contents of a machine tool coolant tank.

7.C4 Find the volume of:

(a) a square steel bar, 15 mm side and 12 cm long;
(b) a round brass bar, 4.5 cm diameter and 24 cm long;
(c) a piece of plate, $45\,cm \times 32\,cm \times 2.5\,cm$.

7.C5 The rectangular base of a garage is $10\,ft \times 22\,ft$. If concrete of thickness 3 in is to be laid on its surface, what volume of concrete will be required?

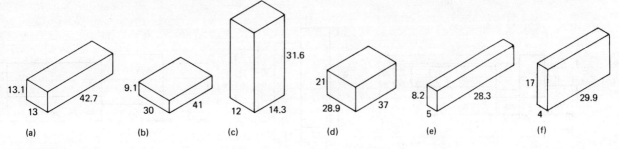

Fig. 7.34

7.C6 The container on a paint spray gun has internal dimensions of 14 cm in diameter and 18 cm long. Determine the amount of paint which could be contained in (*a*) cubic centimetres and (*b*) litres.

7.C7 Wood preservative is sold in cans 15 cm × 7.5 cm × 18 cm. Find the amount of preservative contained in one can in (*a*) cubic centimetres and (*b*) litres.

7.C8 (*a*) Find the volume of a plank having a section of 30 cm × 4 cm and a length of 2.25 m.
(*b*) A scaffold is made of steel tubes having a bore of 3.5 cm and an outside diameter of 5 cm. Each piece of tube is 5 m long. What is the volume of steel in each tube?
(*c*) Find the volume of a ball bearing 15 mm in diameter.
(*d*) A conical road beacon has a base 36 cm in diameter and a vertical height of 65 cm. What is its volume?

7.C9 Figure 7.36 shows a concrete path laid around a garden lawn to a thickness of 10 cm. The path is of a uniform width of 0.5 m. Estimate the volume of concrete that would be required in cubic metres.

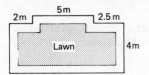

Fig. 7.36

7.C10 Make reference to a car handbook to list the capacities of:
(*a*) the petrol tank;
(*b*) the windscreen-washer reservoir;
(*c*) the engine-oil sump;
(*d*) the back-axle oil;
(*e*) the gearbox oil;
(*f*) the cooling system coolant.

Add all these capacities together and calculate the volume of fluid the car carries when all systems are full.

7.C11 A round steel blank 15 cm diameter × 4 cm thick is forged to the shape of a rectangular block 15 cm long by 9 cm wide. What will be the thickness of the block?

7.C12 Calculate the volume of each of the turned components shown in Fig. 7.37. Give the answer correct to the nearest whole cubic millimetre. (All dimensions in millimetres.)

7.C13 A steel flange is 130 mm in diameter and 15 mm thick. The flange is drilled with six holes 12 mm in diameter. Calculate the volume of the flange (*a*) before drilling and (*b*) after drilling.

7.C14 The component shown in Fig. 7.38 consists of a turned diameter of 30 mm and a 12 mm square portion

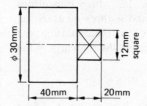

Fig. 7.38

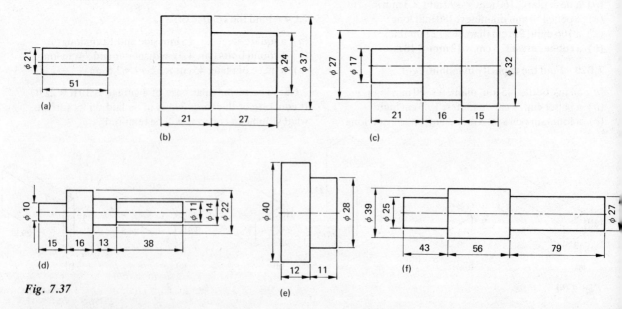

Fig. 7.37

milled at one end. Calculate the volume of the component.

7.C15 Calculate the volume in cubic millimetres of each of the cast-iron bushes shown in Fig. 7.39. (Dimensions in millimetres.)

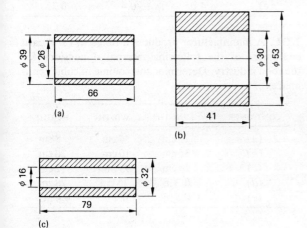

Fig. 7.39

7.C16 Determine the volume of metal in six steel washers having an outside diameter of 70 mm, a bore of 36 mm and a thickness of 10 mm.

7.C17 (a) Find the volume of a gas cylinder 4 m in diameter and 7 m long.
(b) A bar of 30 mm diameter is turned down on a lathe to 15 mm diameter for a length of 10 mm. What volume of metal is removed?

7.C18 Determine the volume of a beam 2 m long having the cross-section shown in Fig. 7.40.

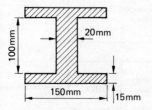

Fig. 7.40

7.C19 Find the capacity in litres of:

(a) an electric water heater, 0.4 m diameter and 0.5 m high;
(b) an electric kettle, 18 cm diameter and 15 cm high;
(c) a machine coolant tank, 1.5 m × 0.6 m × 0.2 m;
(d) an oil dashpot, 50 mm diameter by 75 mm.

7.C20 Find the volume contained in cubic millimetres of:

(a) an electrician's toolbox of internal dimensions 200 mm × 300 mm × 420 mm;
(b) a 0.5 m length of metal trunking of internal dimensions 85 mm × 55 mm;
(c) a moulding box of internal dimensions 350 mm × 250 mm × 175 mm.

7.C21 Calculate the volume of metal removed in machining the tee-slot shown in Fig. 7.41, in a cast-iron block 300 mm long.

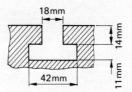

Fig. 7.41

7.C22 Determine the volume of the steel wedge shown in Fig. 7.42.

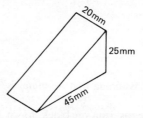

Fig. 7.42

7.C23 Use the densities given in Table 7.1 in Section 7.8 and calculate the mass in kilogrammes of each of the following:

(a) a steel block, 9 cm × 5 cm × 12 cm;
(b) a 1 m length of brass bar, 10 mm square;
(c) an aluminium tube, 60 mm outside diameter, 50 mm inside diameter and 450 mm long;
(d) a steel shaft, 80 mm diameter and 0.5 m long.

7.C24 Determine the mass of each of the steel components shown in Fig. 7.43. (All dimensions in millimetres.)

7.C25 (a) Calculate the mass in kilogrammes of a steel strip of rectangular section 20 mm by 5 mm and length 1.4 m.
(b) Determine the length of aluminium bar 14 mm in diameter that could be extruded from a billet 30 mm in

Volume, Mass and Density 109

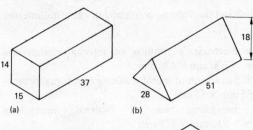

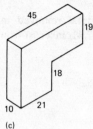 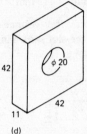

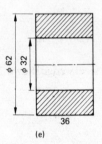

Fig. 7.43

diameter and 150 mm long, assuming that no metal is lost in the process.

(c) Calculate the mass of the cast iron vee block shown in Fig. 7.44. (Dimensions in centimetres.)

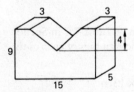

Fig. 7.44

Exercises 7.D General Manufacturing and Processing

7.D1 Sketch and dimension each of the following rectangular crates and find its volume in cubic metres for shipping purposes.

CRATE	LENGTH (m)	WIDTH (m)	HEIGHT (m)
(a)	3.2	1.4	0.4
(b)	1.9	0.8	0.5
(c)	0.7	0.55	0.3
(d)	2.4	1.8	1.35
(e)	4.05	0.4	0.25

7.D2 A manufacturer produces a range of cardboard containers of the following internal dimensions for the food industry. Determine the volume in cubic centimetres of each container.

CONTAINER	LENGTH	WIDTH	HEIGHT
(a)	16 cm	4 cm	8 cm
(b)	35 cm	10 cm	20 cm
(c)	18 cm	95 mm	12 cm
(d)	0.3 m	15 cm	28 cm
(e)	0.5 m	0.1 m	0.4 m

7.D3 Calculate the volume in cubic metres of the load compartment of a delivery van having internal dimensions 2.8 m by 2.4 m by 4.1 m.

7.D4 Figure 7.45 shows the floor plan of a warehouse, the height of each room is 6.5 m. Calculate the volume in cubic metres of:

(a) the small parts store;
(b) the finished goods store;
(c) the raw materials store;
(d) the records office;
(e) the transport office;
(f) the gangway.

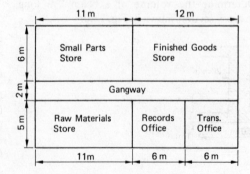

Fig. 7.45

7.D5 Find the volume in cubic metres of:

(a) a large processing vat, 2.3 m diameter and 4.6 m high;
(b) a rectangular chemical storage tank, 1.9 m by 1.3 m by 0.8 m;

(c) a cylindrical boiler, 3.2 m diameter and 5.05 m long.

7.D6 A firm produces a range of plastic extrusions of the cross-section shown in Fig. 7.46. Calculate the volume in cubic millimetres of a length of 300 mm of each shape. (Dimensions are in millimetres.)

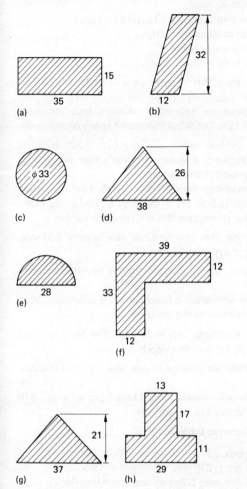

Fig. 7.46

7.D7 Calculate the volume of:

(a) a funnel which is conical in shape and has a base of 25 cm radius and a length of 40 cm;
(b) a rectangular fish tank of dimensions 60 cm by 30 cm by 25 cm.

7.D8 (a) A gas balloon is 2.6 m in diameter, what is its volume in cubic metres?
(b) Calculate the volume of a chemical transporter which is cylindrical in shape with hemispherical ends. The radius of the cylindrical portion is 2.2 m and its length is 20 m.

7.D9 State the units of volume which would be used to measure the following (litres, ml, cm^3):

(a) a bottle of medicine;
(b) a bottle of eye drops;
(c) a churn of milk;
(d) a tank of heating oil;
(e) a can of lubricating oil.

7.D10 (a) A showcase has internal measurements of 2 m × 2 m × 1 m. What volume of storage space is available?
(b) A vacuum flask is cylindrical in shape and has internal dimensions of 8 cm diameter and 32 cm length. What volume of liquid can be contained? Give the answer in (i) cubic centimetres and (ii) litres.

7.D11 (a) The diameter of a gas pipe is 5 cm and the pipe has a length of 3 m. What volume of gas under normal pressure can be contained in the pipe?
(b) A coal hopper is rectangular in cross-section and measures 3 m wide, 4 m high and 6 m long. It contains water to a depth of 0.5 m before a load of coal is tipped into it. After the tipping the water has risen to a depth of 2.3 m. What volume of coal was tipped into the hopper?

7.D12 (a) A coal truck is 6 m long and 3 m wide. Its capacity is 72 m^3. What is its height?
(b) Thermoplastic pellets are stored in a hopper and are delivered to a machine in exact quantities to fill a mould. The mould is rectangular in section and measures 16 cm × 7 cm and is 10 cm long. A full hopper will make 180 of these pieces. What is the capacity of the hopper in cubic centimetres?

7.D13 (a) A hosepipe is 27 m long and will hold 35.64 litres of water. What is the diameter of the pipe?
(b) A concrete paving flag has a volume of 18 250 cm^3. It is 28 cm wide and 12 cm thick. How long is it?

7.D14 A feeder tank delivers chemicals to a processing plant at the rate of 22 litres per minute. It is in use for 15 h per day and is filled three times during that time.

(a) What is the capacity of the tank in litres?
(b) The tank is cylindrical with a 1 m diameter base. What is its height?

7.D15 (a) Fluid flows from a chemical plant at the rate of 50 m^3 per minute. What volume of fluid flows in (i) 1 h and (ii) 1 day?
(b) A fish box is 2.2 m long, 1.5 m wide and 1.2 m high. What volume of fish will it hold?

7.D16 A water channel is 4 m long and has the cross-section shown in Fig. 7.47. Calculate:

(a) the volume of the channel in cubic centimetres;
(b) the quantity of water in litres that could be contained by the channel.

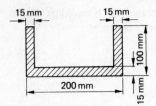

Fig. 7.47

7.D17 A buoy is made in the shape of a sphere 0.78 m in diameter.

(a) Calculate the volume of the buoy in cubic metres.
(b) How many litres of water would be displaced by the buoy if it was fully submerged?

7.D18 A lorry is loaded with 4 m³ of sand. If the mass of the lorry is 15 t, what is the total mass of the lorry and its load? (Density is given in Table 7.1 in Section 7.8.)

7.D19 A load of fish weighs 35 cwt, give its weight in (a) pounds, (b) kilogrammes, (c) tons and (d) tonnes.

7.D20 Each of the boat parts shown in Fig. 7.48 is made from bronze having a density of 8.7 g/cm³. Calculate the mass of each part. (Dimensions in millimetres.)

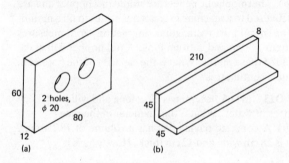

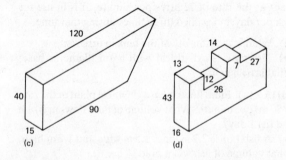

Fig. 7.48

7.D21 Calculate the total mass in kilogrammes of 24 lengths of deck planking 30 cm wide, 8 cm thick and 8 m long. (Density is given in Table 7.1 in Section 7.8.)

Exercises 7.E Services to People I: Community Services

7.E1 State the units of volume which would be used to measure each of the following (litres, ml, cm³):

(a) a flask of fruit juice;
(b) a swimming pool;
(c) a water container for a touring caravan;
(d) a bottle of cleaning fluid;
(e) a tin of floor polish;
(f) a tankful of petrol.

7.E2 (a) Find the volume of a tennis ball 7 cm in diameter.
(b) An enclosure around a shower unit measures 2.5 m × 2.5 m × 3 m, what volume of space is enclosed?

7.E3 (a) A hen cabin is 7 m × 10 m × 3 m high. If each hen requires a space of 50 cm³, how many hens could be housed in the cabin?
(b) A swimming pool 25 yd × 15 yd has a sloping bottom from 15 ft to 3 ft. If the pool is filled to the brim, what volume of water will it hold in cubic yards?

7.E4 A path 30 m long and 2 m wide is to be laid with concrete to a depth of 40 cm.

(a) Calculate the volume of concrete required in cubic metres.
(b) Use the information from Table 7.1 in Section 7.8 to calculate the mass of the concrete.

7.E5 An allotment 7 m wide and 20 m long is to be covered with top soil to a depth of 0.3 m.

(a) Calculate the volume of top soil required in cubic metres.
(b) Use the information from Table 7.1 in Section 7.8 to calculate the mass of the top soil.

7.E6 Determine the volume of:

(a) A wig box, 23 cm square and 40 cm deep;
(b) a gas meter, 210 mm × 230 mm × 320 mm;
(c) a paving stone, 1.2 m square and 90 cm thick;
(d) a toolbox, 25 cm × 45 cm × 35 cm.

7.E7 Calculate the volume of air in cubic metres contained in each of the following:

		WIDTH (m)	LENGTH (m)	HEIGHT (m)
(a)	Sports hall	17	28	8
(b)	Telephone kiosk	0.9	0.9	2.4
(c)	Hairdressing salon	5.3	7.2	4.9
(d)	Changing room	3.6	8.4	3.5
(e)	Steam room	4.6	4.8	3.3

7.E8 Figure 7.49 shows the plan of a garden pool which has a uniform depth of 0.5 m and is full of water. Determine the quantity of water contained in the pool in (a) cubic metres and (b) litres.

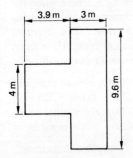

Fig. 7.49

7.E9 Use the density given in Table 7.1 in Section 7.8 to calculate the mass of earth that was excavated to make the garden pool shown in Fig. 7.49.

7.E10 The table gives the dimensions of rectangular petrol tanks. In each case find the capacity of the tank in litres.

TANK	LENGTH	WIDTH	DEPTH
(a)	0.8 m	0.6 m	0.3 m
(b)	75 cm	50 cm	40 cm
(c)	80 cm	65 cm	55 cm
(d)	0.9 m	72 cm	30 cm
(e)	0.7 m	0.5 m	45 cm

7.E11 A treatment for hair care is sold in a 200-ml container which is sufficient to treat 15 customers.

(a) How much of the solution is used per treatment?
(b) If 200 ml costs £3.27 what would one treatment cost?

7.E12 A cylindrical water dispenser contains 16 litres and has a height of 30 cm. Calculate its diameter.

7.E13 A gymnasium is 72 m long and 48 m wide. If the volume of the gymnasium is 52 000 m³, what is its height?

7.E14 A cylindrical silage container is 10 m in diameter and 5 m high and is filled to the top. The contents are transferred to a 14 m diameter container, how high does the silage come up the larger container?

7.E15 Figure 7.50 shows the plan of the barber's shop in a large department store. The ceiling has a uniform height of 5.2 m. Determine the volume in cubic metres of:

(a) booth 1;
(b) booth 2;
(c) booth 3;
(d) the waiting room;
(e) the whole area.

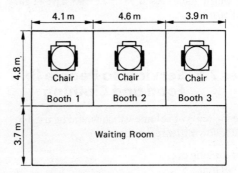

Fig. 7.50

7.E16 Figure 7.51 shows the cross-section of a rabbit hutch. The hutch is 3 m long. What volume of air can be contained in the hutch?

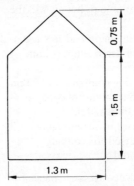

Fig. 7.51

7.E17 A farmer digs a rectangular cross-sectional ditch around a field. The ditch measures 1.2 m deep and 0.8 m wide. The field is 100 m long and 75 m wide. After a storm the ditch is three-quarters full with water. Estimate:

(a) the amount of water in the ditch;
(b) the mass of earth removed to make the ditch.

7.E18 A market gardener has a plastic-covered greenhouse of semi-circular section. 8 m radius and 16 m long. What volume of air does each greenhouse contain?

7.E19 A swimming pool is 25 yd by 20 yd and has a flat bottom giving a depth of 7 ft. If a gallon of water is equal to 0.16 ft³, how many gallons does the pool hold?

7.E20 (a) The hole in a putting green is 4 in in diameter and 6 in deep. A golf ball is 1.062 5 in in diameter.

If a golf ball is dropped into the hole, what volume of space is left?

(b) Determine the capacity of a dustbin 1.5 m tall and 1 m in diameter in (i) cubic metres and (ii) litres.

(c) How many gallons of water can be contained in a water tank which measures $6.5\,\text{ft} \times 2.5\,\text{ft} \times 2\,\text{ft}$. (1 gal of water = $0.16\,\text{ft}^3$.)

Exercises 7.F Services to People II: Food and Clothing

7.F1 State the units of volume which would be used to measure the following (litres, ml, cm^3):

(a) a can of cooking oil;
(b) a bottle of milk;
(c) a cup of coffee;
(d) an urn of tea;
(e) a kettle of water;
(f) a bottle of vanilla essence.

7.F2 (a) The inside of a gas oven measures $60\,cm \times 55\,cm \times 50\,cm$. What volume of hot air will the gas oven hold?

(b) A saucepan is cylindrical in shape and has a diameter of 30 cm, the pan is 18 cm deep. What volume of liquid will it hold?

(c) A deep fat fryer has a cooking compartment 20 cm in diameter and 20 cm deep. What volume of cooking oil will it hold if it is to be no more than half full?

(d) The oil in the chip fryer rises to seven-eighths full when a load of chips are immersed for cooking. What volume of chips does the fryer hold?

7.F3 (a) A chest model deep freezer measures $2\,\text{ft} \times 4\,\text{ft} \times 3\,\text{ft}$ high. What volume of space would it occupy in a kitchen?

(b) Inside the freezer is a section for housing the freezer unit which measures $2\,\text{ft} \times 1\,\text{ft} \times 1\,\text{ft}$. What volume of space does the freezer unit occupy?

7.F4 An electric washing machine has a cylindrical rotary drum which, in order to allow for the water to flow, must not be filled to more than two-thirds of its capacity with clothes. The drum measures 45 cm in diameter and 60 cm deep. What is the maximum volume of clothing that can be washed at any one time?

7.F5 A rectangular roasting tin measures $42\,cm \times 32\,cm \times 8\,cm$. What volume will the tin hold?

7.F6 (a) A microwave oven has a compartment measuring $35\,cm \times 85\,cm \times 75\,cm$. The surrounding switch gear and cabinet make the overall size three times as large as the compartment. What is the overall size of the microwave oven?

(b) Calculate:
(i) the volume of the compartment in cubic centimetres;
(ii) the volume of the oven in cubic metres.

7.F7 Water flowing into a washing machine when the machine is rinsing the clothes, drains a tank measuring $30\,cm \times 30\,cm \times 45\,cm$ for each rinse. The machine rinses three times for every washing cycle. How many litres of water does the machine use for four washings?

7.F8 A tumble drier must be filled not more than half full to allow the air to circulate. The drum is 35 cm in diameter and 43 cm deep and has five ridges inside which take up one-eighth of the capacity. What capacity of clothing will the tumble drier hold?

7.F9 A bread bin has a cross-section as shown in Fig. 7.52. The bin is 60 cm long. What is its capacity in cubic centimetres?

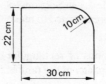

Fig. 7.52

7.F10 Calculate the capacity of each of the following cooking pans (i) in cubic centimetres and (ii) in litres.

PAN	DIAMETER	DEPTH
(a)	24 cm	20 cm
(b)	15 cm	15 cm
(c)	30 cm	23 cm
(d)	350 mm	165 mm
(e)	195 mm	135 mm

7.F11 Figure 7.53 shows the plan of a snack bar. The ceiling has a uniform height of 4 m. Calculate the volume in cubic metres of (a) the seating area, (b) the serving area and (c) the kitchen.

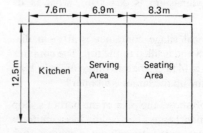

Fig. 7.53

7.F12 The dimensions of a boutique sales-room are 9.2 m wide, 12.6 m long and 4.2 m high. It contains 30 display racks 1.3 m wide, 2 m long and 1.4 m high. What volume of space in cubic metres is left free?

7.F13 Calculate the volume of:

(a) a decorative light in the shape of a sphere, 30 cm diameter;
(b) a changing cubicle, 1.8 m by 1.9 m by 2.4 m;
(c) a hat box, 42 cm diameter and 32 cm deep;
(d) a pattern book, 220 mm × 300 mm × 95 mm;
(e) a video display monitor, 0.4 m × 0.38 m × 28.5 cm;
(f) a roll of cloth, 50 cm diameter and 1.5 m long.

7.F14 Figure 7.54 shows a wooden rule used for measuring cloth. Determine:

(a) its volume in cubic centimetres;
(b) its mass in kilogrammes. (Density is given in Table 7.1 in Section 7.8.)

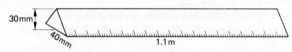

Fig. 7.54

7.F15 Calculate the volume of each of the following rolls of material (answers in cubic centimetres):

ROLL	DIAMETER	LENGTH
(a)	30 cm	2.1 m
(b)	53 cm	1.7 m
(c)	240 mm	0.6 m
(d)	0.4 m	1.85 m
(e)	18 cm	0.45 m

7.F16 A scent bottle is in the shape of a cube of 22 mm side. Determine the capacity of the bottle in millilitres.

7.F17 (a) Find the total mass of the following delivery of greengroceries:

15 kg potatoes
2.5 kg tomatoes
5 kg carrots
500 g grapes
1 kg onions
250 g cherries

(b) A recipe for treacle toffee requires the following ingredients:

440 g barbados sugar
400 g black treacle
160 g butter

Rewrite the list of ingredients to show the correct masses if only 300 g of black treacle is available.

7.F18 Calculate the mass of a wooden bread-board 52 cm × 36 cm × 22 mm. (Density is given in Table 7.1 in Section 7.8.)

7.F19 A funnel is in the shape of a cone and has a base of 12 cm and a vertical height of 21 cm. Calculate the capacity of the funnel in litres.

7.F20 Figure 7.55 shows the plan view of different shapes of moulds for hand-made chocolates. Calculate the volume of each mould in cubic centimetres if the depth is constant at 2 cm.

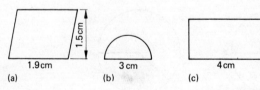
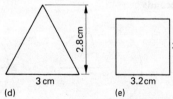
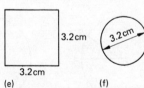

Fig. 7.55

8 Angles and Shapes

8.1 Measurement of Angle

The basic unit of angle is the **degree** which is defined as $\frac{1}{360}$th of one complete rotation of a circle. In Fig. 8.1 the angle x is the amount of rotation of the line OA to reach the position OB. As OA rotates about O to occupy the position OB it moves through an angle of x degrees. Hence,

360 degrees = 1 full revolution

usually written as

360° = 1 rev

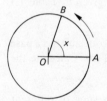

Fig. 8.1

Figure 8.2 shows angles of varying numbers of degrees within 1 revolution. For an accurate measurement of the angle the degree may be divided into smaller portions called **minutes** and **seconds**.

1 degree = 60 minutes

usually written as

1° = 60′

1 minute = 60 seconds

usually written as

1′ = 60″

Thus an angle of 73 degrees 42 minutes 29 seconds would be written as 73°42′29″.

To convert degrees to minutes we multiply by 60, and to convert minutes to seconds we multiply by 60. Thus,

degree × 60 = minutes

$$\frac{\text{minutes}}{60} = \text{degrees}$$

minutes × 60 = seconds

$$\frac{\text{seconds}}{60} = \text{minutes}$$

Example 8.1 Convert the angle 19°35′ to minutes.

Angle in minutes = (19 × 60) + 35
= 1 140 + 35
= 1 175′ (*Ans.*)

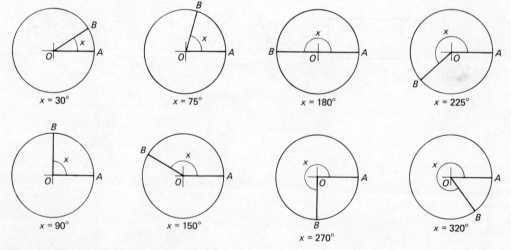

Fig. 8.2

Example 8.2 Convert 7°28′ to seconds.

Angle in minutes = (7 × 60) + 28
= 420 + 28
= 428′
Angle in seconds = 428 × 60
= 26 880″ (*Ans.*)

Example 8.3 Convert 1 354 minutes to degrees.

Angle in degrees = $\frac{1\,354}{60}$ = 22 rem 34
= 22°34′ (*Ans.*)

Sometimes we may require to state an angle in decimal points of a degree.

Example 8.4. Convert 8.27 degrees to an angle in degrees, minutes and seconds.

The 8° remains unchanged. To convert the decimal portion to minutes we multiply by 60

0.27 × 60 = 16.2′

To convert the decimal minutes to seconds

0.2 × 60 = 12″
∴ angle 8.27° = 8°16′12″ (*Ans.*)

Example 8.5 Convert the angle 12°24′36″ to a decimal.
The 12° remains unchanged. First convert the seconds to minutes

$\frac{36}{60} = 0.6′$

thus

24′36″ = 24.6′

Now convert the minutes to degrees

$\frac{24.6}{60} = 0.41°$
∴ angle 12°24′36″ = 12.41° (*Ans.*)

8.2 Types of Angles

Figure 8.3 shows the names given to angles depending on their magnitude:

(*a*) an **acute** angle—an angle less than 90°;
(*b*) a **right** angle—and angle of 90°;
(*c*) an **obtuse** angle—an angle between 90° and 180°; and
(*d*) a **reflex** angle—an angle between 180° and 360°.

When angles add up to 90° they are called **complementary** angles; in Fig. 8.4 angles *x* and *y* are complementary angles. When angles add up to 180° they are called **supplementary** angles; in Fig. 8.5 angles *x* and *y* are supplementary angles.

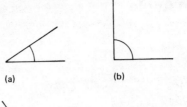
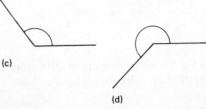

Fig. 8.3

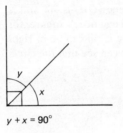

y + *x* = 90°

Fig. 8.4

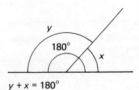

y + *x* = 180°

Fig. 8.5

Example 8.6 Find the angle complementary to 38°23′.
Let *x* = angle to be found. Complementary angles add up to 90°.

∴ *x* + 38°23′ = 90°
x = 90° − 38°23′
x = 51°37′ (*Ans.*)

Example 8.7 Find the angle supplementary to 103°17′49″.

Let *x* = angle to be found. Supplementary angles add up to 180°.

∴ *x* + 103′17′14″ = 180°
x = 180° − 103°17′49″
x = 76°42′11″ (*Ans.*)

8.3 The Right Angle

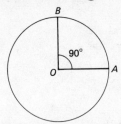

Fig. 8.6

When line *OA* in Fig. 8.6 rotates to position *OB* it moves through one-quarter of a revolution

1 rev = 360°
¼ rev = 90°

It can be seen from Fig. 8.6 that *OB* is perpendicular or 'square' to *OA*. Thus, the angle between a horizontal line (*OA*) and a perpendicular line (*OB*) is 90° and is given the name **right angle**. In an engineer's square, for instance, the perpendicular blade is said to be at right angles to the base. In Fig. 8.6 we can see that the circle contains four right angles.

∴ circle = 360° = 4 right angles.

Fig. 8.7

A right angle may be indicated by either of the methods shown in Fig. 8.7. Angles in a right angle are complementary, i.e. they add up to 90° (Fig. 8.8).

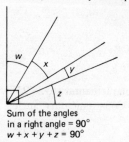

Sum of the angles
in a right angle = 90°
$w + x + y + z = 90°$

Fig. 8.8

Example 8.8   Find the angle x in Fig. 8.9.

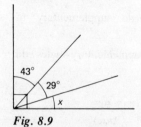

Fig. 8.9

The sum of the angles in a right angle = 90°.

$$x + 29° + 43° = 90°$$
$$x + 72° = 90°$$
$$x = 90° - 72°$$
$$x = 18° \quad (Ans.)$$

Example 8.9   Find the angle x in Fig. 8.10.

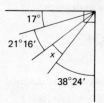

Fig. 8.10

$$17° + 21°16' + x + 38°24' = 90°$$
$$76°40' + x = 90°$$
$$x = 90° - 76°40'$$
$$x = 13°20' \quad (Ans.)$$

8.4 Angles in a Circle

The sum of the angles in a circle is 360° (Fig. 8.11).

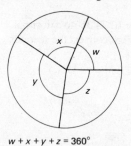

$w + x + y + z = 360°$

Fig. 8.11

Example 8.10   Find the angle x in Fig. 8.12.

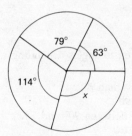

Fig. 8.12

The sum of the angles = 360°.

$$63° + 79° + 114° + x = 360°$$
$$x = 360° - 256°$$
$$x = 104° \quad (Ans.)$$

8.5 Angles in a Semi-circle

The sum of the angles in a semi-circle is 180° (Fig. 8.13).

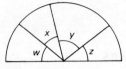

$w + x + y + z = 180°$

Fig. 8.13

Example 8.11   Find the angle x in Fig. 8.14.

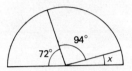

Fig. 8.14

The sum of the angles = 180°.

$$x + 94° + 72° = 180°$$
$$x + 166° = 180°$$
$$x = 180° - 166°$$
$$x = 14° \quad (Ans.)$$

8.6 Angles and Straight Lines

Figure 8.15 shows the intersection of two straight lines. The angles on opposite sides of the intersection are said to be **vertically opposite** angles and are equal.

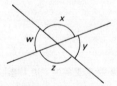

x and z are vertically opposite angles
$x = z$
w and y are vertically opposite angles
$w = y$

Fig. 8.15

Example 8.12   Find the angles x, y and z in Fig. 8.16.

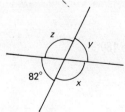

Fig. 8.16

The angles 82° and z are supplementary angles.

$$82° + z = 180°$$
$$z = 180° - 82°$$
$$z = 98°$$

The angles x and z are vertically opposite angles,

$$\therefore x = z$$
$$x = 98°$$

The angle y and 82° are vertically opposite angles,

$$\therefore y = 82°$$

thus

$$x = 98°, y = 82°, z = 98° \quad (Ans.)$$

To check this answer, the sum of the angles gives a full circle about the point of intersection

$$\therefore x + y + z + 82° = 360°$$
$$98° + 82° + 98° + 82° = 360°$$
$$360° = 360°$$

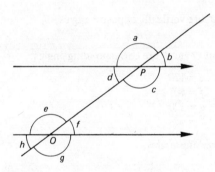

Fig. 8.17

Figure 8.17 shows two parallel lines (indicated by arrows) cut at points O and P by a straight line called a *transversal*. Each angle about point P has a *corresponding* angle about point O which is equal.

Corresponding angles are equal:

$a = e$
$b = f$
$c = g$
$d = h$

The angles on alternate sides of the transversal are called **alternate** angles and are equal.

Alternate angles are equal:

$c = e$
$d = f$

There is also a number of pairs of supplementary angles, e.g.

$a + b = 180°$
$f + g = 180°$

Example 8.13 Find all the missing angles in Fig. 8.18.

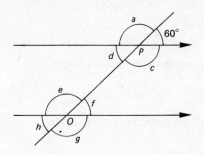

Fig. 8.18

Angles a and $60°$ are supplementary angles,

$$\therefore a + 60° = 180°$$
$$a = 180° - 60° = 120°$$

Angles a and c are vertically opposite angles,

$$\therefore a = c$$
$$c = 120°$$

Angles d and $60°$ are vertically opposite angles,
$$\therefore d = 60°$$

The angles about O and P are corresponding angles:

$a = e$	$\therefore e = 120°$
$60° = f$	$f = 60°$
$c = g$	$g = 120°$
$d = h$	$h = 60°$

Listing all the missing angles:

$a = 120°$
$c = 120°$
$d = 60°$
$e = 120°$
$f = 60°$
$g = 120°$
$h = 60°$ (*Ans.*)

Example 8.14 Determine the angles x and y in Fig. 8.19.

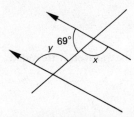

Fig. 8.19

Supplementary angles

$$x + 69° = 180°$$
$$x = 180° - 69°$$
$$x = 111°$$

Alternate angles

$$y = x$$
$$y = 111°$$

$$\therefore x = 111°, y = 111° \quad (Ans.)$$

8.7 Angles in Triangles

The angles A, B and C in the triangle shown in Fig. 8.20 are called **interior** angles. The sum of the interior angles of any triangle is $180°$.

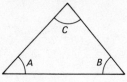

$A + B + C = 180°$

Fig. 8.20

Example 8.15 Determine the angle x in the triangle shown in Fig. 8.21.

Fig. 8.21

The sum of the angles = $180°$

$$\therefore x + 83° + 65° = 180°$$
$$x + 148° = 180°$$
$$x = 180° - 148°$$
$$x = 32° \quad (Ans.)$$

Example 8.16 Determine the angles x, y and z in Fig. 8.22.

Fig. 8.22

Supplementary angles,

$$x + 118° = 180°$$
$$x = 180° - 118°$$
$$x = 62°$$

The sum of angles in a triangle = $180°$.

$$x + 85° + y = 180°$$
$$62° + 85° + y = 180°$$
$$y + 147° = 180°$$
$$y = 180° - 147°$$
$$y = 33°$$

Supplementary angles,

$$y + z = 180°$$
$$33° + z = 180°$$
$$z = 180° - 33°$$
$$z = 147°$$

$x = 62°, y = 33°, z = 147°$ (Ans.)

8.8. Useful Angle Theorems

Figure 8.23 shows the main components of a circle.

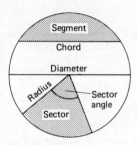

Fig. 8.23

Theorem 1 *The angle made at the centre of a circle by an arc or chord is twice the angle made at the circumference* (Fig. 8.24).

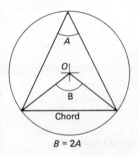

Fig. 8.24

Example 8.17 Determine the angle x in Fig. 8.25.

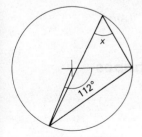

Fig. 8.25

The angle at the centre = 2 × angle at the circumference.

$$112° = 2 \times x$$
$$x = \frac{112°}{2}$$
$$x = 56°\ (Ans.)$$

Theorem 2 *All angles made by an arc or chord at the circumference in the same sector of a circle are equal* (Fig. 8.26).

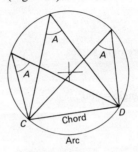

Fig. 8.26

Example 8.18 Determine the angles x and y in Fig. 8.27.

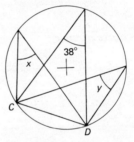

Fig. 8.27

All angles are made by the same chord *CD* and therefore are equal.

$$x = 38° = y$$
$$\therefore x = 38°, y = 38°\ (Ans.)$$

Theorem 3 *The angle made by the diameter at the circumference of a circle is* $90°$, *i.e. angles in semi-circles are right angles* (Fig. 8.28).

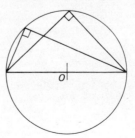

Fig. 8.28

Angles and Shapes

Example 8.19 Determine the angles x and y in Fig. 8.29, where AB is the diameter of the circle.

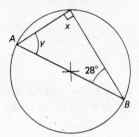

Fig. 8.29

The angles in semi-circles are right angles.

$\therefore x = 90°$

The sum of the angles of a triangle = 180°.

$\therefore x + y + 28° = 180°$
$90° + y + 28° = 180°$
$y = 180° - 118°$
$y = 62°$

$x = 90°, y = 62°$ (*Ans.*)

Theorem 4 *The shaded area in Fig. 8.30 is a sector of a circle, and for any sector*

$$\text{area of sector} = \frac{\text{sector angle}}{360} \times \text{area of circle}$$

$$= \frac{x}{360} \times \pi r^2.$$

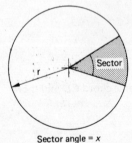

Sector angle = x

Fig. 8.30

Example 8.20 Calculate the area of the shaded portion of the circle shown in Fig. 8.31.

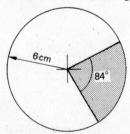

Fig. 8.31

Radius = 6 cm
Sector angle = 84°

$$\text{Area of sector} = \frac{\text{sector angle}}{360} \times \text{area of circle}$$

$$= \frac{84}{360} \times \frac{22}{7} \times 6 \times 6$$

$$= 26.4 \text{ cm}^2 \quad (Ans.)$$

Theorem 5 *A quadrilateral is any four-sided figure. The sum of the interior angles of a quadrilateral is 360°* (Fig. 8.32).

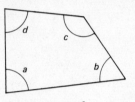

$a + b + c + d = 360°$

Fig. 8.32

Example 8.21 Determine the angles x, y and z in Fig. 8.33.

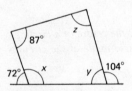

Fig. 8.33

Supplementary angles,

$x + 72° = 180°$
$x = 180° - 72° = 108°$
$y + 104° = 180°$
$y = 180° - 104° = 76°$

The sum of the angles in a quadrilateral = 360°.

$x + y + z + 87° = 360°$
$108° + 76° + z + 87° = 360°$
$z + 271 = 360°$
$z = 360° - 271 = 89°$

$\therefore x = 108°, y = 76°, z = 89°$ (*Ans.*)

8.9 Properties of Shapes

1. RECTANGLE

The **rectangle** (Fig. 8.34) is a quadrilateral or four-sided figure containing four right angles. Opposite sides are equal in length and parallel.

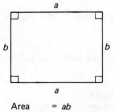

Area = ab
Perimeter = 2(a + b)

Fig. 8.34

2. SQUARE

The **square** (Fig. 8.35) is a rectangle in which all four sides are equal in length. The diagonals of a square intersect at right angles.

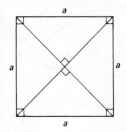

Area = a × a = a^2
Perimeter = 4a

Fig. 8.35

3. PARALLELOGRAM

The **parallelogram** (Fig. 8.36) is a quadrilateral whose opposite sides are equal and parallel but it does not contain a right angle.

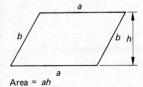

Area = ah

Fig. 8.36

4. POLYGON

A **polygon** is a plane figure bounded by straight lines. Names are given to polygons according to their number of sides.

NUMBER OF SIDES	NAME OF POLYGON
3	Triangle
4	Quadrilaterial
5	Pentagon
6	Hexagon
7	Heptagon
8	Octagon

When the sides have the same length the polygon is said to be a regular polygon (Fig. 8.37).

Regular Hexagon

Regular Octagon

Fig. 8.37

5. CIRCLE

A **circle** (Fig. 8.38) is a plane figure which is bounded by a line which is at a constant distance or radius from its centre.

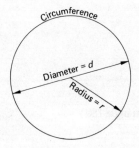

Fig. 8.38

Diameter = 2 × radius
$d = 2r$
Circumference = $2\pi r$ or πd
Area = πr^2 or $\frac{1}{4}\pi d^2$

Angles and Shapes 123

6. TRIANGLE

Triangles are three-sided figures which are classified according to the types of angle that they contain or the relationship between their sides (Fig. 8.39):

(*a*) a **right-angled** triangle has one angle equal to 90°;
(*b*) an **acute-angled** triangle has all angles less than 90°;
(*c*) an **obtuse-angled** triangle has one angle greater than 90°;
(*d*) an **equilateral** triangle has all its sides and all its angles equal, each angle is 60°;
(*e*) an **isosceles** triangle has two sides equal and two angles equal. The equal angles lie opposite to the equal sides; and
(*f*) a **scalene** triangle has all three sides of different length.

Exercises 8.A Core

8.A1 Use a protractor to measure each of the angles marked x in Fig. 8.41.

8.A2 Convert to minutes:
(*a*) 5°   (*b*) 3°12′   (*c*) 14°32′   (*d*) 0.5°   (*e*) 2.5°
(*f*) 8°47′   (*g*) 27°19′

8.A3 Convert to degrees and minutes:
(*a*) 90′   (*b*) 110′   (*c*) 275′   (*d*) 700′   (*e*) 134′

8.A4 Convert to seconds:
(*a*) 4′   (*b*) 2′13″   (*c*) 15′45″   (*d*) 1°8′22″
(*e*) 10°32′16″

8.A5 Convert:
(*a*) 200″ to minutes and seconds;
(*b*) 90.5′ to degrees, minutes and seconds;

     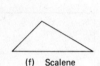

(a) Right-angled   (b) Acute   (c) Obtuse   (d) Equilateral   (e) Isosceles   (f) Scalene

Fig. 8.39

Example 8.22 Find the angles x and y in the isosceles triangle shown in Fig. 8.40.

Fig. 8.40

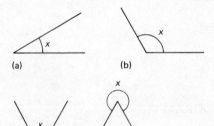

Fig. 8.41

The sum of the angles in a triangle 180°.

$$\therefore x + y + 24° = 180°$$
$$x + y = 180° - 24°$$
$$x + y = 156°$$

x and y are equal angles

$$\therefore x = \frac{156°}{2} \text{ and } y = \frac{156°}{2}$$

$x = 78°, y = 78°$ (*Ans.*)

(*c*) 16 000″ to degrees, minutes and seconds;
(*d*) 350′ to degrees and minutes.

8.A6 Find the value of:
(*a*) 33°15′ + 17°58′ + 9°26′;
(*b*) 175°23′ − 86°44′;
(*c*) 12°38′45″ + 9°41′28″;
(*d*) 28°19′14″ − 13°22′48″.

8.A7 Convert to degrees and minutes:
(*a*) 14.5°   (*b*) 9.3°   (*c*) 121.9°   (*d*) 7.28°   (*e*) 38.16°

8.A8 Convert the following angles to decimals:
(*a*) 2°36′   (*b*) 43°15′   (*c*) 19°42′   (*d*) 150′   (*e*) 13°30′

8.A9 Complete the following table by placing a tick in the correct column.

ANGLE	ACUTE	OBTUSE	REFLEX
35°			
195°			
310°			
85°			
105°			
30°			
245°			

8.A10 (a) State the complementary angle to:
(i) 33° (ii) 78° (iii) 43°35' (iv) 21°12'
(b) State the supplementary angle to:
(i) 104° (ii) 65° (iii) 40°20' (iv) 139°16'

8.A11 Determine the angle x in Fig. 8.42.

(a) (b) (c)

(d) (e) (f)

Fig. 8.42

8.A12 Determine the angle x in Fig. 8.43.

(a) (b)

(c) (d)

 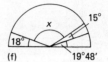
(e) (f)

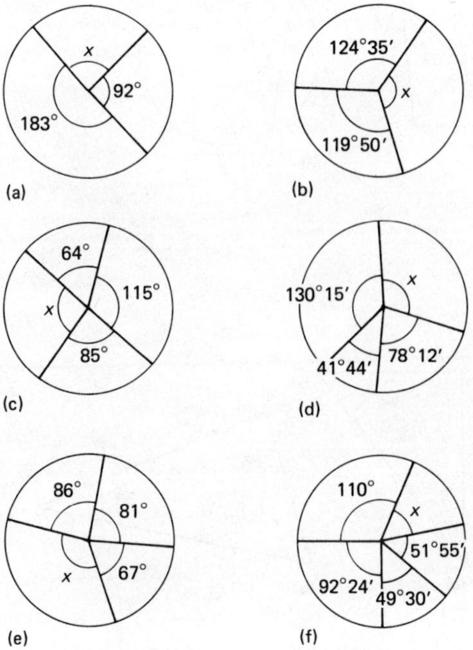

Fig. 8.43

8.A13 Determine the angle x in Fig. 8.44.

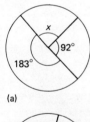

Fig. 8.44

8.A14 Find the angles x, y and z in Fig. 8.45.

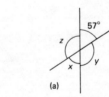
(a) (b)

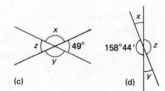
(c) (d)

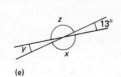 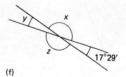
(e) (f)

Fig. 8.45

Angles and Shapes 125

8.A15 Determine all the missing angles in Fig. 8.46.

8.A18 Determine the angles x, y and z in Fig. 8.49.

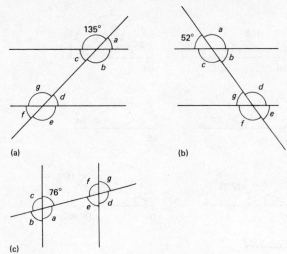
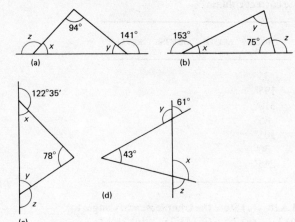

Fig. 8.46

Fig. 8.49

8.A16 Determine the angles x and y in Fig. 8.47.

8.A19 Determine the angle x in Fig. 8.50.

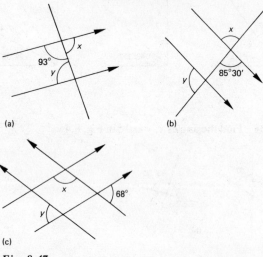
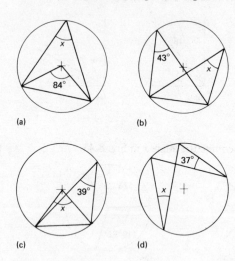

Fig. 8.47

Fig. 8.50

8.A17 Find the angle x in each of the triangles shown in Fig. 8.48.

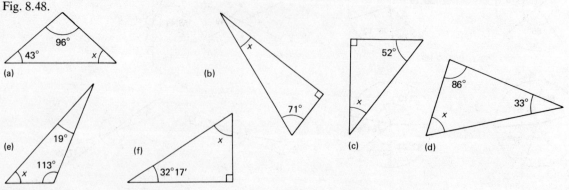

Fig. 8.48

126 **Spotlight on Numeracy**

8.A20 Determine the angles x and y in Fig. 8.51.

8.A22 Calculate the area of a 48° sector of a circle of 100 mm diameter.

8.A23 Determine the angles x, y and z in Fig. 8.53.

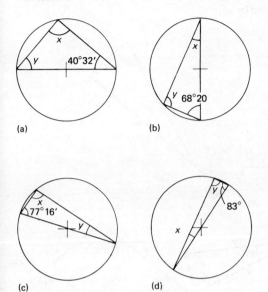

Fig. 8.51

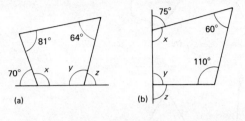

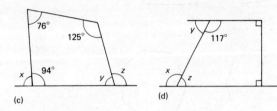

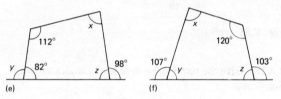

Fig. 8.53

8.A21 Calculate the area of each of the shaded sectors in Fig. 8.52.

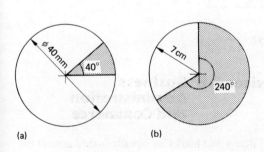

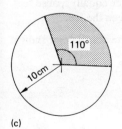

Fig. 8.52

8.A24 Figure 8.54 shows a square cut by two diagonals.
(a) Calculate (i) the area of the square and (ii) the perimeter of the square.
(b) Give the value of (i) angle x and (ii) angle y.
(c) What fraction of the area is shown shaded?

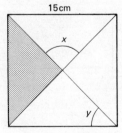

Fig. 8.54

8.A25 Calculate the area of:
(a) a square, 35 mm side;
(b) a parallelogram, 5 cm base, 15 cm vertical height;
(c) a circle, 21 cm radius;
(d) a triangle, 50 mm base, 75 cm vertical height.

Angles and Shapes 127

8.A26 Give the name of each of the polygons shown in Fig. 8.55.

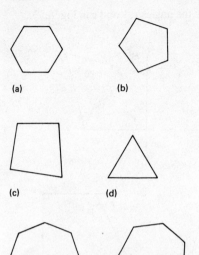

Fig. 8.55

8.A27 Give the name of each of the triangles shown in Fig. 8.56.

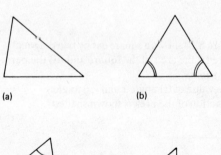

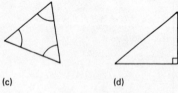

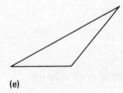

Fig. 8.56

8.A28 Determine the angles x and y in each of the isosceles triangles shown in Fig. 8.57 (the equal sides are indicated by a).

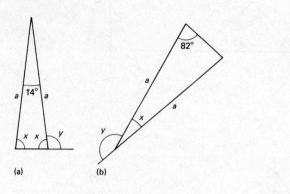

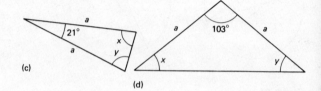

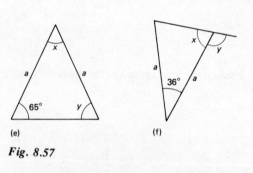

Fig. 8.57

Exercises 8.B Business, Administration and Commerce

8.B1 Fifteen coat hooks are equally spaced around the top of a circular coat stand. What is the angle at the centre of the stand, between any two adjacent hooks?

8.B2 A typist's swivel chair has tree equally spaced feet at its base. What is the angle between the feet?

8.B3 Determine the angle x of the pencil point shown in Fig. 8.58.

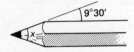

Fig. 8.58

8.B4 Figure 8.59 shows the funnel used to fill a spirit duplicator, determine the angle x.

Fig. 8.59

8.B5 What is the approximate angle between the fingers of a clock at the following times?
(a) 12.30 (b) 12.15 (c) 5.45 (d) 11.10 (e) 2.15
(f) 12.05 (g) 8.50 (h) 3.00 (i) 3.40 (j) 7.45
(k) 2.35 (l) 5.55

8.B6 A blotting pad is in the shape of a square of 35 cm side.
(a) Sketch the pad and dimension its sides.
(b) Find the perimeter.
(c) Find the area.
(d) Draw two diagonals.
(e) Insert the value of the angle between the diagonals.
(f) Insert the value of the angle between a diagonal and the side of the blotter.

8.B7 The following shapes of self-adhesive labels are used in an office to identify files:
(a) a square marked L;
(b) a circle marked R;
(c) an octagon marked K;
(d) a triangle marked Q;
(e) a pentagon marked S;
(f) a hexagon marked P.
Sketch each label and show its marking.

8.B8 (a) The circular label in Exercise 8.B7 has a radius of 18 mm. What is (i) its diameter, (ii) its circumference and (iii) its area?
(b) Calculate the area of a self-adhesive label having the shape of a 125° sector of a circle of 4 cm diameter.

8.B9 A pencil eraser has the shape shown in Fig. 8.60.

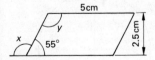

Fig. 8.60

(a) Give the name of the shape.
(b) Find the area.
(c) Determine the values of the angles x and y.

8.B10 A safe dial is circular and has seven equally spaced markings. Find the angle between the markings in:
(a) degrees and minutes (to nearest minute);
(b) degrees (to 2 decimal places).

8.B11 Figure 8.61 shows the side view of a spirit duplicating machine. Handle A rotates the roller and handle B adjusts the printing pressure in three positions, 1, 2 and 3.

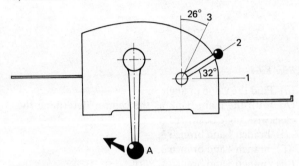

Fig. 8.61

(a) How many revolutions will the roller make when handle A is turned through 900°?
(b) What fraction of a revolution will the roller make when handle A is turned through 160°?
(c) Through what angle does handle B move to change the pressure setting from 2 to 3?
(d) Through what angle does handle B move to change the pressure setting from 3 to 1?

8.B12 Figure 8.62 shows the side view of a video display monitor. Find the values of the angles x, y and z.

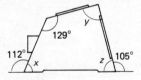

Fig. 8.62

8.B13 (a) How many sides has a 50p coin?
(b) What is the name given to this shape?
(c) Sketch a design for a non-circular £2 coin and name the shape used.

8.B14 Figure 8.63 shows the side view of a word processor.
(a) Give the name of the shape.
(b) Find the values of the angles x, y and z.

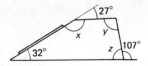

Fig. 8.63

8.B15 The map shown in Fig. 8.64 gives the locations of a town centre bank and five of its branches numbered 1 to 5.

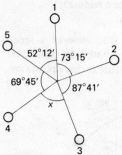

Fig. 8.64

(a) Find the value of angle *x*.
(b) With the main bank as the centre, determine the **clockwise** angle between:
  (i) branch 1 and branch 3;
  (ii) branch 4 and branch 2;
  (iii) branch 3 and branch 5.

Exercise 8.C Technical Services: Engineering and Construction

8.C1 Determine the angle *x* on each of the components shown in Fig. 8.65.

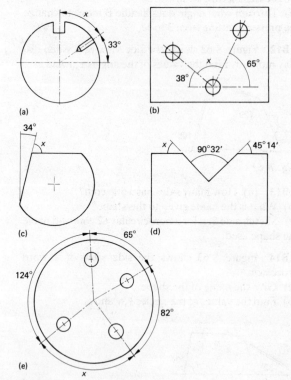

Fig. 8.65

8.C2 Determine the angles *x* and *y* in Fig. 8.66.

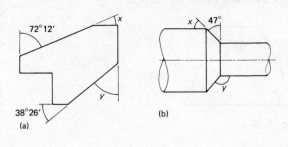

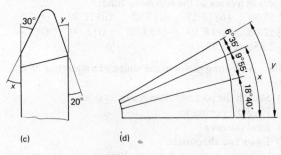

Fig. 8.66

8.C3 Determine the angle at the centre of the flange between the equally spaced holes in each of the drilled flanges shown in Fig. 8.67.

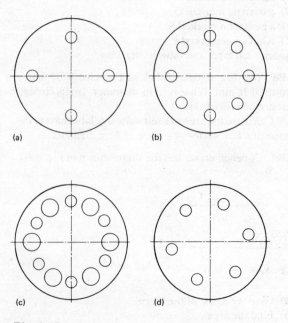

Fig. 8.67

130 Spotlight on Numeracy

8.C4 (*a*) Find the angle between the keyways in the shaft shown in Fig. 8.68.

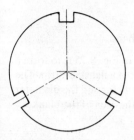

Fig. 8.68

(*b*) Find the angle *x* between the holes A and B in the drilled plate shown in Fig. 8.69.

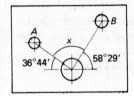

Fig. 8.69

8.C5 Measurements of the included angle *x* of six taper shafts are shown in Fig. 8.70. Determine the mean value of the angle.

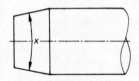

Shaft	Angle *x*
1	30°06'
2	29°58'
3	30°17'
4	30°07'
5	29°42'
6	30°14'

Fig. 8.70

8.C6 Figure 8.71 gives the names of angles on a cutting tool. In each of the exercises give the correct name and the value of the angle marked.

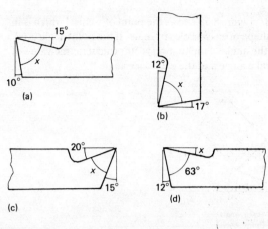

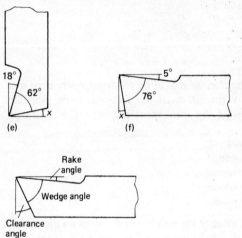

Fig. 8.71

8.C7 Figure 8.72 gives the names of the cutting angles at the point of a chisel. Determine:
(*a*) the value of the rake angle when the point angle is 60° and the clearance angle is 10°;
(*b*) the value of the clearance angle when the rake angle is 15° and the point angle is 63°;
(*c*) the value of the point angle when the rake angle is 25° and the clearance angle is 15°.

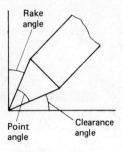

Fig. 8.72

Angles and Shapes 131

8.C8 Figure 8.73 shows the point of a chisel which is in the shape of an isosceles triangle. If the point angle is 60° and the angle of inclination is 50°, calculate the value of the rake angle and the clearance angle.

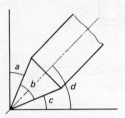

a = Rake angle
b = Point angle
c = Clearance angle
d = Angle of inclination

Fig. 8.73

8.C9 The circular dial of an electrical instrument has 60 equally spaced divisions. Determine the angular movement of the instrument pointer when it moves through:
(a) $\frac{1}{2}$ revolution;
(b) 1 division;
(c) 10 divisions;
(d) $\frac{1}{4}$ revolution;
(e) 33 divisions.

8.C10 Calculate the number of divisions on the dial of the electrical instrument in Exercise 8.C9 which corresponds to an angular movement of 114° by the pointer.

8.C11 Figure 8.74 shows a number of cast concrete slabs. Determine the angles x, y and z.

8.C12 A rectangular bench top, $2\,m \times 1.2\,m$, has rounded corners of a radius of 20 cm. A protection strip is to be attached all round the edge of the bench. Calculate:
(a) the length of the strip;
(b) the surface area of the bench top.

8.C13 The metal blank shown in Fig. 8.75 is to form a funnel and has been stamped out of a flat steel strip. The hole in the centre is 10 mm in diameter and the diameter of the blank is 15 cm. Calculate the area of the blank.

Fig. 8.75

8.C14 Figure 8.76 shows the face of a punch which is used in conjunction with a die to pierce a plate, 8 mm thick, on a power press. Determine:
(a) the surface area of the hole to be punched;

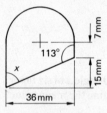

Fig. 8.76

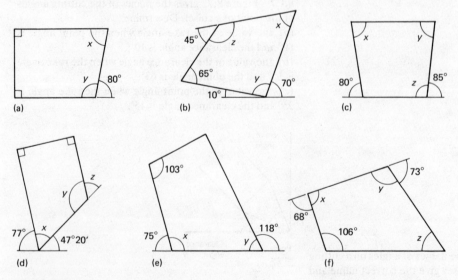

Fig. 8.74

(b) the volume of metal which would be removed from a large plate when seven holes have been punched through it;
(c) the value of angle x.

8.C15 (a) The end ring of the rotor of an induction motor has 22 equally spaced fins to give air circulation. Calculate the angle between the fins correct to the nearest minute.
(b) The coupling on the motor shaft has five equally spaced holes. Calculate the angle between the holes.
(c) Figure 8.77 shows the section of the vee-belt used on the motor driving pulley. Calculate its cross-sectional area.

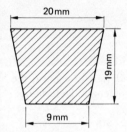

Fig. 8.77

Exercises 8.D General Manufacturing and Processing

8.D1 Determine the angle x on each of the plastic parts shown in Fig. 8.78.

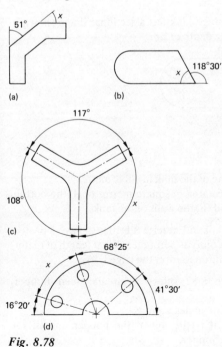

Fig. 8.78

8.D2 Figure 8.79 shows a bell crank lever in the operating mechanism of a process plant. Determine the angles x and y.

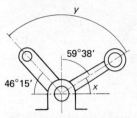

Fig. 8.79

8.D3 A maintenance section stocks various drilled flanges which are used to connect high-pressure steam and water pipes. In use the flanges are bolted together through a number of equally spaced drilled holes. Complete the table showing the angle between the holes on different types of flanges.

FLANGE	NUMBER OF HOLES	ANGLE BETWEEN HOLES
(a)	4	
(b)	6	
(c)	8	
(d)	12	
(e)	18	

8.D4 A large rectangular field is divided by intersecting fences as shown in Fig. 8.80. Determine the value of: (a) angle x, (b) angle y and (c) angle z.

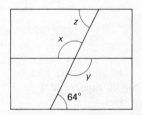

Fig. 8.80

8.D5 A drainage ditch having the section shown in Fig. 8.81 is dug down the length of a field 30 m long.
(a) Determine the value of angle x.
(b) Calculate the cross-sectional area of the ditch.
(c) Calculate the volume of earth which will be removed.

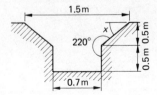

Fig. 8.81

8.D6 Figure 8.82 shows the layout of flower beds in a large garden. Calculate the area of each flower bed.

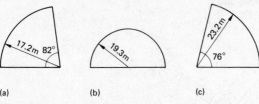 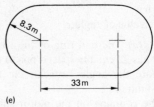

Fig. 8.82

8.D7 Figure 8.83 shows the plan of a roundabout at the junction of five roads on an industrial estate. Determine:
(a) the angle x;
(b) the **clockwise** angle between the roads leading to:
 (i) the town centre and the docks;
 (ii) the railway station and the power station;
 (iii) the docks and the town centre;
 (iv) the power station and the steel works.

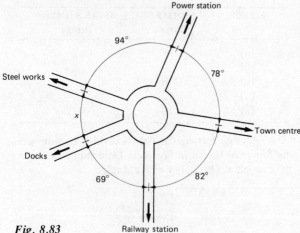

Fig. 8.83

8.D8 The map given in Fig. 8.84 shows the angular locations of five tyre-fitting depots numbered 1 to 5 around a central store in a large town.

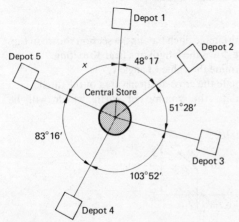

Fig. 8.84

(a) Find the value of angle x.
(b) With the central store as the centre, determine the **clockwise** angle between:
 (i) depot 1 and depot 4;
 (ii) depot 3 and depot 5;
 (iii) depot 2 and depot 5;
 (iv) depot 3 and depot 1.

8.D9 Figure 8.85 shows the end view of a barn, the sloping sides of the roof have equal length. Determine:
(a) the values of the angle x and y;
(b) the cross-sectional area of the barn.

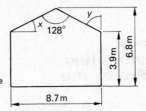

Fig. 8.85

8.D10 A storage tank has a sectional shape shown in Fig. 8.86 and a depth of 2 m.

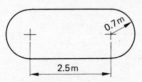

Fig. 8.86

Determine:
(a) the volume of the tank in cubic metres;
(b) the surface area in square metres of (i) the bottom of the tank and (ii) the walls of the tank.

8.D11 The tank in Exercise 8.D10 is to have a protective edge built onto its perimeter. What length of material will be required to form the edge?

8.D12 Figure 8.87 shows a circular feed hopper. Determine:
(a) the value of angles x and y;
(b) the area at (i) the top of the hopper and (ii) the bottom of the hopper.

134 Spotlight on Numeracy

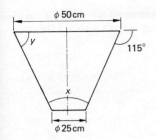

Fig. 8.87

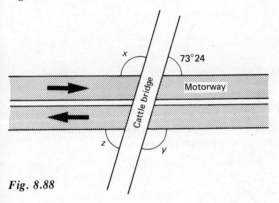

Fig. 8.88

8.D13 Figure 8.88 shows the plan of a cattle bridge crossing over a motorway. Determine the angles (a) x, (b) y and (c) z.

8.D14 A plastics manufacturer produces large plant pots which have a number of drain holes in their base, as shown in Fig. 8.89. Determine the angle between the holes on (a) circle A, (b) circle B and (c) circle C.

8.D15 Determine the angles x, y and z on each of the wooden wedges shown in Fig. 8.90.

Exercises 8.E Services to People I: Community Services

8.E1 Figure 8.91 shows the angular locations of houses, marked 1 to 5, which receive a meals-on-wheels service from a central kitchen. Determine:
(a) the value of angle x;
(b) the clockwise angle between:
 (i) houses 1 and 4;
 (ii) houses 5 and 2;
 (iii) houses 1 and 3;
 (iv) houses 3 and 5.

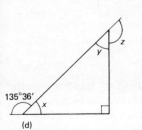

Fig. 8.89

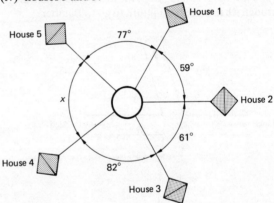

Fig. 8.91

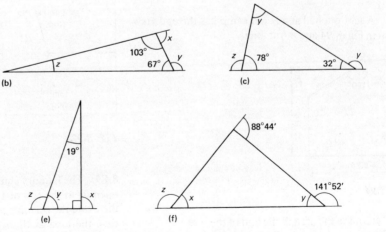

Fig. 8.90

8.E2 State the approximate angle between the hands of a nurse's watch when it shows the following times:
(a) 1.15 (b) 6.00 (c) 4.30 (d) 5.20 (e) 2.05 (f) 7.15 (g) 1.35 (h) 8.10

8.E3 Figure 8.92 shows the measured angles at three corners of an indoor tennis court, determine the angle x at the fourth corner.

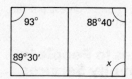

Fig. 8.92

8.E4 Figure 8.93 shows a junction of major and minor roads encountered on a journey by a delivery van.
(a) If the van is coming from the direction of Leyland, through what angle must it turn to go to Bamber?
(b) If the van is coming from the direction of Bamber, through what angle must it turn to go to Grange?
(c) If the van is coming from the direction of Winckley, through what angle must it turn to go to Bamber?

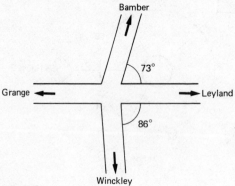

Fig. 8.93

8.E5 A tent pitched at a Scout camp has the end view shown in Fig. 8.94 and is 6 m long.

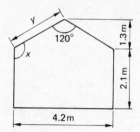

Fig. 8.94

(a) Determine the area of the end of the tent.
(b) Determine the value of the angle x.

8.E6 A large letter C is painted on the side of a bus to the dimensions shown in Fig. 8.95. Calculate the area covered by the letter.

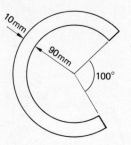

Fig. 8.95

8.E7 Figure 8.96 shows a piece of fine weave material used to make a hairdresser's cape. Calculate the area of the material in square metres.

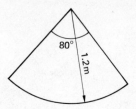

Fig. 8.96

8.E8 Figure 8.97 shows a swing door between the salon and the washroom at a hairdresser's shop. The movement of the door is restricted by limit stops on either side. Determine the value of the angles x, y and z.

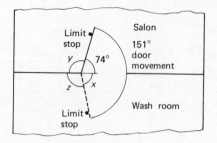

Fig. 8.97

8.E9 The laundry chute in a hospital ward has a square opening at the top of 1.2 m side. At the bottom end of the chute the opening is circular and its area is 20% less than the area at the square end. Find the area of the chute at (a) the top and (b) the bottom.

8.E10 Figure 8.98 shows a snooker table with three balls (white, pink and black) all touching the cushion. Make a sketch of the layout of the table and give the values of all the missing angles: a, b, c, d, e, f and g.

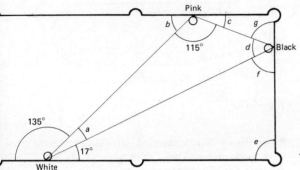

Fig. 8.98

8.E11 Figure 8.99 shows a circular garden pond with three garden ornaments equally spaced around its circumference. Determine the angles x and y.

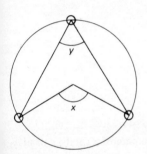

Fig. 8.99

8.E12 Figure 8.100 shows a pedestrian bridge over a railway line. Determine the angles x, y and z.

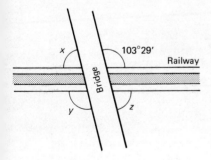

Fig. 8.100

8.E13 (*a*) Calculate the area of the serving tray shown in Fig. 8.101.
(*b*) Estimate the percentage of its surface area that would be covered by a teapot having a circular base 14 cm in diameter.

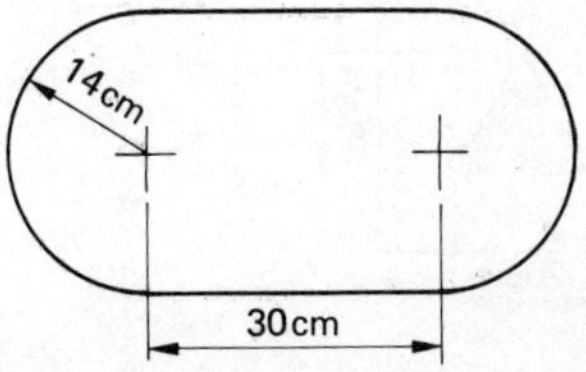

Fig. 8.101

8.E14 A marker flag used in a car rally is in the shape of a square of 40 cm side. The square is divided by diagonal lines into four equal triangles which are coloured red, white, green and black.
(*a*) Sketch the flag and insert the value of the angle between the diagonals.
(*b*) What area of the flag is coloured red?
(*c*) What area of the flag is not coloured black?

8.E15 In an orienteering event, five check points are marked by white squares of 40 cm, each bearing a different red shape. The angular locations of the check points are shown in Fig. 8.102 and the markers are shown in Fig. 8.103.
(*a*) With marker 1 as the centre, determine the clockwise angle between:
  (i) markers 4 and 5;
  (ii) markers 2 and 4;
  (iii) markers 3 and 2.

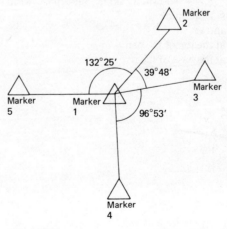

Fig. 8.102

(b) Calculate the area not covered by the red shape on each of the markers shown in Fig. 8.103.
(c) Sketch other markers available for use bearing the following shapes: (i) hexagon, (ii) pentagon and (iii) octagon.

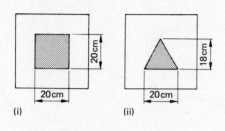
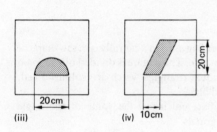

Fig. 8.103

Exercises 8.F Services to People II: Food and Clothing

8.F1 Each of the following light fittings has a number of lamps equally spaced around a circle. Determine the angle between the lamps at the centre of the fitting.
(a) 3 lamps   (b) 4 lamps   (c) 5 lamps   (d) 6 lamps

8.F2 Figure 8.104 shows a small saucepan with tapering sides. Determine:
(a) angles x and y;
(b) the area at the top of the pan;
(c) the area at the base of the pan.

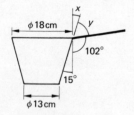

Fig. 8.104

8.F3 Figure 8.105 shows the top face of an ironing board. Determine:
(a) the value of angle x;
(b) the area of the shaded portion.

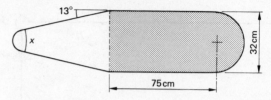

Fig. 8.105

8.F4 Figure 8.106 shows a circular swimming pool 42 m in diameter. Three swimmers, A, B and C, are shown at the circumference and a fourth swimmer, D, at the centre of the pool. Determine:
(a) the circumference of the pool;
(b) the area of the pool;
(c) the value of angle x.

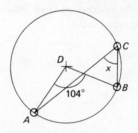

Fig. 8.106

8.F5 Figure 8.107 shows the plan of a rotating display unit in a boutique. The unit has seven radial arms at equal spacing. Calculate the angle between:
(a) any two adjacent arms;
(b) arm 1 and arm 4 in a clockwise direction;
(c) arm 6 and arm 1 in an anti-clockwise direction.
(Give answers correct to the nearest minute.)

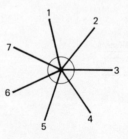

Fig. 8.107

8.F6 A circular pie block is shown in Fig. 8.108. Determine:
(a) the value of angle x;
(b) the surface area at the top of the block;
(c) the surface area at the bottom of the block.

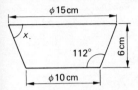

Fig. 8.108

8.F7 What is the approximate angle between the hands of a department-store clock at the following times?
(a) 9.15 (b) 10.10 (c) 12.25 (d) 7.20 (e) 8.30
(f) 2.15

8.F8 A sponge cake is 24 cm in diameter and is iced on the top surface. The cake is cut into 10 equal pieces. Determine:
(a) the angle at the centre of the cake contained in one piece;
(b) the area of icing on one piece;
(c) the volume of one piece if the cake is 5 cm thick.

8.F9 A cylindrical pressure cooker is 400 mm in diameter and 180 mm deep. It has five equally spaced dividers to allow different vegetables to be cooked at the same time. Determine:
(a) the angle between the dividers at the centre of the cooker;
(b) the volume of one compartment.

8.F10 (a) An umbrella has 15 radial spokes, what is the angle between the spokes?
(b) A kitchen ladle is in the shape of a hemisphere 60 cm in diameter, what volume of soup will it contain?

8.F11 Figure 8.109 shows a kitchen knife. Determine the values of the angles x, y and z.

8.F12 Figure 8.110 shows a mould for producing chocolates of different shapes. State the correct name of each shape (a) to (l).

8.F13 A pastry cutter is 60 mm in diameter and has 30 serrations on its edge. Calculate:
(a) the circumference of the pastry cutter;
(b) the angle between the serrations;
(c) the number of serrations in a 120° arc of the cutting edge.

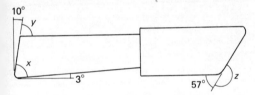

Fig. 8.109

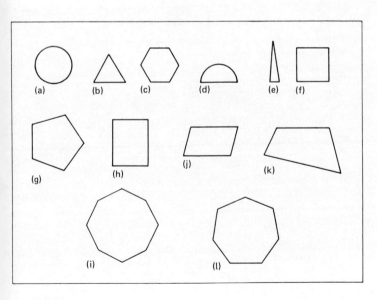

Fig. 8.110

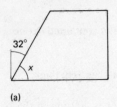 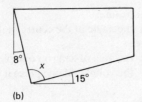 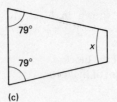 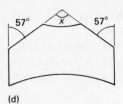

(a) (b) (c) (d)

Fig. 8.111

8.F14 Determine the angle x on each of the pattern pieces shown in Fig. 8.111.

8.F15 A piece of cloth is cut to the shape shown in Fig. 8.112. Determine:
(*a*) the angles x and y;
(*b*) the area of the cloth.

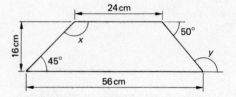

Fig. 8.112

9 Practical Algebra

9.1 Constructing Simple Formulae

Letters or symbols are used in algebra in place of numbers so that general formulae can be readily constructed which will apply not just to one particular problem but to all similar situations.

Figure 9.1 shows a rectangle of length 8 cm and breadth 6 cm. The area of the rectangle is given by the product of its length and breadth.

Area of rectangle = 8 cm × 6 cm
= 48 cm^2

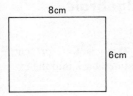

Fig. 9.1

By representing the sides by letters we can obtain a general formula which will apply to rectangles of all sizes.

Let length = l
let breadth = b
let area of rectangle = A

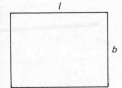

Fig. 9.2

Area of rectangle = length × breadth
$A = l \times b$
usually written as
$A = lb$

Example 9.1 Find the area of a rectangle of length 30 mm and breadth 25 mm.

Using the general formula $A = lb$:

$l = 30$ mm
$b = 25$ mm
$A = lb$
$= 30 \times 25$
$A = 750$ mm^2 (*Ans.*)

Example 9.2 Construct a general formula to convert dimensions in inches to dimensions in millimetres, given that 1 in = 25.4 mm.

Let dimensions in inches = x
let dimensions in millimetres = y
then $y = 25.4 \times x$
$y = 25.4x$ (*Ans.*)

Example 9.3 Construct a formula to find the total cost of producing an engineering component by welding, drilling and fitting, given that welding costs are £20 per hour, drilling costs are £14 per hour and fitting costs are £16 per hour.

Let x = welding time in hours
y = drilling time in hours
z = fitting time in hours
C = total cost in £s

Total cost = welding cost + drilling cost + fitting cost
C = £20 × welding time + £14 × drilling time + £16 × fitting time
$C = 20x + 14y + 16z$ (*Ans.*)

Example 9.4 Use the formula constructed in Example 9.3 to find the cost of producing a component which requires 3 h of welding time, $1\frac{1}{2}$ h of drilling time and 2 h of fitting time.

$x = 3$
$y = 1\frac{1}{2}$
$z = 2$
$C = 20x + 14y + 16z$
$= 20 \times 3 + 14 \times 1\frac{1}{2} + 16 \times 2$
$= 60 + 21 + 32$
$= 113$

Total cost = £113 (*Ans.*)

9.2 Like and Unlike Terms

When two or more quantities are multiples of the same symbol, such as $5x$ and $3x$, they are said to be **like** terms and may be added together or subtracted.

$5x + 3x = 8x$
$5x - 3x = 2x$

Two quantities such as $5x$ and $3y$ are not multiples of the same symbols, these are **unlike** terms and cannot be added together or subtracted.

Example 9.5 $13x + 9x + 2x + 5x$.
These are **like** terms and may be added:

$13x + 9x + 2x + 5x = 29x$ (Ans.)

Example 9.6 $43y - 17y$.
These are **like** terms and may be subtracted:

$43y - 17y = 26y$ (Ans.)

Example 9.7 $14x + 3y$.
These are **unlike** terms and cannot be added:

$14x + 3y$ (Ans.)

Example 9.8 Simplify $3x + 9y + 2y + 7x - 5x - 3y$.
The **like** terms may be added and subtracted:

$(3x + 7x - 5x) + (9y + 2y - 3y)$
$5x + 8y$ (Ans.)

9.3 Brackets

In the last section we used **brackets** to collect the like terms together. Brackets containing unlike terms may also occur in algebraic expressions and must be removed by multiplication before simplifying the expression.

Example 9.9 Simplify $4(a + b) + 7a$.
The bracket is removed by multiplying each individual term inside the bracket by 4:

$4(a + b) + 7a = 4a + 4b + 7a$
$= 11a + 4b$ (Ans.)

Example 9.10 Simplify $8(3x + 2y) + 2(5x + y)$.

$8(3x + 2y) + 2(5x + y) = 24x + 16y + 10x + 2y$
$= 34x + 18y$ (Ans.)

In some problems where negative values occur the rules for the multiplication of **directed numbers** must be observed. When multiplying **like** signs the product is **positive**:

$+ \times + = +$ (plus times plus = plus)
$- \times - = +$ (minus times minus = plus)

When multiplying **unlike** signs the product is **negative**:

$+ \times - = -$ (plus times minus = minus)
$- \times + = -$ (minus times plus = minus)

Example 9.11 Simplify $6(4a + 3b) - 4(2a - 5b)$.

$6(4a + 3b) - 4(2a - 5b) = 24a + 18b - 8a + 20b$
$= 16a + 38b$ (Ans.)

Example 9.12 Evaluate $3(9 - 4) - 2(3 + 1)$.
In this case the quantity inside the bracket can be worked out before removing the bracket:

$3(9 - 4) - 2(3 + 1) = 3(5) - 2(4)$
$= 15 - 8$
$= 7$ (Ans.)

Example 9.13 Evaluate $4(5 - 2) - 2(4 - 7)$.

$4(5 - 2) - 2(4 - 7) = 4(3) - 2(-3)$
$= 12 + 6$
$= 18$ (Ans.)

9.4 Substitution in Algebraic Expressions

An algebraic expression uses symbols to represent numbers. Numbers can be substituted back into the expression in order to evaluate it.

Example 9.14 Evaluate $7x - 4y$, when $x = 4$ and $y = 3$.

$7x - 4y = 7 \times 4 - 4 \times 3$
$= 28 - 12$
$= 16$ (Ans.)

Example 9.15 Find the value of $\dfrac{3x + 12y}{x + y}$, when $x = 4$ and $y = 5$.

$\dfrac{3x + 12y}{x + y} = \dfrac{3 \times 4 + 12 \times 5}{4 + 5}$

$= \dfrac{12 + 60}{9}$

$= \dfrac{72}{9} = 8$ (Ans.)

Example 9.16 Evaluate $4(x + y) - 3(7x + 4z)$, when $x = -2$, $y = 4$ and $z = 3$.

$4(x + y) - 3(7x + 4z) = 4(-2 + 4) - 3(7 \times -2 + 4 \times 3)$
$= 4(2) - 3(-14 + 12)$
$= 8 - 3(-2)$
$= 8 + 6 = 14$ (Ans.)

An alternative method is to simplify the expression before substituting the values:

$$4(x+y) - 3(7x+4z) = 4x + 4y - 21x - 12z$$
$$= 4 \times -2 + 4 \times 4 - 21 \times -2 - 12 \times 3$$
$$= -8 + 16 + 42 - 36$$
$$= 58 - 44$$
$$= 14 \quad (Ans.)$$

9.5 Squares

The **square** of a number is simply the result of multiplying the number by itself. Thus,

the square of $5 = 5 \times 5 = 25$

usually written as

$5^2 = 25$

Example 9.17 Draw up a table of squares for whole numbers from 12 to 20.

NUMBER		SQUARE
12	12×12	144
13	13×13	169
14	14×14	196
15	15×15	225
16	16×16	256
17	17×17	289
18	18×18	324
19	19×19	361
20	20×20	400 (Ans.)

Example 9.18 Evaluate $9^2 + 4^2 - 6^2$.

$$9^2 + 4^2 - 6^2 = 81 + 16 - 36$$
$$= 61 \quad (Ans.)$$

Example 9.19 Find the value of $\dfrac{3^2 \times 6^2}{9^2}$.

$$\frac{3^2 \times 6^2}{9^2} = \frac{9 \times 36}{81}$$
$$= \frac{324}{81} = 4 \quad (Ans.)$$

Algebraic terms can also be squared, thus

$a \times a = a^2$
$x \times x = x^2$
$y \times y = y^2$

Example 9.20 Evaluate $4x^2 - 2y^2$, when $x = 5$ and $y = 3$.

$$4x^2 - 2y^2 = 4 \times 5 \times 5 - 2 \times 3 \times 3$$
$$= 100 - 18 = 82 \quad (Ans.)$$

Example 9.21 Find the value of $\dfrac{8a^2 + 6b^2}{2c^2}$, when $a = 3$, $b = 6$ and $c = 10$.

$$\frac{8a^2 + 6b^2}{2c^2} = \frac{8 \times 3 \times 3 + 6 \times 6 \times 6}{2 \times 10 \times 10}$$
$$= \frac{72 + 216}{200}$$
$$= \frac{288}{200} = 1.44 \quad (Ans.)$$

Example 9.22 Find the value of $3a^2 + b^2$, when $a = 4$ and $b = -3$.

$$3a^2 + b^2 = 3 \times 4 \times 4 + (-3) \times (-3)$$
$$= 48 + 9 = 57 \quad (Ans.)$$

9.6 Square Roots

Finding the **square root** of a number is the reverse of the process of finding the square. The square root of a number is that value which when multiplied by itself equals the number. Thus,

the square root of $9 = 3$ because $3 \times 3 = 9$

usually written as

$\sqrt{9} = 3$

Example 9.23 Find the square root of (a) 64, (b) 144 and (c) 400.

(a) By observation $8 \times 8 = 64$

$\therefore \sqrt{64} = 8 \quad (Ans.)$

(b) By observation $12 \times 12 = 144$

$\therefore \sqrt{144} = 12 \quad (Ans.)$

(c) By observation $20 \times 20 = 400$

$\therefore \sqrt{400} = 20 \quad (Ans.)$

Example 9.24 Find the square root of (a) 225, (b) 324 and (c) 196.

To find the square root of these numbers we can use the table of squares drawn up in Example 9.17

(a) $\sqrt{225} = 15$ because $15 \times 15 = 225$

$\sqrt{225} = 15 \quad (Ans.)$

(b) $\sqrt{324} = 18$ because $18 \times 18 = 324$

$\sqrt{324} = 18 \quad (Ans.)$

(c) $\sqrt{196} = 14$ because $14 \times 14 = 196$

$\sqrt{196} = 14 \quad (Ans.)$

To find squares and square roots of numbers we can use published tables; this is discussed further in Chapter 10.

Algebraic terms can also have square roots

$$a \times a = a^2$$
$$\therefore \sqrt{a^2} = a$$
$$\sqrt{y^2} = y$$

Example 9.25 Evaluate $3\sqrt{x} + y$, when $x = 49$ and $y = 20$.

$$3\sqrt{x} + y = 3\sqrt{49} + 20$$
$$= 3 \times 7 + 20$$
$$= 41 \quad (Ans.)$$

Example 9.26 Find the value of $2\sqrt{a} + 3b^2$, when $a = 100$ and $b = 4$.

$$2\sqrt{a} + 3b^2 = 2\sqrt{100} + 3 \times 4^2$$
$$= 2 \times 10 + 3 \times 16$$
$$= 68 \quad (Ans.)$$

9.7 Cubes

To find the **cube** of a number we multiply the number by itself **twice**. Thus,

cube of $3 = 3 \times 3 \times 3 = 27$

usually written as

$$3^3 = 27$$

It can be seen that the cube is also given by multiplying the square by the number

$$\therefore \text{cube of } 3 = 3^2 \times 3 = 9 \times 3 = 27$$

Example 9.27 Draw up a table of cubes of numbers from 1 to 10.

NUMBER	CUBE	
1	$1 \times 1 \times 1$	1
2	$2 \times 2 \times 2$	8
3	$3 \times 3 \times 3$	27
4	$4 \times 4 \times 4$	64
5	$5 \times 5 \times 5$	125
6	$6 \times 6 \times 6$	216
7	$7 \times 7 \times 7$	343
8	$8 \times 8 \times 8$	512
9	$9 \times 9 \times 9$	729
10	$10 \times 10 \times 10$	1 000 (Ans.)

Example 9.28 Evaluate $9^3 - 17^2$.

Using the table in Example 9.27 $9^3 = 729$. Using the table in Example 9.17 $17^2 = 289$

$$\therefore 9^3 - 17^2 = 729 - 289$$
$$= 440 \quad (Ans.)$$

Example 9.29 Find the value of $5a^3 + 3b^2$, when $a = 4$ and $b = 5$.

$$a^3 = 4 \times 4 \times 4 = 64$$
$$b^2 = 5 \times 5 = 25$$
$$5a^3 + 3b^2 = 5 \times 64 + 3 \times 25$$
$$= 395 \quad (Ans.)$$

Example 9.30 The formula for the volume of a sphere is

$$V = \frac{4}{3}\pi r^3$$

Find the volume of a sphere having a radius of 14 cm.

$$r^3 = 14 \times 14 \times 14 = 2\,744$$
$$V = \frac{4}{3} \times \frac{22}{7} \times 2\,744$$
$$= 11\,498.7 \text{ cm}^3 \quad (Ans.)$$

9.8 Simple Equations and Transposition

A statement showing that one quantity is equal to another is called an **equation**. Examples of equations are:

1 in = 25.4 mm
1 cm = 10 mm
Diameter = 2 × radius
$$V = \frac{4}{3}\pi r^3$$
$$A = lb$$

In studies at this stage equations will contain only one unknown which is represented by a symbol such as x. This type of equation is called a **simple equation** and does not contain any powers of the unknown such as x^2, x^3, etc. It is required to solve the equation in order to find the value of x. For example,

$$2x = 12$$

the **solution** of the equation is $x = 6$.

To assist in solving equations we can use the rules of **transposition** to manipulate the equation in order to obtain the unknown quantity by itself on one side of the equation.

The rules of transposition are as follows:

Rule 1 If a quantity is connected to one side of the equation by a plus or minus sign, it may move to the other side by **changing its sign**, so that plus becomes minus and minus becomes plus.

L.H.S. = R.H.S.
+ ↔ −
− ↔ +

Rule 2 If a quantity multiplies or divides on one side of the equation it may move to the other side by **changing its operation,** so that multiplication becomes division and division becomes multiplication

L.H.S. = R.H.S.
$$\times \leftrightarrow \div$$
$$\div \leftrightarrow \times$$

Example 9.31 Solve $x + 3 = 7$.

To obtain the unknown x by itself on one side of the equation, we must move or transpose the value $+3$. Using Rule 1, the value $+3$ moves from the R.H.S. to the L.H.S. of the equation and becomes -3:

$$x + 3 = 7$$
$$x = 7 - 3$$
$$x = 4 \quad (Ans.)$$

The solution $x = 4$ can be checked by substituting the value 4 for x in the equation

$$x + 3 = 7$$
$$4 + 3 = 7$$
$$7 = 7$$

Therefore solution $x = 4$ is correct.

Example 9.32 Solve $x - 16 = 28$.
Using Rule 1,
$$x - 16 = 28$$
$$x = 28 + 16$$
$$x = 44 \quad (Ans.)$$

Example 9.33 Solve $6x = 42$.
Using Rule 2,
$$6x = 42$$
$$x = \frac{42}{6}$$
$$x = 7 \quad (Ans.)$$

Example 9.34 Solve $\frac{x}{5} = 6$.
Using Rule 2,
$$\frac{x}{5} = 6$$
$$x = 6 \times 5$$
$$x = 30 \quad (Ans.)$$

Example 9.35 Solve $3x + 4 = 31$.
Using Rule 1,
$$3x + 4 = 31$$
$$3x = 27$$

Using Rule 2,
$$x = \frac{27}{3}$$
$$x = 9 \quad (Ans.)$$

Example 9.36 Solve $\frac{x-3}{6} = 7$.
Using Rule 2,
$$\frac{x-3}{6} = 7$$
$$x - 3 = 42$$
Using Rule 1,
$$x = 45 \quad (Ans.)$$

Example 9.37 Solve $\frac{4}{5}k - 8 = \frac{1}{5}k + 4$.
Using Rule 1,
$$\frac{4}{5}k - 8 = \frac{1}{5}k + 4$$
$$\frac{4}{5}k - \frac{1}{5}k = 4 + 8$$
$$\frac{3}{5}k = 12$$

Using Rule 2,
$$k = 12 \times \frac{5}{3}$$
$$k = 20 \quad (Ans.)$$

Check:
$$\frac{4}{5}k - 8 = \frac{1}{5}k + 4$$
$$\frac{4}{5} \times 20 - 8 = \frac{1}{5} \times 20 + 4$$
$$16 - 8 = 4 + 4$$
$$8 = 8$$

Therefore solution $k = 20$ is correct.

Example 9.38 Solve $3(x - 2) = 9$.
$$3(x - 2) = 9$$
Removing brackets
$$3x - 6 = 9$$
Using Rule 1,
$$3x = 15$$
Using Rule 2,
$$x = 5 \quad (Ans.)$$

9.9 Transposition of Formulae

A **formula** is simply an equation and can be manipulated using the same **rules of transposition**. The formula for the area of a rectangle is:

$A = lb$

where A = area,
l = length and
b = breadth.

In its present form A is said to be the **subject** of the formula, i.e. the quantity which would be found when using the formula. However, the formula can be rearranged by transposition to make any other quantity the subject.

To make b the subject,

$A = lb$

$\dfrac{A}{l} = b$

$b = \dfrac{A}{l}$

To make l the subject,

$A = lb$

$\dfrac{A}{b} = l$

$l = \dfrac{A}{b}$

Example 9.39 Transpose the formula $N = \dfrac{1\,000V}{\pi d}$ to make d the subject.

$N = \dfrac{1\,000V}{\pi d}$

$Nd = \dfrac{1\,000V}{\pi}$

$d = \dfrac{1\,000V}{\pi N}$ (Ans.)

Example 9.40 Given that $V = E - IR$ transpose the formula to give (a) E, (b) I and (c) R.

(a) $V = E - IR$
$V + IR = E$
$E = V + IR$ (Ans.)

(b) $V = E - IR$
$V + IR = E$
$IR = E - V$
$I = \dfrac{E - V}{R}$ (Ans.)

(c) $V = E - IR$
$V + IR = E$
$IR = E - V$
$R = \dfrac{E - V}{I}$ (Ans.)

Example 9.41 The area of circle is given by $A = \pi r^2$. Transpose the formula to find r.

$A = \pi r^2$

$\dfrac{A}{\pi} = r^2$

$r^2 = \dfrac{A}{\pi}$

Take square root of both sides:

$\sqrt{r^2} = \sqrt{\dfrac{A}{\pi}}$

$r = \sqrt{\dfrac{A}{\pi}}$ (Ans.)

Example 9.42 Given that $g = \dfrac{fv}{t}$:

(a) transpose the formula to give v;
(b) find the value of v when $g = 200$, $f = 25$ and $t = 0.5$.

(a) $g = \dfrac{fv}{t}$

$gt = fv$

$v = \dfrac{gt}{f}$ (Ans.)

(b) $v = \dfrac{200 \times 0.5}{25}$

$v = 4$ (Ans.)

Exercises 9.A Core

9.A1 Write an algebraic expression to replace each of the following statements. Use the symbols:

x = first number
y = second number
z = third number

(a) The first number plus the second number.
(b) The first number minus the second number.
(c) The second number minus the first number.
(d) The sum of the three numbers.
(e) The product of the first number and the second number.
(f) The product of the second number and the third number.
(g) The first number divided by the second number.

(h) The product of the first number and the second number divided by the third number.
(i) Twice the first number plus three times the second number.
(j) Three times the first number minus twice the third number.
(k) Four times the first number plus twice the second number plus five times the third number.
(l) The square of the first number plus the square root of the third number.
(m) The cube of the second number minus the square of the third number.

9.A2 Construct formulae for:
(a) the area of a triangle;
(b) the perimeter of a rectangle;
(c) the volume of a cube;
(d) the area of a parallelogram.

9.A3 (a) Construct a formula to find angle a in the triangle shown in Fig. 9.3.
(b) Transpose the formula to give angle b.

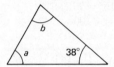

Fig. 9.3

9.A4 Write an algebraic expression to give the area of each of the shapes shown in Fig. 9.4.

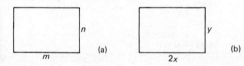

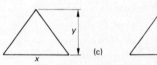

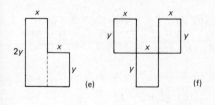

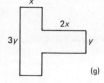

Fig. 9.4

9.A5 (a) Construct a formula to find the volume of the solid cylinder shown in Fig. 9.5.
(b) Transpose the formula to make L the subject.

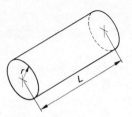

Fig. 9.5

9.A6 Construct a formula to find the total labour cost of building an office extension given that a brick-layer's wages are £2 per hour, a plasterer's £2.50 per hour and an electrician's £3 per hour.

9.A7 Use the formula constructed in Exercise 9.A6 to find the total labour cost if the bricklayer works 30 h, the plasterer 20 h and the electrician 10 h.

9.A8 Figure 9.6 shows the angles on the face of a lathe tool. Construct a formula to give the trail angle in terms of the other two angles.

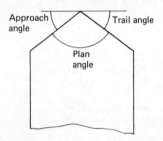

Fig. 9.6

9.A9 Construct a formula to convert:
(a) kilometres to metres;
(b) millimetres to metres;
(c) metres to centimetres;
(d) grammes to kilograms;
(e) millimetres to inches.

9.A10 Simplify the following expressions where possible:
(a) $4x + 9x$ (b) $13y + 8y$ (c) $5x + 4y$
(d) $32x - 17x$ (e) $3x + 9x + 4x$ (f) $6x + 5x + 3y$
(g) $3a + 4b + 5a$ (h) $2a + 3b + c$
(i) $7x + 3y + 14x + y$ (j) $10x + 6y - 4x + 3y$

9.A11 Simplify:
(a) $3 \times 2b$ (b) $2x^2 + 3x^2$ (c) $4x \times 2x$
(d) $3x + 2x^2 + 5x + 3x^2$ (e) $6a^2 + 3a + 5a + a^2$
(f) $3a \times a$ (g) $4b^2 + 3b - b^2 + 6b$ (h) $3b \times 2b$

9.A12 Simplify:
(a) $2x + \dfrac{x}{2} + \dfrac{3x}{2}$ (b) $\dfrac{30ab}{2a}$ (c) $\dfrac{24xy}{3y}$
(d) $\dfrac{15x^2 y}{3x}$ (e) $\dfrac{18a^2}{9a}$

9.A13 Expand:
(a) $3(a+b)$ (b) $7(x+y)$ (c) $2(x-y)$
(d) $2(3x+4)$ (e) $6(2k-1)$ (f) $10(2x+3y)$
(g) $5(3x^2+2x)$ (h) $9(4b^2+3b+4)$

9.A14 Find the value of:
(a) $8+7-6$ (b) $10(7-3)$
(c) $4(13-6)+5(18-9)$ (d) $5(6-1)-(7-9)$
(e) $13(8-5)-2(5-8)$ (f) $4(3+11)+(6-12)$
(g) $12(8+4-7)+2(3+5-12)$
(h) $8(2-9+14)-3(7+6-8)$

9.A15 Given that $a=1$, $b=2$ and $c=3$, evaluate:
(a) $3a+4b+9c$ (b) $15a+2b-8c$
(c) $3(a+b+c)$ (d) $2(a+b)+4c$
(e) $c+(b-a)$ (f) $\dfrac{2a+4c}{b}$
(g) $\dfrac{1}{a}+\dfrac{4}{b}+\dfrac{12}{c}$ (h) $6a+\dfrac{14}{b}-2c$

9.A16 Given that $x=2$, $y=-2$ and $z=4$, evaluate:
(a) $x+y+z$ (b) $2x-y$ (c) $y+2z$
(d) xyz (e) $3yz$ (f) $x(y+z)$
(g) $\dfrac{xy}{z}$ (h) $\dfrac{8xz}{y}$

9.A17 Given that $m=3$, $n=2$ and $p=4$, evaluate:
(a) m^2+n^2 (b) $3p^2$ (c) $\sqrt{p}$ (d) $2m^2+p$
(e) p^2-n^2 (f) $2n^3$ (g) $6m^2-n^2$ (h) $\dfrac{mp^2}{n}$
(i) $(m+n)^2$ (j) $m^3+n^2+p^3$ (k) $\dfrac{p^2}{n^2}$
(l) $4m+6n^2+3p^3$

9.A18 Given that $R=100$, $S=0.4$ and $T=1.2$, evaluate:
(a) $\dfrac{RT}{1000}$ (b) $RS+T$
(c) $ST+R$ (d) $\dfrac{T}{R}+S$
(e) $0.3R+10S$ (f) $2(S+T)+R$
(g) $\dfrac{3R}{4}-5T$ (h) $\dfrac{30S+5T}{R}$

9.A19 Draw up a table of squares for whole numbers from 15 to 30.

9.A20 Use the table in Exercise 9.A19 to evaluate:
(a) 20^2-15^2 (b) 16^2+28^2 (c) $15^2+17^2+19^2$
(d) 27^2-24^2 (e) $3(22^2-18^2)$
(f) $\dfrac{23^2-17^2}{10}$

9.A21 Draw up a table of cubes of whole numbers from 7 to 15.

9.A22 Use the tables in Exercises 9.A19 and 9.A21 to evaluate:
(a) 16^2+7^3 (b) 13^3-9^3 (c) $2(8^3-17^2)$
(d) 15^3-19^2 (e) $20^2+16^2-7^3$

9.A23 State the square root of:
(a) 49 (b) 400 (c) 81 (d) 121 (e) 169

9.A24 Find the value of:
(a) $4^2-\sqrt{4}$ (b) $5^2+\sqrt{100}$ (c) $\sqrt{36}+7^3$
(d) $8^2+\sqrt{25}+2^3$ (e) $10(\sqrt{144}+5^3)$

9.A25 Simplify:
(a) $a \times a \times a$ (b) $b \times b$ (c) $a \times a + b \times b$
(d) $2b \times b$ (e) $a \times 2a \times 3a$
(f) $K \times K \times K - T \times T$ (g) $(\sqrt{a})^2$ (h) $\sqrt{a^2}$

9.A26 Solve:
(a) $x+7=21$ (b) $x+3=9$ (c) $x-2=8$
(d) $x-14=6$ (e) $x+\tfrac{1}{2}=2$ (f) $x-\tfrac{1}{4}=\tfrac{1}{2}$
(g) $x+0.3=1.1$ (h) $x-1.2=5$
(i) $x+2.76=13.09$ (j) $x+11.165=20.598$
(k) $x-0.075=0.295$ (l) $x+3^2=\sqrt{100}$

9.A27 Solve:
(a) $4x=12$ (b) $10x=30$ (c) $7x=21$ (d) $4x=32$
(e) $8x=48$ (f) $2x=\sqrt{16}$ (g) $10x=1.5$
(h) $0.7x=1.4$

9.A28 Solve:
(a) $\dfrac{x}{4}=3$ (b) $\dfrac{x}{10}=1$
(c) $\dfrac{x}{4}=13$ (d) $\dfrac{x}{9}=6$
(e) $\dfrac{x}{2}=1.3$ (f) $\dfrac{x}{6}=\sqrt{9}$
(g) $\dfrac{x}{100}=0.06$ (h) $\dfrac{p}{0.2}=13$

9.A29 Solve:
(a) $4x+2=18$ (b) $2x+1=7$ (c) $3x-8=16$
(d) $5p-1=9$ (e) $7m-12=72$ (f) $5x-16=34$
(g) $2x-9=\sqrt{49}$ (h) $4x+6=7$

9.A30 Solve:
(a) $\tfrac{1}{2}x+8=12$ (b) $\dfrac{1}{3}x+2=17$
(c) $\dfrac{1}{7}x-1=9$ (d) $\dfrac{2}{3}m+6=12$
(e) $\tfrac{3}{4}k-1=5$ (f) $\dfrac{4}{3}t+0.6=1$
(g) $\tfrac{1}{2}x+6=\sqrt{144}$ (h) $\dfrac{x}{10}-1.2=3.15$

9.A31 Solve:
(a) $\dfrac{x-4}{5}=2$ (b) $\dfrac{x+6}{2}=9$
(c) $\dfrac{x-11}{3}=1$ (d) $\dfrac{2x+4}{10}=3$
(e) $\dfrac{7x+6}{5}=4$ (f) $\dfrac{3x+2}{4}=\sqrt{64}$
(g) $\dfrac{\sqrt{9x+5}}{2}=\sqrt{100}$ (h) $\dfrac{100x+0.3}{0.1}=17$

9.A32 Solve:
(a) $3(x+2)=18$ (b) $2(x+3)=24$
(c) $7(x-6)=14$ (d) $4(2x+1)=28$
(e) $10(3x+5)=290$ (f) $5(6x-12)=60$
(g) $3(4x-1)=57$ (h) $2(x+\tfrac{1}{2})=6$

9.A33 Transpose each of the following to make x the subject of the formula:
(a) $L=rtx$ (b) $f=gx+4$
(c) $S=px-3t$ (d) $n=\dfrac{f}{x}$
(e) $Q=\dfrac{rp}{x}$ (f) $\dfrac{m}{n}=\dfrac{y}{x}$
(g) $y=\dfrac{a}{x+2}$ (h) $A=\pi x^2$

9.A34 Transpose:
(a) $R=mxy$ for x (b) $g=mh$ for m
(c) $s=\pi rh$ for r (d) $I=PRT$ for T
(e) $A=\pi r^2 v$ for v (f) $V=\dfrac{\pi d^2 l}{4}$ for l
(g) $n=pm-q$ for m (h) $A=\pi r^2 v$ for r

9.A35 Transpose:
(a) $T=\dfrac{x}{p+q}$ for x (b) $S=p(n-2)$ for n
(c) $C=100(k-a)$ for k (d) $y=a(x+b)$ for x
(e) $F=\dfrac{9C}{5}+32$ for C (f) $P=I^2 R$ for I
(g) $t=r+s(n-1)$ for n (h) $G=\dfrac{a}{p+q}$ for a
(i) $x=\dfrac{0.9}{b+d}$ for b (j) $A=4\pi r^2$ for r

9.A36 The volume of a cone is found from the formula
$$V=\dfrac{\pi r^2 h}{3}$$
where V = volume,
  r = radius of base and
  h = vertical height.

(a) Determine the volume of a cone having a base 18 cm in diameter and a vertical height of 28 cm.
(b) Transpose the formula to (i) make h the subject and (ii) make r the subject.

Exercises 9.B Business, Administration and Commerce

9.B1 Construct formulae for:
(a) the area of an envelope, length l and width w;
(b) the volume of a box file, length l, breadth b and depth d;
(c) the volume of a duplicator roller, diameter D and length L;
(d) the volume of air contained in an office.

9.B2 Construct formulae for the area of each of the gummed labels shown in Fig. 9.7.

(a)

(b)

(c)

(d)

(e)

(f)

(g)

(h)

Fig. 9.7

9.B3 Figure 9.8 shows the dimensions of a three-drawer filing cabinet. Construct formulae to find (a) height H, (b) width W, (c) length L and (d) volume V.

9.B4 The filing cabinet in Exercise 9.B3 has the following dimensions: $a=3$ cm, $b=28$ cm, $c=3$ cm, $d=37$ cm and $l=54$ cm. Calculate its (a) height H, (b) width W, (c) length L and (d) volume V.

9.B5 The total cost of producing an advertising leaflet is made up of writer's costs at £30 per hour, artist's costs at £40 per hour and printing costs at £20 per hour. Construct a formula to find the total cost.

9.B6 Use the formula constructed in Exercise 9.B5 to find the total cost of producing a leaflet, when writer's time is 14 h, artist's time is 9 h and printing time is 45 h.

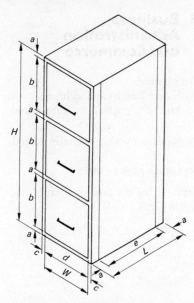

Fig. 9.8

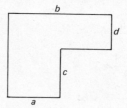

Fig. 9.9

9.B12 The formula for calculation of simple interest is

$$I = \frac{PRT}{100}$$

where P = principal,
R = rate per cent and
T = time in years.

(a) Transpose the formula to give (i) P, (ii) R and (iii) T.
(b) Use the formula to calculate the simple interest when P = £500, R = 11% and T = 3 years.

9.B13 The following formula is used when calculating compound interest: $A = P\left(1 + \dfrac{R}{100}\right)^n$

where A = money accruing after n years investment, P = principal, R = rate per cent and n = number of years of investment. Use the formula to calculate the amount of money which has accrued when £500 has been invested for 2 years at an interest rate of 10%.

9.B14 Construct a formula to give the area of the envelope shown in Fig. 9.10.

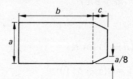

Fig. 9.10

9.B15 The following formula is used when calculating the profit per cent earned by the sale of goods:

$$p = \frac{s - c}{c} \times 100$$

where p = profit per cent, s = selling price and c = cost price. Use the formula to calculate the profit per cent made by a shopkeeper who buys an article for £5 and sells it for £7.

9.B16 A personal saver deposits a sum of £x in the bank each week for a full year. The second year he increases his saving to £y.
(a) Write an algebraic expression to show the amount of money in the saver's account (neglecting interest) (i) at the end of the first year and (ii) at the end of the second year.

9.B7 (a) A car hire company operates a fleet consisting of Ford cars costing £4 800 each, Vauxhall cars costing £5 600 each and Leyland cars costing £3 700 each. Construct a formula to find the total cost of the fleet.
(b) Use the formula to find the total cost of a fleet of 12 Fords, 8 Vauxhalls and 10 Leylands.

9.B8 (a) A typewriter mechanic is paid £3 for each working hour plus £2 for each travelling hour and a subsistence allowance of £15 per week. Construct a formula to calculate his total salary in any week.
(b) Use the formula to calculate his total salary in a week in which he claims for 38 working hours and 12 travelling hours.

9.B9 Construct a formula for:
(a) converting paper sizes given in inches to centimetres;
(b) converting distances given in miles to kilometres;
(c) converting storage space given in cubic metres to cubic feet.
(Use 1 in = 2.54 cm; 1 mile = 1.61 km; 1 f^3 = 0.028 m^3.)

9.B10 Figure 9.9 shows the car-parking area at the rear of a bank.
(a) Construct a formula to find the area.
(b) Use the formula to find the area when a = 11 m, b = 20 m, c = 10 m and d = 12 m.

9.B11 Construct a formula for the perimeter of the car-parking space shown in Fig. 9.9.

(b) If the total amount saved at the end of the second year is £312, and the weekly amount saved in the second year was twice that saved in the first year, find the value of (i) x and (ii) y.

9.B17 The weekly wages of five typists in an office are $£x + 3$, $£x + 1$, $£x$, $£x + 6$ and $£x - 2$.
(a) Write an algebraic expression to show the total weekly wage bill.

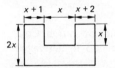

Fig. 9.11

(b) If the total weekly wage bill is £508, what is the value of x?
(c) State the weekly wages of each of the five typists.

9.B18 Figure 9.11 shows the floor plan of a small office. Write an algebraic expression for the floor area.

9.B19 A bank clerk pays income tax on his monthly salary at the rate of $£x + 12$, $£x + 15$ and $£x + 13$ over a 3-month period.
(a) Write an algebraic expression for the total tax paid in the 3-month period.
(b) If the total tax paid was £190 how much tax was paid each month?

9.B20 (a) Punched holes in computer print-out paper are x mm apart and the same distance from the folds in each section. Write an algebraic expression for the length of a section containing 52 holes.
(b) From a ball of string x m long, 15 lengths are cut each b m long. What is the remaining length of string on the ball?
(c) A business has a turnover of $£x$ per month and an expenditure of $£y$ per month. How much profit will be made in 1 year?
(d) A rectangular box containing duplicating paper has the dimensions $(x + 20)$ mm long, $(x + 10)$ mm wide and $(x - 2)$ mm depth. What is the volume of the box?

Exercises 9.C Technical Services: Engineering and Construction

9.C1 Construct a formula to find the dimension x on each of the shafts shown in Fig. 9.12.

9.C2 Construct a formula to find the dimension x on each of the components shown in Fig. 9.13.

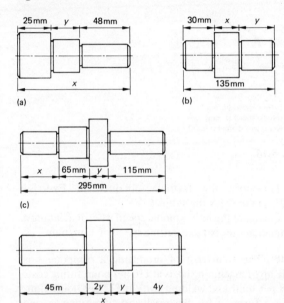

Fig. 9.12

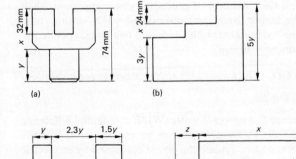

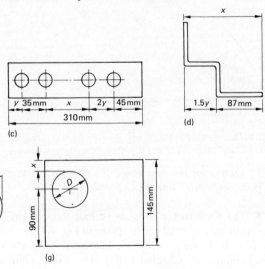

Fig. 9.13

9.C3 Write an algebraic expression for the volume of metal removed when machining the tee-slot shown in Fig. 9.14 in a block of metal 200 mm long.

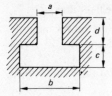

Fig. 9.14

9.C4 Figure 9.15 shows an instrument panel having a punched hole to accommodate an instrument dial. Write an algebraic expression for its area.

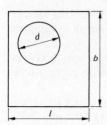

Fig. 9.15

9.C5 Use the expression in Exercise 9.C4 to find the area of an instrument panel having the following dimensions: $d = 90$ mm, $l = 150$ mm and $b = 130$ mm.

9.C6 Figure 9.16 shows the angles at the point of a chisel. Construct a formula to give the rake angle in terms of the other two angles.

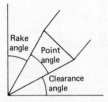

Fig. 9.16

9.C7 Determine the rake angle of a chisel point when the point angle is 60° and the clearance angle is 10°.

9.C8 (*a*) Construct a formula to find the current I flowing in the electrical circuit shown in Fig. 9.17.
(*b*) Use the formula to find the current flowing when $R_1 = 18$ ohms, $R_2 = 30$ ohms, $E = 24$ V. (Use Ohm's Law $E = IR$.)

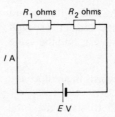

Fig. 9.17

9.C9 (*a*) Figure 9.18 shows a turning operation. Construct a formula to find the cutting speed V in metres per minute.

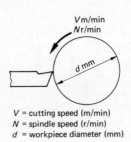

V = cutting speed (m/min)
N = spindle speed (r/min)
d = workpiece diameter (mm)

Fig. 9.18

(*b*) Transpose the formula obtained in Exercise 9.C9(*a*) to make N the subject.
(*c*) Calculate the lathe spindle speed when machining a 28 mm diameter bar at a cutting speed of 30 m/min.

9.C10 The total cost of producing a component is made up of machining costs at £10 per hour, fitting costs at £8 per hour and welding costs at £12 per hour. Construct a formula for the total cost of producing one component.

9.C11 Use the formula obtained in Exercise 9.C10 to find the total cost of producing one component when machining time per component = $1\frac{1}{2}$ h, fitting time per component = 2 h and welding time per component = 30 min.

9.C12 The formula for electrical power can be written:

$P = EI$

where P = power in watts (W), E = potential difference in volts (V) and I = current in amperes (A). Use the formula to calculate the power consumed by an electric fire drawing a current of 8 A from a 250 V supply.

9.C13 (*a*) Transpose the formula given in Exercise 9.C12 to make I the subject.
(*b*) Calculate the current required by a 120-W electric lamp connected to a power supply of 240 V.

9.C14 Construct a formula to find the area of each of the concrete foundations shown in Fig. 9.19.

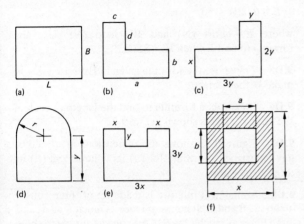

Fig. 9.19

9.C15 The total resistance of two electrical resistors connected in parallel can be found from the formula:

$$\frac{1}{R} = \frac{1}{R_1} + \frac{1}{R_2}$$

where R = total resistance (ohms), R_1 = first resistor (ohms) and R_2 = second resistor, (ohms). Use the formula to find the total resistance of two resistors of 6 ohms and 12 ohms connected in parallel.

9.C16 Temperature in degrees Celsius may be converted to degrees Fahrenheit by using the formula:

$$F = \frac{9}{5}C + 32$$

Convert to degrees Fahrenheit the melting point of a brazing spelter given as 900° Celsius.

9.C17 (a) Transpose the formula given in Exercise 9.C16 to produce a formula for converting degrees Fahrenheit to degrees Celsius.
(b) Convert (i) 68°F to °C and (ii) 100°C to °F.

9.C18 (a) Construct a formula to find the length of the pipe shown in Fig. 9.20.

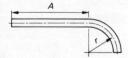

Fig. 9.20

(b) Find the length of the pipe when $A = 300$ mm and $r = 140$ mm.

9.C19 Use Ohm's Law $E = IR$ to complete the following table:

E (VOLTS)	I (AMPERES)	R (OHMS)
	5	10
240	6	
12		8
	3	35
30	0.6	
12	2	
9		45
110		21
240	12	

9.C20 (a) Construct a formula to find the surface speed in metres per minute of a grinding wheel of d mm diameter, when rotating at N r/min.
(b) Use the formula to find the surface speed of a grinding wheel having a diameter of 200 mm and a spindle speed of 2 800 r/min.

9.C21 Given that $V = E - IR$ transpose the formula to find (a) E, (b) I and (c) R.

9.C22 Use the formulae given in Exercise 9.C21 to find:
(a) V when $E = 3$, $I = 1$ and $R = 0.7$;
(b) I when $V = 5.1$, $E = 6$ and $R = 0.6$;
(c) R when $V = 9.6$, $E = 12$ and $I = 3$;
(d) E when $V = 7.2$, $I = 2.5$ and $R = 0.75$.

Exercises 9.D General Manufacturing and Processing

9.D1 Figure 9.21 shows a belt drive to a factory conveyor system. The length of the belt can be found from:

$$\text{length} = \left(\frac{\text{circumference of driving pulley}}{2}\right)$$

$$+ \left(\frac{\text{circumference of drive pulley}}{2}\right)$$

$$+ (2 \times \text{distance between centres})$$

Use the symbols on the diagram to construct a formula for the length of the belt.

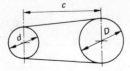

Fig. 9.21

9.D2 Use the formula constructed in Exercise 9.D1 to find the length of the belt when the diameter of the driving pulley = 30 cm, the diameter of the driven pulley = 50 cm and the distance between the centres = 1.8 m.

9.D3 Cardboard boxes for individual packaging of rectangular bars of soap are made to the following dimensions: length = $l + 5$ mm, width = $w + 4$ mm and depth = $t + 3$ mm, where l, w and t are the dimensions of the soap bar in millimetres.
(*a*) Construct a formula for the volume of the box.
(*b*) Calculate the volume of the box for soap bars of dimensions 100 mm × 70 mm × 25 mm.

9.D4 The effort required to raise a screw-operated pit prop is related to the load on the prop by the formula:

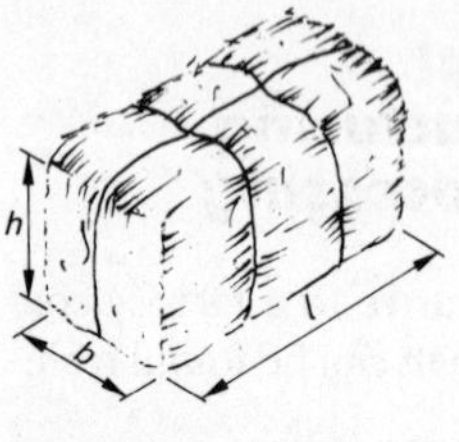

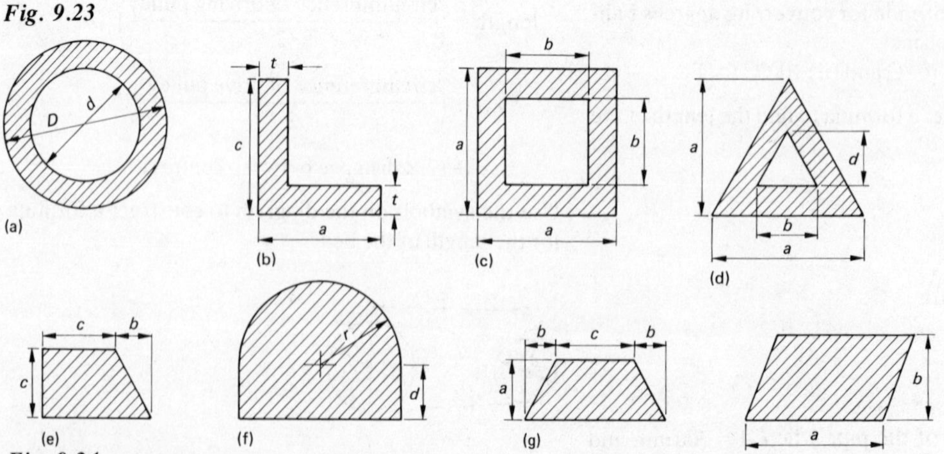

$$E = 0.02W + 50$$

where E = effort (N) and W = load (N). Use the formula to find E when $W = 8500$.

9.D5 Transpose the formula given in Exercise 9.D4 to make W the subject.

9.D6 Construct a formula to find the length of each of the piping sections shown in Fig. 9.22.

9.D7 Figure 9.23 shows the dimensions of a bale of hay, construct a formula for (*a*) its volume and (*b*) its total surface area.

9.D8 A chemical mixture is made up of three constituents A, B and C. It contains twice as much B as A, and three times as much C as B. Construct a formula to show the composition of the mixture.

9.D9 A farmer pays £A for a load of feedstuff and an extra £b for delivery. He is allowed a discount of 10% on the total cost of the feedstuff plus the delivery charge. Construct a formula to find the final cost to the farmer.

9.D10 Construct a formula to find the volume of each of the following processing tanks:
(*a*) a cylindrical tank, x m diameter and y m high;
(*b*) a rectangular tank, x m long, y m wide and z m high;
(*c*) a cylindrical tank with hemispherical top, x m diameter and y m high to the top of the cylindrical portion;
(*d*) a cube-shaped tank, x m side.

9.D11 Construct a formula to find the area of each of the sections of plastic extrusions shown in Fig. 9.24.

9.D12 Calculate the cross-sectional area of each of the extrusions shown in Fig. 9.24, when $a = 48$ mm, $b = 18$ mm, $c = 35$ mm, $d = 23$ mm, $r = 30$ mm, $t = 12$ mm, and $D = 33$ mm.

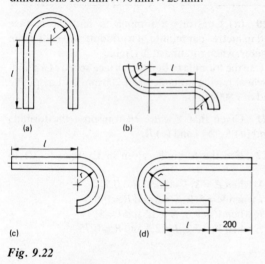

Fig. 9.22

Fig. 9.23

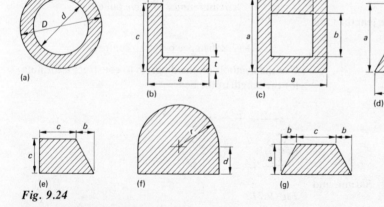

Fig. 9.24

9.D13 Construct a formula to give the mass of a lead net weight of diameter d cm, thickness t cm and density m g/cm^3.

9.D14 Calculate the mass of the net weight in Exercise 9.D13 when the diameter = 80 mm, the thickness = 25 mm and the density = 11 g/cm^3.

9.D15 A market gardener makes deliveries which involve an average travelling time of t min each way and an unloading time of s min.
(a) Construct a formula to give the total time in minutes for making the delivery and returning.
(b) Modify the formula to give the total time taken in hours.

9.D16 An irrigation ditch is x m wide, y m deep and z m long.
(a) Write an equation for the volume (V) of water that could be contained in the ditch.
(b) Determine the volume contained when $x = 0.5$ m, $y = 0.6$ m and $z = 105$ m.

Exercises 9.E Services to People I: Community Services

9.E1 A roll of carpet is L m long. Four pieces of length x m are cut from the roll. Write an algebraic expression for the length of carpet remaining on the roll.

9.E2 The swept volume of one engine cylinder of a delivery van is given by the formula:
$$V = \pi r^2 h$$
where V = swept volume (mm^3), r = radius of piston (mm) and h = stroke of piston (mm).
(a) Calculate the swept volume if the piston diameter is 90 mm and the piston stroke is 100 mm.
(b) Transpose the formula to make r the subject.

9.E3 (a) A sports field is L m long and W m wide. Construct a formula to find its (i) area and (ii) perimeter.
(b) Calculate the area and perimeter when $L = 110$ m and $W = 63$ m.

9.E4 A cast-iron training weight is d mm in diameter and t mm thick. Construct a formula to find the mass of the weight taking the density of the metal as m g/cm^3.

9.E5 Calculate the mass of the training weight in Exercise 9.E4, given that $d = 240$ mm, $t = 50$ mm and $m = 7.3$ g/cm^3.

9.E6 A small delivery van has a 12 V electrical system. Two headlamp bulbs are each 60 W and the sidelights and both tail lights are each 12 W. Use the formula watts = volts × amperes to find the current flowing in amperes when driving at night with headlamps lit.

9.E7 The load compartment of a delivery van has the dimensions, a m long, b m wide and c m high.
(a) Construct a formula to find the volume of the load compartment.
(b) Given that the volume of the load compartment is 9 m^3, $a = 3$ m and $c = 1.5$ m, what is its width?

9.E8 An ambulance has a mass of M kg and carries equipment having a mass of N kg. It carries two ambulance men each having a mass of P kg.
(a) Construct a formula to give the total mass of the vehicle and its contents.
(b) The ambulance is required to cross a temporary bridge which is restricted to vehicles having a maximum mass of S kg. What is the maximum additional load that could be carried across the bridge in the ambulance?

9.E9 A hot-water storage tank at a hospital is filled by two pipes having cross-sectional areas of x and y. The volume of water (V m^3) in the tank after a period of time (T min) can be found from the formula
$$V = \frac{(6x + 9y)}{5} T$$
Calculate the volume of water in the tank after 5 min if $x = 6$ and $y = 7$.

9.E10 Construct a formula to find the area of each of the fitted carpets shown in Fig. 9.25.

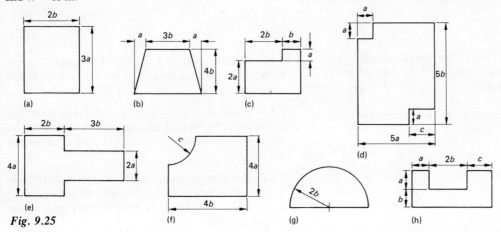

Fig. 9.25

9.E11 Calculate the area of each of the fitted carpets shown in Fig. 9.25, given that $a = 3.2$ m, $b = 4.1$ m and $c = 5.15$ m.

9.E12 State the perimeter of each of the fitted carpets shown in Fig. 9.25 in algebraic form.

9.E13 Figure 9.26 shows the dimensions of a running track. Construct a formula to find the distance run by a competitor in N laps.

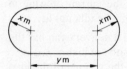

Fig. 9.26

9.E14 Calculate the length of the running track shown in Fig. 9.26 if $x = 35$ m and $y = 120$ m.

9.E15 At a country show a rectangular area of ground is roped off as shown in Fig. 9.27.
(a) Construct a formula to find (i) the length of rope required and (ii) the area of the enclosed ground.

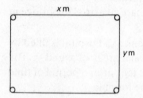

Fig. 9.27

(b) If the length of rope is 30 m and $x = 2y$, what is the area of the enclosed ground?

9.E16 (a) The distance round a circular traffic island is given by

$C = \pi d$

where C = circumference (m) and d = diameter (m).
(i) Transpose the formula to make d the subject.
(ii) Find the diameter of the island when the distance round is 121 m.
(b) The area of a manhole cover is given by

$$A = \frac{\pi d^2}{4}$$

where A = area (cm^2) and d = diameter (cm).
(i) Transpose the formula to make d the subject.
(ii) Find the area of the manhole cover when the diameter is 42 cm.

Exercises 9.F Services to People II: Food and Clothing

9.F1 Construct a formula to find the area of each of the pattern pieces shown in Fig. 9.28.

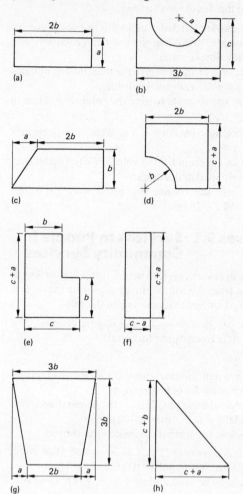

Fig. 9.28

9.F2 Calculate the area of each of the pattern pieces shown in Fig. 9.28, given that $a = 2.5$ m, $b = 3.4$ m and $c = 5.2$ m.

9.F3 A roll of cloth is x m long. Three pieces y m long and two pieces z m long are cut from the roll. Write an algebraic expression for the length of cloth remaining on the roll.

9.F4 A cotton thread is wound round a wooden bobbin x number of times. If the diameter of the bobbin is d mm, construct a formula to find the length of the thread in metres.

9.F5 A patterned fabric is sold in three qualities, A, B and C, at prices of x, y and z £s per metre length. Write algebraic expressions for each of the following:
(a) total cost of 3 m of A, 5 m of B and 10 m of C;
(b) total cost of $2\frac{1}{2}$ m of A, 1.3 m of B and 4 m of C;
(c) the difference in cost between 6 m of A and 4 m of C;
(d) the difference in cost between $8\frac{1}{2}$ m of A plus 3 m of B and 6 m of B plus 2 m of C.

9.F6 Give the answers to Exercise 9.F5(a)–(d) in £s, given that x = £4.80, y = £3.20 and z = £2.50.

9.F7 A sandwich bar has a turnover of £x each day of a 6-day working week. The total expenses of the sandwich bar amount to £y per week. How much profit is made (a) per day and (b) per week?

9.F8 A bakehouse employs 12 bakers who receive a basic weekly wage plus a bonus depending on the weekly output of the bakery. In one week the bonus amounted to £390 to be equally divided between the 12 bakers. If the weekly basic wage was £x, how much did each baker receive?

9.F9 If the total amount paid to the 12 bakers in Exercise 9.F8 was £1 345.80, what was their basic wage?

9.F10 (a) Decorative stitches are made around the hem of a skirt at intervals of x mm. If 76 of these stitches are made, how long is the hem on the skirt?
(b) Write an algebraic expression for the total weekly wage bill at a boutique employing four sales assistants who are paid £x + 2.50 per week and a manageress who is paid £$6x$ + 8 per month.

9.F11 The total weekly wage bill for six workers at a café amounts to £462. The workers receive different amounts as follows: (a) £x, (b) £x + 4, (c) £x + 9, (d) £x + 12, (e) £x + 7 and (f) £x − 2. Determine the amount of money received by each worker.

9.F12 Construct a formula to find the volume of a rectangular shoe box x + 75 mm long, x + 15 mm wide and x − 10 mm depth.

9.F13 A rectangular roasting tin has an area (A) found by multiplying its length (L) by its width (W).
(a) Write an equation for the area.
(b) Show how the length can be expressed in terms of the width and the area.
(c) If the depth of the roasting tin is D, write an equation for the volume.
(d) Determine the volume of the roasting tin when L = 44.72 cm, W = 51.96 cm and D = 32.15 cm.

9.F14 A list of some of the ingredients required to make a fruit cake is:

a raisins
b glacé cherries
c mixed peel
d currants
e sultanas

(a) Write an expression for the total mass of the listed ingredients.
(b) What quantity of glacé cherries would be required to make four cakes?
(c) Write an expression for the difference in mass of raisins and sultanas.
(d) Determine the total mass of the listed ingredients when a = 100, b = 75, c = 125, d = 425 and e = 110.

9.F15 A fish and chip shop sold x fish and y bags of chips in one evening. A fish cost three times as much as a bag of chips.
(a) Write an equation for the evenings takings in £s, if the price of one fish was 90p.
(b) Use the equation to determine the evenings takings if 120 fish and 450 bags of chips were sold.

Practical Algebra 157

10 Use of Tables, Graphs and Diagrams

In every occupation there is an amount of job-related information which is used in carrying out the work. This information may be in the form of instructions, tables of values, readings of measurements or any other type of data which is used when performing specific tasks. Some examples of job-related information are:

- Price list
- Fuel-gauge reading
- Recipe
- Menu
- Railway timetables
- Gearbox speed plate
- Currency conversion table
- Temperature chart
- Speedometer reading
- Radar plot

There are many methods of displaying job-related information, and the choice of method used in a particular job will depend on its convenience and suitability. Some common methods are shown in this chapter.

10.1 Linear and Circular Scales

Graduated scales are used for making measurements of various quantities in all kinds of occupations. The scales may be a **linear** type, such as rules and measuring tapes, or a **circular** type, such as the dial of a clock or an instrument. Some examples of these types are shown in Fig. 10.1 and in the following table.

LINEAR SCALES	CIRCULAR SCALES
Thermometer tube	Ammeter
Vernier	Boiler pressure gauge
Tyre-tread gauge	Car speedometer
Plimsoll load-lines	Oven thermostat
Graduated flask	Shop scales
Surveyor's tape	Gas meter

Example 10.1   State the reading shown on each of the linear scales in Fig. 10.2.

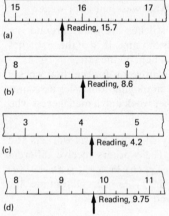

Fig. 10.2

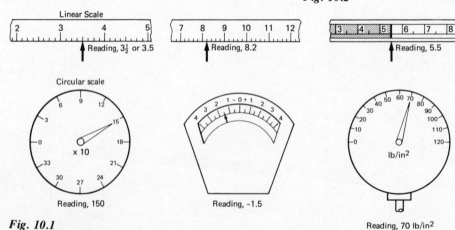

Fig. 10.1

Example 10.2 State the reading shown on each of the circular scales shown in Fig. 10.3.

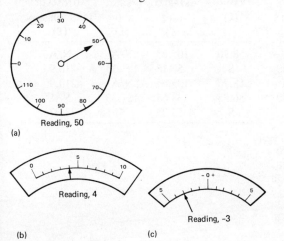

Fig. 10.3

10.2 Tables

Tables are a quick and easy-to-use method of displaying information. The table must indicate clearly the unit value of the listed quantities, e.g. miles, pence, kilogrammes, metres.

Example 10.3 Table 10.1 shows the cost of short-length computer cassettes when ordering in quantity.

TABLE 10.1

LENGTH OF CASSETTE (min)	COST IN PENCE PER CASSETTE		
	10–99 CASSETTES	100–199 CASSETTES	200–500 CASSETTES
C10	28	25	23
C12	30	27	25
C15	35	32	30
C20	40	37	35

Find the cost of:

(a) 50 C12 cassettes;
(b) 180 C20 cassettes;
(c) 480 C10 cassettes;
(d) 3 doz C12 cassettes.

(a) Total cost = 50 × 30
  = 1 500p = £15 (*Ans.*)
(b) Total cost = 180 × 37
  = 6 660p = £66.60 (*Ans.*)
(c) Total cost = 480 × 23
  = 11 040 = £110.40 (*Ans.*)
(d) Total cost = 36 × 30
  = 1 080p = £10.80 (*Ans.*)

Example 10.4 Table 10.2 shows the basic price and the cost of extras for a range of cars. The on-the-road costs are found by adding on the road tax, insurance and delivery charges. (The insurance quotations are for the same driver.) Find the on-the-road cost of:

(a) De Luxe 1.1L with cloth trim and stereo tape;
(b) Coupe 1.3L with leather trim, sports wheels and sun roof;
(c) Estate 1.3L with radio and automatic transmission;
(d) Sports 1.6L with leather trim, stereo tape and automatic transmission;
(e) Super 1.3L with cloth trim, radio and sun roof.

(a) De Luxe 1.1L

	£
Basic price	4 653.00
Cloth trim	44.30
Stereo tape	105.65
Road tax	90.00
Insurance	76.20
Delivery charge	74.00
	5 043.15 (*Ans.*)

(b) Coupe 1.3L

	£
Basic price	5 380.00
Leather trim	139.70
Sports wheels	164.80
Sun roof	121.33
Road tax	90.00
Insurance	121.80
Delivery charge	74.00
	6 091.63 (*Ans.*)

(c) Estate 1.3L

	£
Basic price	5 064.00
Radio	58.70
Automatic transmission	210.90
Road tax	90.00
Insurance	83.50
Delivery charge	74.00
	5 581.10 (*Ans.*)

(d) Sports 1.6L

	£
Basic price	6 255.00
Stereo tape	73.25
Automatic transmission	335.76
Road tax	90.00
Insurance	182.40
Delivery charge	74.00
	7 010.41 (*Ans.*)

TABLE 10.2

MODEL	BASIC PRICE (£)	INSURANCE GROUP[a]	CLOTH TRIM (£)	LEATHER TRIM (£)	SPORTS WHEELS (£)	RADIO (£)	STEREO TAPE (£)	SUN ROOF (£)	AUTOMATIC TRANSMISSION (£)
De Luxe 1.1L	4 653	1	44.30	154.10	N/A[c]	58.70	105.65	N/A	N/A
Super 1.3L	4 917	3	S[c]	128.60	109.20	S	53.15	121.33	210.90
Estate 1.3L	5 064	2	N/A	174.38	N/A	58.70	105.65	N/A	210.90
Coupe 1.3L	5 380	4	S	139.70	164.80	86.25	138.64	121.33	N/A
Sports 1.6L	6 255	6	N/A	S	S	S	73.25	S	335.76

Notes
N/A = Not available
S = Standard cost
Road tax = £90 per year on all models
Delivery charge = £74 on all models

[a]Insurance cost per year:
Group 1 £76.20
Group 2 £83.50
Group 3 £106.50
Group 4 £121.80
Group 5 £159.30
Group 6 £182.40

(e) Super 1.3L

	£
Basic price	4 917.00
Sun roof	121.33
Road tax	90.00
Insurance	106.50
Delivery charge	74.00
	5 308.83 (*Ans.*)

Example 10.5 Table 10.3 shows the multiplication of numbers from 15 to 20.

TABLE 10.3

×	15	16	17	18	19	20
15	225	240	255	270	285	300
16	240	256	272	288	304	320
17	255	272	289	306	323	340
18	270	288	306	324	342	360
19	285	304	323	342	361	380
20	300	320	340	360	380	400

Use the table to find the value of:

(a) $(16 \times 19) + (18 \times 17)$;
(b) $(15 \times 20) - (16 \times 17)$;
(c) $19^2 - 16^2$;
(d) $(19 \times 18) + 17^2$.

(a) $(16 \times 19) + (18 \times 17) = 304 + 306$
$= 610$ (*Ans.*)
(b) $(15 \times 20) - (16 \times 17) = 300 - 272$
$= 28$ (*Ans.*)
(c) $19^2 - 16^2 = 361 - 256$
$= 105$ (*Ans.*)
(d) $(19 \times 18) + 17^2 = 342 + 289$
$= 631$ (*Ans.*)

Example 10.6 Use Table 10.3 to find the cost of:

(a) 18 stamps at 16p each;
(b) 15 chocolate bars at 19p each;
(c) 20 dinner-dance tickets at £16 each.

(a) $18 \times 16 = 288p = £2.88$ (*Ans.*)
(b) $15 \times 19 = 285p = £2.85$ (*Ans.*)
(c) $20 \times 16 = £320$ (*Ans.*)

10.3 Conversion Tables

The use of **conversion tables** is a simple and rapid method of converting values from one unit to another, e.g. yards to metres, Centigrade to Fahrenheit. In the next chapter some simple computer programs are given to produce conversion tables.

Example 10.7 Table 10.4 shows the conversion of inches to centimetres over the range 0 to 39 in.

Use the table to convert to centimetres: (a) 9 in, (b) 16 in, (c) 28 in, (d) 21 in and (e) 250 in.

(a) 9 in = 22.86 cm (*Ans.*)
(b) 16 in = 40.64 cm (*Ans.*)
(c) 28 in = 71.12 cm (*Ans.*)
(d) 21 in = 53.34 cm (*Ans.*)
(e) 25 in = 63.50 cm
250 in = 63.50 × 10 = 635 cm (*Ans.*)

Example 10.8 Determine the area in square centimetres of a rectangle 13 in by 24 in.

Using Table 10.4:
13 in = 33.02 cm
24 in = 60.96 cm
Area of rectangle = $33.02 \times 60.96 = 2\,013\,cm^2$
(*Ans.*)

TABLE 10.4

	INCHES									
	0	1	2	3	4	5	6	7	8	9
INCHES					CENTIMETRES					
0	0	2.54	5.08	7.62	10.16	12.70	15.24	17.78	20.32	22.86
10	25.40	27.94	30.48	33.02	35.56	38.10	40.64	43.18	45.72	48.26
20	50.80	53.34	55.88	58.42	60.96	63.50	66.04	68.58	71.12	73.66
30	76.20	78.74	81.28	83.82	86.36	88.90	91.44	93.98	96.52	99.06

10.4 Tables of Squares and Square Roots

To find the square of large numbers or numbers having a decimal portion, four-figure **tables of squares** may be used. A portion of a squares tables is shown in Table 10.5. Referring to Table 10.5, the value of 5.8^2 is found on the row of values marked 5.8 and in the column headed 0, thus,

$5.8^2 = 33.64$

The value of 6.37^2 is found on the row marked 6.3 and in the column headed 7, thus,

$6.37^2 = 40.58$

The value of 6.184^2 is found from the value on the row marked 6.1 and in the column headed 8, plus the mean difference value in the column headed 4, thus,

$6.184^2 = 38.19$
$ + 5$
$ \overline{38.24}$

Note that the place value of the figure in the mean difference column is the same as that of the last figure in the body of the table.

Example 10.9 Use Table 10.5 to evaluate (a) 5.9^2, (b) 6.38^2 and (c) 6.049^2.

(a) $5.9^2 = 34.81$ (Ans.)
(b) $6.38^2 = 40.70$ (Ans.)
(c) $6.049^2 = 36.48$
$ + 11$
$ \overline{36.59}$ (Ans.)

The four-figure tables of squares are usually given for values from 1.000 to 9.999, but can be used for values outside this range by re-positioning the decimal point in the answer. This can be done by making a rough estimate of the order or size of the answer before referring to the tables.

Example 10.10 Find the value of 61.3^2.

Make a rough estimate by squaring an easy value which is close to 61.3, say 60

$60^2 = 60 \times 60 = 3600$
$\therefore 61.3^2$ must be in the order of 3600

Referring to Table 10.5 and ignoring the place values of the number, the value required is on the row marked 61 and under the column headed 3, giving 37.58. From the

TABLE 10.5 Squares

	0	1	2	3	4	5	6	7	8	9	MEAN DIFFERENCES								
											1	2	3	4	5	6	7	8	9
5.5	30.25	30.36	30.47	30.58	30.69	30.80	30.91	31.02	31.14	31.25	1	2	3	4	6	7	8	9	10
5.6	31.36	31.47	31.58	31.70	31.81	31.92	32.04	32.15	32.26	32.38	1	2	3	5	6	7	8	9	10
5.7	32.49	32.60	32.72	32.83	32.95	33.06	33.18	33.29	33.41	33.52	1	2	3	5	6	7	8	9	10
5.8	33.64	33.76	33.87	33.99	34.11	34.22	34.34	34.46	34.57	34.69	1	2	4	5	6	7	8	9	11
5.9	34.81	34.93	35.05	35.16	35.28	35.40	35.52	35.64	35.76	35.88	1	2	4	5	6	7	8	10	11
6.0	36.00	36.12	36.24	36.36	36.48	36.60	36.72	36.84	36.97	37.09	1	2	4	5	6	7	9	10	11
6.1	37.21	37.33	37.45	37.58	37.70	37.82	37.95	38.07	38.19	38.32	1	2	4	5	6	7	9	10	11
6.2	38.44	38.56	38.69	38.81	38.94	39.06	39.19	39.31	39.44	39.56	1	3	4	5	6	8	9	10	11
6.3	39.69	39.82	39.94	40.07	40.20	40.32	40.45	40.58	40.70	40.83	1	3	4	5	6	8	9	10	11
6.4	40.96	41.09	41.22	41.34	41.47	41.60	41.73	41.86	41.99	42.12	1	3	4	5	6	8	9	10	12

rough estimate the answer is known to be in the order of 3 600

$$\therefore 61.3^2 = 3758 \quad (Ans.)$$

Example 10.11   Find the value of 0.6132^2.
   Rough estimate

$$0.6^2 = 0.6 \times 0.6 = 0.36$$

From Table 10.5, 0.6132^2 gives

$$\begin{array}{r} 37.58 \\ + \quad 2 \\ \hline 37.60 \end{array}$$

Re-positioning the decimal point to the order of the rough estimate

$$0.6132^2 = 0.3760 \quad (Ans.)$$

Two four-figure **tables of square roots** are provided to cover numbers 1 to 10 and numbers from 10 to 100. Portions of these tables are shown in Tables 10.6 and 10.7. The tables of square roots are used in the same manner as the tables of squares.

Example 10.12   Evaluate (*a*) $\sqrt{5.964}$ and (*b*) $\sqrt{59.64}$.

(*a*) Using Table 10.6 (numbers 1 to 10):

$$\begin{array}{r} \sqrt{5.964} = 2.441 \\ + \quad 1 \\ \hline 2.442 \quad (Ans.) \end{array}$$

(*b*) Using Table 10.7 (numbers 10 to 100):

$$\begin{array}{r} \sqrt{59.64} = 7.720 \\ + \quad 3 \\ \hline 7.723 \quad (Ans.) \end{array}$$

To find the square root of numbers less than 1 or numbers greater than 100, a rough estimate is made to position the decimal point in the answer. This estimate is also used to decide which of the two square root tables gives the correct value.

Example 10.13   Evaluate $\sqrt{600}$.
   Rough estimate $25^2 = 625$, therefore the answer is in the order of 25. Table 10.6 gives a value of 2.449 which

TABLE 10.6   Square roots from 1 to 10.

	0	1	2	3	4	5	6	7	8	9	MEAN DIFFERENCES								
											1	2	3	4	5	6	7	8	9
5.5	2.345	2.347	2.349	2.352	2.354	2.356	2.358	2.360	2.362	2.364	0	0	1	1	1	1	1	2	2
5.6	2.366	2.369	2.371	2.373	2.375	2.377	2.379	2.381	2.383	2.385	0	0	1	1	1	1	1	2	2
5.7	2.387	2.390	2.392	2.394	2.396	2.398	2.400	2.402	2.404	2.406	0	0	1	1	1	1	1	2	2
5.8	2.408	2.410	2.412	2.415	2.417	2.419	2.421	2.423	2.425	2.427	0	0	1	1	1	1	1	2	2
5.9	2.429	2.431	2.433	2.435	2.437	2.439	2.441	2.443	2.445	2.447	0	0	1	1	1	1	1	2	2
6.0	2.449	2.452	2.454	2.456	2.458	2.460	2.462	2.464	2.466	2.468	0	0	1	1	1	1	1	2	2
6.1	2.470	2.472	2.474	2.476	2.478	2.480	2.482	2.484	2.486	2.488	0	0	1	1	1	1	1	2	2
6.2	2.4900	2.492	2.494	2.496	2.498	2.500	2.502	2.504	2.506	2.508	0	0	1	1	1	1	1	2	2
6.3	2.510	2.512	2.514	2.516	2.518	2.520	2.522	2.524	2.526	2.528	0	0	1	1	1	1	1	2	2
6.4	2.530	2.532	2.534	2.536	2.538	2.540	2.542	2.544	2.546	2.548	0	0	1	1	1	1	1	2	2

TABLE 10.7   Square roots from 10 to 100.

	0	1	2	3	4	5	6	7	8	9	MEAN DIFFERENCES								
											1	2	3	4	5	6	7	8	9
55	7.416	7.423	7.430	7.436	7.443	7.450	7.457	7.463	7.470	7.477	1	1	2	3	3	4	5	5	6
56	7.483	7.490	7.497	7.503	7.510	7.517	7.523	7.530	7.537	7.543	1	1	2	3	3	4	5	5	6
57	7.550	7.556	7.563	7.570	7.576	7.583	7.589	7.596	7.603	7.609	1	1	2	3	3	4	5	5	6
58	7.616	7.622	7.629	7.635	7.642	7.649	7.655	7.662	7.668	7.675	1	1	2	3	3	4	5	5	6
59	7.681	7.688	7.694	7.701	7.707	7.714	7.720	7.727	7.733	7.740	1	1	2	3	3	4	4	5	6
60	7.746	7.752	7.759	7.765	7.772	7.778	7.785	7.791	7.797	7.804	1	1	2	3	3	4	4	5	6
61	7.810	7.817	7.823	7.829	7.836	7.842	7.849	7.855	7.861	7.868	1	1	2	3	3	4	4	5	6
62	7.874	7.880	7.887	7.893	7.899	7.906	7.912	7.918	7.925	7.931	1	1	2	3	3	4	4	5	6
63	7.937	7.944	7.950	7.956	7.962	7.969	7.975	7.981	7.987	7.994	1	1	2	3	3	4	4	5	6
64	8.000	8.006	8.012	8.019	8.025	8.031	8.037	8.044	8.050	8.056	1	1	2	2	3	4	4	5	6

can be adjusted in accordance with the rough estimate to 24.49 (i.e. in the order of 25). Table 10.7 gives a value of 7.746 which cannot be adjusted to give a value in the order of the rough estimate and hence is incorrect.

$$\therefore \sqrt{600} = 24.49 \quad (Ans.)$$

Example 10.14 Evaluate $\sqrt{0.55}$.

Rough estimate $0.7^2 = 0.49$, therefore the answer is in the order of 0.7. Table 10.6 gives 2.345 which cannot be adjusted to the order of 0.7 and hence is incorrect. Table 10.7 gives 7.416 which can be adjusted to 0.7416

$$\therefore \sqrt{0.55} = 0.7416 \quad (Ans.)$$

10.5 Applications of Squares and Roots

The tables of squares and square roots are used in many practical problems in which the solution of a right-angled triangle is required. In any **right-angled** triangle the length of one side can be found when the length of the other two sides are known by the application of **Pythagoras' Theorem**.

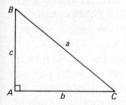

Fig. 10.4

In the triangle shown in Fig. 10.4 the angle at A is a right angle, i.e. $A = 90°$. The longest side of the triangle is called the **hypotenuse** and lies opposite to the right angle. Hence side a is the hypotenuse and the other two sides are b and c.

Pythagoras' Theorem states that *in any right-angled triangle the square on the hypotenuse is equal to the sum of the squares on the other two sides*:

$$(\text{length of hypotenuse } a)^2 = (\text{length of side } b)^2 + (\text{length of side } c)^2$$
$$a^2 = b^2 + c^2$$

Example 10.15 Calculate the length of side a in the right-angled triangle shown in Fig. 10.5.

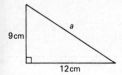

Fig. 10.5

The hypotenuse of the triangle is side a.

$$\therefore a^2 = b^2 + c^2$$
$$= 12^2 + 9^2$$
$$= 144 + 81 = 225$$
$$\therefore a = \sqrt{225} = 15 \text{ cm} \quad (Ans.)$$

Example 10.16 Calculate the length of side x in the triangle shown in Fig. 10.6.

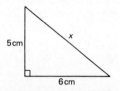

Fig. 10.6

$$x^2 = 6^2 + 5^2$$
$$= 36 + 25$$
$$x = \sqrt{61}$$

Using Table 10.7

$$x = 7.810 \text{ cm} \quad (Ans.)$$

Tables 10.5–10.7 are only portions of the complete tables. The full four-figure tables will be required to answer exercises in this chapter.

Example 10.17 Determine the length of side x in the triangle shown in Fig. 10.7.

Fig. 10.7

The length of the hypotenuse is 10 cm.

$$10^2 = 7^2 + x^2$$
$$100 = 49 + x^2$$
$$x^2 = 100 - 49 = 51$$
$$x = \sqrt{51}$$

Using square root tables,

$$x = 7.141 \text{ cm} \quad (Ans.)$$

Example 10.18 Determine the centre distance C between the holes in the drilled plate shown in Fig. 10.8.

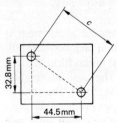

Fig. 10.8

Use of Tables, Graphs and Diagrams 163

Sketching the triangle (Fig. 10.9), the hypotenuse is *C*.

Fig. 10.9

$C^2 = 44.5^2 + 32.8^2$

(using table of squares)

$= 1980 + 1076 = 3056$

(using table of square roots)

$C = \sqrt{3056} = 55.28 \text{ mm}$ (*Ans.*)

10.6 Presenting Information by Graphs

Graphs are a means of presenting large quantities of numerical information in a simple pictorial form, which can be readily assessed and interpreted. A graph can show clearly how one quantity is changing with respect to another and show underlying trends which may be hidden by numerical data. Once the scale of a graph is known its picture of information means the same in any language, and for this reason graphs are much used in international technical and commercial communications.

The table below shows the number of manufactured items of the same type produced in a factory in a production run of 16 consecutive working weeks.

Week	1	2	3	4	5	6
Output	1030	1470	1820	1930	1880	1950

Week	7	8	9	10	11
Output	1520	1110	770	680	840

Week	12	13	14	15	16
Output	1310	1650	1730	1950	1940

The table gives numerical information about what is happening on the factory production line, but it is difficult to understand the information in this form because of the large quantity of numbers. The same information is shown in the form of a graph in Fig. 10.10. The number of manufactured items in each week's production is plotted as a point on the graph, and straight lines are drawn to join the points and give a picture of the information.

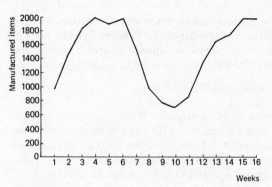

Fig. 10.10

The graph clearly shows that the number of items produced each week increased rapidly in the first four weeks, remained steady at a high level for the next two weeks, but then production dropped rapidly to a very low level in the tenth week. Over the remaining weeks of the run the production rate increased rapidly back to its previous high level and remained constant over the last two weeks. The picture shown by the graph clearly tells us some important facts about the manufacturing process.

1. The process took some weeks after its introduction to build up to a high production rate (weeks 1 to 4).
2. The process was capable of maintaining a high production rate (weeks 5 to 6 and weeks 15 to 16).
3. Some difficulty affected the process (weeks 6 to 10) and reduced production to a very low level. This may have been caused by factors such as shortage of raw materials, failure of machinery or shortage of labour.
4. the difficulty was overcome and production gradually increased back to a high level (weeks 10 to 15).

Example 10.19 The graph given in Fig. 10.11 shows the change in the measured temperature of a small furnace over a period of 7 h.

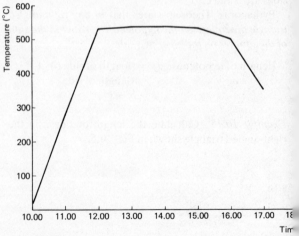

Fig. 10.11

(a) Read from the graph the temperature of the furnace at each hourly interval.
(b) State three items of information that can be interpreted from the graph.

(a) Taking readings of °C from the graph:

Time	10.00	11.00	12.00	13.00	14.00	15.00	16.00	17.00
°C	20	280	570	580	580	570	500	350

(b) (i) The furnace was heated from room temperature (20°C) at a uniform rate for the first 2 h.
  (ii) The furnace was held at its working temperature for a period of 3 h.
  (iii) The furnace was switched off at 15.00 and allowed to cool.

10.7 Procedure for Plotting Graphs

1. Decide which variable is to be represented by the horizontal axis and which by the vertical axis. When plotting equations such as $y = 3x$, the horizontal axis would be used to represent x, hence,
   the horizontal axis is called the x-axis
   the vertical axis is called the y-axis
2. Decide the scale on each axis so that the maximum and minimum values of both variables can be accommodated on the graph. Choose scales which are easy to sub-divide.
3. Draw the axes. The point of intersection of the axes is called the **origin**.
4. Label each axis to show what variable is represented, and sub-divide the axis from the origin according to the scale.
5. Plot each point from the table of values by drawing a vertical line up from the value on the x-axis, to intersect with a horizontal line across from the value on the y-axis. Mark each point with a cross or a ringed dot.
6. Join the points with straight lines or smooth curves.

Example 10.20 Given that the rate of exchange between the pound sterling and the Austrian schilling is 27 schillings = £1, construct a graph to give the value of the schilling in pounds up to £10.

A table is first drawn up to give the number of schillings for the range £0 to £10.

£	0	1	2	3	4	5	6	7	8	9	10
Schillings	0	27	54	81	108	135	162	189	216	243	270

The horizontal axis will represent pounds and be numbered from 0 to 10. The vertical axis will represent schillings and be numbered from 10 to at least 270. Figure 10.12 shows the graph.

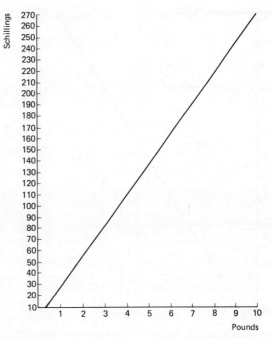

Fig. 10.12

Example 10.21 Construct a graph to find the circumference of circles over a range of diameters from 0 to 50 cm.

A table is drawn up to give the circumference C over a range of diameters d from 0 to 40 cm, using $C = \pi d$.

d (cm)	0	10	20	30	40
C (cm)	0	31.4	62.8	94.2	125.6

Figure 10.13 shows the graph.

Example 10.22 Use the graph shown in Fig. 10.13 to determine (a) the circumference of a circle 35 cm in diameter and (b) the diameter of a circle of 85 cm circumference.

(a) Referring to Fig. 10.13, a vertical line (the dashed line) is drawn up from the value of $d = 35$ on the horizontal axis until it cuts the curve of the graph. From this intersection a horizontal line is drawn across to the vertical axis and the value of C is read from the scale:

$C = 110$ cm   (*Ans.*)

(b) The same method is used in reverse (the dashed and dotted line) to find the value of d:

$d = 27$ cm   (*Ans.*)

The process shown in Example 10.22 of finding values within the graph is called **interpolation**.

Example 10.23 Draw the graph of $y = x^3$ for values of x from 0 to 5. By interpolation from the graph, obtain

Use of Tables, Graphs and Diagrams 165

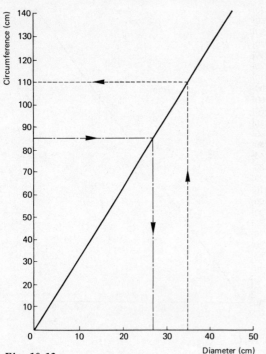

Fig. 10.13

(*a*) the value of *y* when *x* = 4.5 and (*b*) the value of *x* when *y* = 60.

x	0	1	2	3	4	5
$y = x^3$	0	1	8	27	64	125

The graph is shown in Fig. 10.14. From Fig. 10.14:

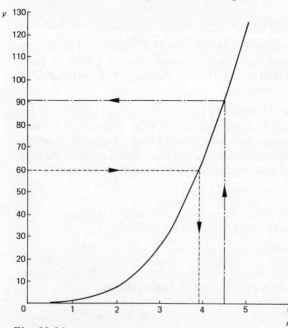

Fig. 10.14

(*a*) when $x = 4.5$, $y = 91$ (*Ans.*)
(*b*) when $y = 60$, $x = 3.9$ (*Ans.*)

10.8 Conversion Graphs

Conversion graphs are often used in industry and commerce as a rapid means of converting from one quantity to another. They have the advantage over conversion tables that all values within the range can be converted by using interpolation. However, the accuracy of making the conversion will depend on the accuracy of drawing and reading the graph.

Example 10.24 Draw a graph to convert inches to millimetres for the range 0 to 5 in. Use the graph to (*a*) convert 4.5 in to millimetres and (*b*) convert 70 mm to inches. (1 in = 25.4 mm.)

Inches	0	1	2	3	4	5
Millimetres	0	25.4	50.8	76.2	101.6	127

Figure 10.15 shows the graph.

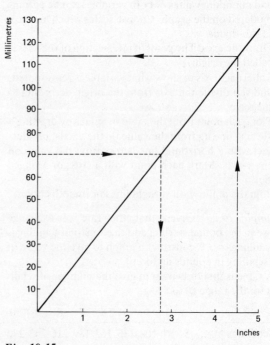

Fig. 10.15

(*a*) 4.5 in = 114.00 mm (*Ans.*)
(*b*) 70 mm = 2.75 in (*Ans.*)

10.9 Bar Charts and Histograms

Bar charts are a common method of visual presentation of information used to show the relative size of quantities in a simple and clear fashion. The size of each quantity is represented by the length of a bar. The bar chart may be drawn horizontally or vertically. The width of the bar is not important and only one axis has a scale.

Example 10.25 A clothing factory has a total labour force of workers made up as follows:

Machinists	24
Cutters	10
Finishers and pressers	16
Maintenance workers	5
Packers	14
Clerical workers	9

Display this information as a horizontal bar chart.
Figure 10.16 shows the bar chart.

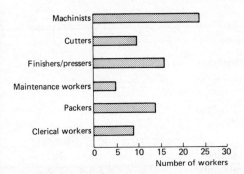

Fig. 10.16

Example 10.26 The table shows the percentage of accidents occurring in a furniture factory which could be attributed to various causes. Display this information on a bar chart.

Moving machinery	15%
Hand tools	18%
Electric shock	6%
Falling objects	9%
Burns	22%
Other causes	30%

The bar chart is shown in Fig. 10.17.

A **histogram** has scales on both axes and is used when the data refer to a continuous variable. In this case the data are represented by the area of the bar and the width of the bar is important.

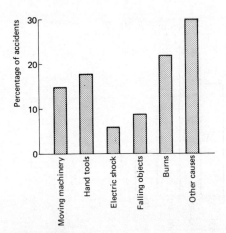

Fig. 10.17

Example 10.27 The table shows the number of small twist drills of standard size in the stock of a machine shop store. Show this information on a histogram.

Drill dia. (mm)	2	3	4	5	6	7	8
Number of drills	12	7	16	20	13	5	24

Figure 10.18 shows the histogram.

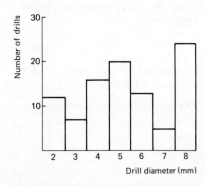

Fig. 10.18

Example 10.28 The table shows the measured life in hours before failure of 200 electric lamps. Show this information in a histogram using the vertical axis for the number of lamps.

LIFE (h)	NUMBER OF LAMPS
400–500	12
500–600	28
600–700	46
700–800	59
800–900	36
900–1 000	19

Use of Tables, Graphs and Diagrams 167

The histogram is shown in Fig. 10.19.

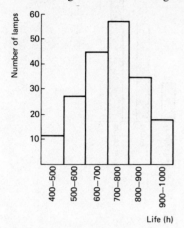

Fig. 10.19

10.10 Pie Charts

Pie charts or **Circular diagrams** are an alternative method of displaying the same information that could be shown on a bar chart. In this diagram the total of the given information is represented by a full circle and each separate piece of information is shown as a sector of the circle. The value of each piece of information is represented by the area of the sector.

Example 10.29 The time taken to produce an engineering component is made up as follows:

Machining	75 min
Fitting	25 min
Heat treatment	60 min
Inspection	20 min

Show this information on a pie chart.

The total time is represented by a full circle of 360°.

Total time = 75 + 25 + 60 + 20 = 180 min
∴ 180 min = 360°

$$1 \text{ min} = \frac{360}{180} = 2°$$

A table can now be drawn up showing the sector angle representing each operation

	TIME (mins)			SECTOR ANGLE (degree)
Machining	75	× 2 =		150
Fitting	25	× 2 =		50
Heat treatment	60	× 2 =		120
Inspection	20	× 2 =		40
	180	× 2 =		360

Figure 10.20 shows the pie chart obtained.

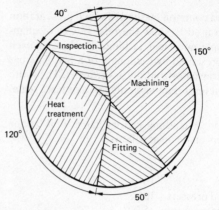

Fig. 10.20

Example 10.30 The causes of breakdown when operating a fleet of hire cars were recorded and shown as percentages of the total number of breakdowns.

Electrical faults	45%
Mechanical faults	32%
Tyres	11%
Fuel system	8%
Other causes	4%

Show this information on a pie chart.

100% = 360°

$$1\% = \frac{360}{100} = 3.6°$$

	PERCENTAGE		DEGREES
Electrical faults	45	× 3.6 =	162
Mechanical faults	32	× 3.6 =	115.2
Tyres	11	× 3.6 =	39.6
Fuel system	8	× 3.6 =	28.8
Other causes	4	× 3.6 =	14.4
	100	× 3.6 =	360

The resulting pie chart is shown in Fig. 10.21.

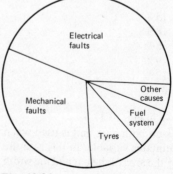

Fig. 10.21

Exercises 10.A Core

10.A1 State the reading on each of the linear scales shown in Fig. 10.22.

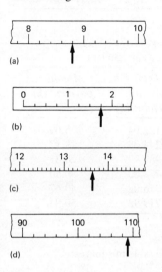

Fig. 10.22

10.A2 State the readings on each of the circular scales shown in Fig. 10.23.

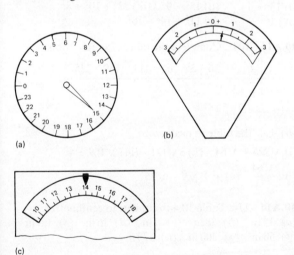

Fig. 10.23

10.A3 Sketch each of the following:
(a) a linear scale marked in centimetres and millimetres showing the reading 16.5 mm;
(b) a linear scale marked in inches and $\frac{1}{8}$ in showing the reading $2\frac{5}{8}$ in;
(c) a full circular scale having 12 divisions showing the reading 8.5 divisions.

10.A4 Explain what is meant by each of the following terms and give a practical example where each may be used:
(a) a linear scale;
(b) a circular scale;
(c) a digital readout.

10.A5 Table 10.8 gives the cost per square metre of grades of floor carpet when ordered in various quantities. Find the cost of:
(a) 25 m² Hardwear; (b) 130 m² Custom;
(c) 15 m² Custom; (d) 60 m² Luxury;
(e) 200 m² Hardwear; (f) 40 m² Boardroom;
(g) 85 m² Luxury; (h) 12 m² Boardroom.

Table 10.8

GRADE OF CARPET	COST (£s per square metre)			
	5 to 20 m²	21 to 80 m²	81 to 140 m²	141 to 200 m²
Custom	4.60	4.40	4.20	4.00
Hardwear	6.20	6.00	5.80	5.60
Luxury	8.70	8.40	8.10	7.80
Boardroom	12.50	12.00	11.80	11.20

10.A6 Draw up a multiplication table for numbers from 20 to 25. Use the table to find the value of:
(a) $(20 \times 21) + (23 \times 25)$ (b) $(22 \times 24) - (21 \times 23)$
(c) $20^2 + 23^2$ (d) $24^2 - 21^2$
(e) $(25 \times 21) - 22^2$ (f) $(20 \times 23) + 24^2$
(g) $22^2 + (20 \times 25) - 24^2$ (h) $0.2 \times 23 \times 21 \times 10$

10.A7 Table 10.9 shows the times of trains between Liverpool and Euston Station.
(a) How long does the 11.25 from Liverpool take to reach Euston?
(b) How long does the 17.20 from Liverpool take between Runcorn and Euston?
(c) What is the latest train I can take from Liverpool to reach Euston by 18.00?
(d) Which train does not stop at Runcorn?
(e) Which train reaches Euston in the shortest time?
(f) Which train takes the longest time?
(g) How many trains reach Euston before 15.00?
(h) Which train takes the longest time between Runcorn and Euston?

10.A8 Table 10.10 is used to calculate monthly payments to repay a loan from a credit company. Determine:
(a) The amount of interest to be paid on a loan of:
 (i) £200 for 2 years;
 (ii) £600 for 1 year;
 (iii) £1 000 for 2 years;
 (iv) £100 for 1 year;
 (v) £1 600 for 2 years.

Use of Tables, Graphs and Diagrams

TABLE 10.9

Liverpool	00.30	06.20	07.00	09.05	10.00	11.25	13.00	14.20	16.00	17.20	19.00
Runcorn	—	06.37	07.17	09.22	10.17	11.42	13.17	14.37	16.17	17.37	19.17
Euston	04.33	09.18	09.42	11.54	12.49	14.20	15.46	17.13	18.21	19.51	21.48

TABLE 10.10

	12 MONTHLY REPAYMENTS			24 MONTHLY REPAYMENTS		
AMOUNT BORROWED (£)	Interest (£)	Total Amount repayable (£)	Monthly payment (£)	Interest (£)	Total amount repayable (£)	Monthly payment (£)
100	11.00	111.00	9.25	22.04	122.04	5.09
200	22.00	222.00	18.50	44.08	244.08	10.17
400	44.00	444.00	37.00	87.92	487.92	20.33
600	66.00	666.00	55.50	132.00	732.00	30.50
800	88.00	888.00	74.00	176.08	976.08	40.67
1 000	110.00	1 110.00	92.50	219.92	1 219.92	50.83

(b) The total amount to be repaid for a loan of:
 (i) £600 for 2 years;
 (ii) £800 for 1 year;
 (iii) £100 for 2 years;
 (iv) £300 for 1 year;
 (v) £1 400 for 2 years.

(c) The monthly payments on a loan of:
 (i) £400 for 2 years;
 (ii) £500 for 2 years;
 (iii) £200 for 1 year;
 (iv) £1 200 for 1 year;
 (v) £2 400 for 2 years.

10.A9 Make out a table similar to Table 10.4 to show the conversion of centimetres to millimetres over the range 0 to 49 cm.

10.A10 Make out a table similar to Table 10.4 to show the conversion of inches to millimetres over the range 40 to 69 in. (1 in = 25.4 mm.)

10.A11 Complete the conversion table (Table 10.11) to show the conversion of miles to kilometres over the range 0 to 29 miles. (Take 1 mile = 1.61 km.)

TABLE 10.11

	MILES									
	0	1	2	3	4	5	6	7	8	9
MILES	KILOMETRES									
0	0	1.61	3.22							
10	16.1									
20	32.2									

10.A12 (a) Complete the table of squares:

x	8	9	10	11	12	13	14	15	16
x^2	64	81							

(b) Use the table to evaluate:
(i) $13^2 + 14^2$ (ii) $15^2 - 9^2$ (iii) $12^2 + 10^2 + 8^2$
(iv) $16^2 - 13^2$ (v) $11^2 + 14^2 - 10^2$

10.A13 (a) Complete the table of square roots:

x	4	9	16	25	64	121	169	225
$\sqrt{x}$	2	3						

(b) Use the table to evaluate:
(i) $\sqrt{225} + \sqrt{64}$ (ii) $3\sqrt{121}$ (iii) $\sqrt{169} + 4^2$
(iv) $\dfrac{\sqrt{64}}{4}$ (v) $9^2 + \sqrt{9} - \sqrt{25}$

10.A14 Use Table 10.4 to convert to centimetres:
(a) 13 in (b) 38 in (c) 9 in (d) 26 in (e) 1.5 in
(f) 50 in (g) 200 in (h) 370 in

10.A15 Use the four-figure table of squares to find the value of:
(a) 1.92^2 (b) 7.63^2 (c) 2.567^2 (d) 50.6^2
(e) 351.8^2 (f) 23.54^2 (g) 0.19^2 (h) 0.852^2

10.A16 Use the four-figure table of square roots to find the value of:
(a) $\sqrt{3.29}$ (b) $\sqrt{12.7}$ (c) $\sqrt{6.518}$ (d) $\sqrt{54.72}$
(e) $\sqrt{130.7}$ (f) $\sqrt{0.65}$ (g) $\sqrt{2387}$ (h) $\sqrt{0.088}$

10.A17 Sketch a right-angled triangle and illustrate Pythagoras' Theorem.

10.A18 Calculate the length of side x in each of the triangles shown in Fig. 10.24.

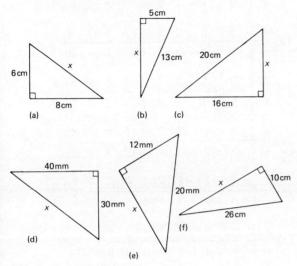

Fig. 10.24

10.A19 Calculate the length of side x in each of the triangles shown in Fig. 10.25.

10.A20 Calculate the length of side x in each of the triangles shown in Fig. 10.26.

10.A21 (*a*) A rectangle is 425 mm by 182 mm, calculate the length of its diagonal.
(*b*) Determine the length of the diagonal of a square with 88 mm side.

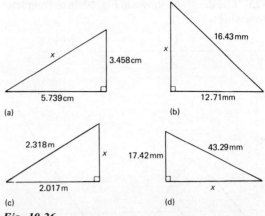

Fig. 10.26

10.A22 Determine the length L of the sloping face on each of the shapes shown in Fig. 10.27.

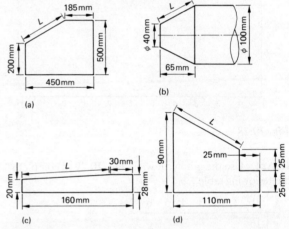

Fig. 10.27

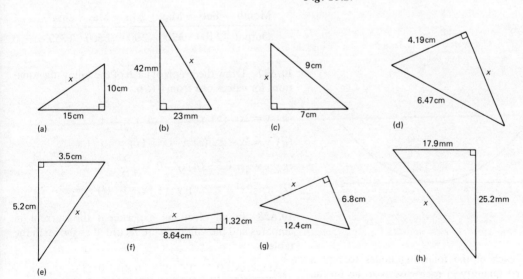

Fig. 10.25

10.A23 Use the graph shown in Fig. 10.28 to complete the following table.

x	0	1	2	3	4	5	6	7	8	9	10	11
y	15											

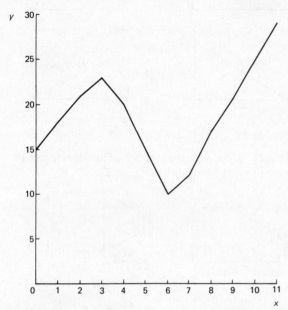

Fig. 10.28

10.A24 Use the graph shown in Fig. 10.29 to complete the following table.

x	0	1	2	3	4	5	6	7
y	16							

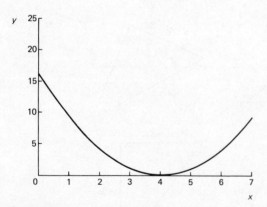

Fig. 10.29

10.A25 Use each of the following tables to plot a graph. The first quantity in each case is to be on the horizontal axis. Use a scale of 1 cm = 1.

(a)
x	0	1	2	3	4	5	6	7	8	9
y	0	1	2	3	4	5	6	7	8	9

(b)
r	0	1	2	3	4	5	6	7	8
s	2	3	4	5	6	7	8	9	10

(c)
M	0	1	2	3	4	5	6	7	8	9
P	9	8	7	6	5	4	3	2	1	0

(d)
x	0	1	2	3	4	5	6	7	8
y	−3	−2	−1	0	1	2	3	4	5

(e)
x	−3	−2	−1	0	1	2	3	4	5
y	0	1	2	3	4	5	6	7	8

(f)
x	0	1	2	3	4	5	6
y	9	4	1	0	1	4	9

10.A26 Each of the tables given below shows the output of a factory over a period of months. Show each set of information on a graph using the horizontal axis for months.

(a)
Month	Jan	Feb	Mar	Apr	May	Jun
Output	300	450	440	460	310	280

(b)
Month	May	Jun	Jul	Aug	Sep	Oct	Nov	Dec
Output	110	122	48	63	95	119	135	120

(c)
Month	Sep	Oct	Nov	Dec	Jan
Output	2 600	2 800	2 750	1 650	1 400

Month	Feb	Mar	Apr	May	Jun
Output	2 100	3 200	2 900	2 500	1 350

10.A27 Draw the graph of each of the following equations for values of x from 0 to 6.

(a) $y = 2x$ (b) $y = \dfrac{3x}{2}$ (c) $y = 2x + 1$

(d) $y = 3x - 2$ (e) $y = x^2$ (f) $y = x^2 + x$

(g) $y = \dfrac{x}{2} + 0.5$ (h) $y = 0.3x + \dfrac{x}{2}$ (i) $y = \dfrac{3x}{2} - 1.2$

(j) $y = 2x^2 - 4$ (k) $y = 1.5x + 1$ (l) $y = x^2 - 2x$

10.A28 In an electrical experiment the current in amperes and the voltage were recorded as shown in the table:

Amperes	0.2	0.4	0.6	0.8	1.0	1.2
Volts	12	24	35	49	60	73

(a) Show these results on a graph with volts on the vertical axis.
(b) Use the graph to estimate (i) the voltage for a current of 0.5 A and (ii) the current in amperes for a voltage of 55 V.

10.A29 Given that the rate of exchange between the pound sterling and the West German Deutschemark is DM3.8 = £1, construct a graph to give the value of the DM in pounds up to £20. The horizontal scale is to be used for pounds.

10.A30 (a) Construct a graph to find the cubes of numbers from 0 to 8.
(b) Use the graph to find (i) 1.7^3, (ii) 5.6^3 and (iii) $6.9^3 - 4.2^3$.

10.A31 (a) Construct a graph to convert yards to metres for the range 0 to 6 yd.
(b) Use the graph to (i) convert 3.8 yd to metres and (ii) convert 4.3 m to yards (take 1 yd = 0.91 m).

10.A32 A sales and service garage has the following staff:
  Mechanics   6
  Apprentices   4
  Storemen   3
  Clerical workers 2
  Salesmen   3
Show this information on (a) a bar chart and (b) a pie chart.

10.A33 In a car park there are 72 cars having the following colours:
  Red   24
  White 18
  Green   8
  Blue   16
  Black   6
Show this information on (a) a bar chart and (b) a pie chart.

Exercises 10.B Business, Administration and Commerce

10.B1 (a) State the reading on the room thermometer shown in Fig. 10.30.

Fig. 10.30

(b) Make a sketch to show a thermometer reading of 72°F.

10.B2 (a) State the reading on the safe dial shown in Fig. 10.31.
(b) Make a sketch showing a dial reading of 23 divisions.

Fig. 10.31

10.B3 State the reading in millimetres on each of the metric rules shown in Fig. 10.32.

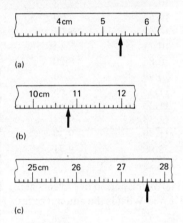

(a)

(b)

(c)

Fig. 10.32

10.B4 Give **one** example of each of the following types of scale that may be seen in a bank or an office: (a) linear, (b) circular and (c) digital readout.

10.B5 Table 10.12 shows foreign exchange rates against the pound sterling.

TABLE 10.12

COUNTRY	RATE OF EXCHANGE (£)
France	11.82 Francs
West Germany	3.85 Deutschemarks
Italy	2 375 Lire
Spain	220 Pesetas
Switzerland	3.09 Francs
U.S.A.	1.42 Dollars

(a) A tourist on holiday in Spain takes £350 to spend and returns home with £27. How many pesetas did he spend in Spain?
(b) A woman buys a coat while on holiday in Switzerland. The coat cost 180 francs, how much was this in pounds sterling?

Use of Tables, Graphs and Diagrams

(c) A hitch-hiker exchanges 726 French francs to West German currency. How many Deutschemarks did he receive?

10.B6 Draw up a table of the present rate of exchange of foreign currency and compare with the rates given in Table 10.12 to find whether the value of the pound sterling has risen or fallen.

10.B7 Table 10.13 shows the calendar monthly repayments on a loan of £1 000 over a number of years.

TABLE 10.13

YEARS	11.75% p.a. (£)	7.95% p.a. (£)
25	10.45	7.78
24	10.53	7.89
23	10.62	8.01
22	10.73	8.14
21	10.85	8.20
20	10.99	8.46
19	11.15	8.65
18	11.33	8.87
17	11.54	9.11
16	11.79	9.39
15	12.08	9.71

(a) A man borrows £1 000 over a period of 17 years at a rate of interest of 11.75% p.a. What is the amount of his monthly repayments?
(b) If the loan had been taken at an interest rate of 7.95%, what would be the difference in monthly repayments?
(c) How much would be saved by the lower interest rate over the 17 years period of the loan?

(d) State the monthly repayments for each of the following loans:
  (i) £3 000 for 15 years at 11.75%;
  (ii) £2 500 for 18 years at 11.75%;
  (iii) £6 000 for 20 years at 7.95%;
  (iv) £4 500 for 25 years at 11.75%;
  (v) £2 750 for 21 years at 7.95%;
  (vi) £10 000 for 23 years at 7.95%.

10.B8 Table 10.14 shows the amount payable on the maturity of an endowment insurance policy which provides life insurance cover for 12 years, together with a cash sum at the end of the 12-year term. The cash sum is made up of a guaranteed sum plus bonus. How much would be paid on maturity to each of the following clients:
(a) a man joining at 27 years old and paying £10 per month;
(b) a man joining at 31 years old and paying £10 per month;
(c) a woman joining at 33 years old and paying £10 per month;
(d) a woman joining at 38 years old and paying £10 per month;
(e) a man joining at 34 years old and paying £10 per month;
(f) a man joining at 35 years old and paying £50 per month;
(g) a woman joining at 35 years old and paying £50 per month;
(h) a woman joining at 40 years old and paying £50 per month;
(i) a woman joining at 32 years old and paying £50 per month;
(j) a man joining at 25 years old and paying £50 per month;

TABLE 10.14 12-Year Cover Plus Bonus (Guaranteed sum is the minimum amount paid on the policy at maturity. The bonus is estimated as shown.)

AGE		NET MONTHLY PREMIUM OF £10		NET MONTHLY PREMIUM OF £50	
Male	Female	Guaranteed sum (£)	Bonus (£)	Guaranteed sum (£)	Bonus (£)
25	30	1 456	720	7 716	4 104
26	31	1 456	720	7 716	4 104
27	32	1 425	702	7 552	4 001
28	33	1 425	702	7 552	4 001
29	34	1 397	687	7 404	3 916
30	35	1 393	681	7 383	3 882
31	36	1 370	659	7 261	3 756
32	37	1 360	643	7 208	3 665
33	38	1 345	630	7 128	3 591
34	39	1 340	619	7 102	3 528
35	40	1 329	601	7 044	3 426

(k) a man joining at 25 years old and paying £30 per month;
(l) a woman joining at 35 years old and paying £40 per month;

10.B9 Table 10.15 shows the amount returnable when £1 is invested at various rates of simple interest for a number of years. Use Table 10.15 to find the amount returned by each of the following investments:
(a) £400 for 5 years at 10%;
(b) £200 for 14 years at 12%;
(c) £800 for 7 years at 8%;
(d) £1 000 for 10 years at 9%;
(e) £1 200 for 4 years at 6%;
(f) £250 for 10 years at 11%;
(g) £2 500 for 6 years at 7%;
(h) £30 for 2 years at 9%;
(i) £650 for 14 years at 12%;
(j) £125 for 7 years at 5%;
(k) £1 350 for 14 years at 7%;
(l) £95 for 5 years at 12%;
(m) £44 for 3 years at 8%;
(n) £105 for 8 years at 9%;
(o) £276 for 9 years at 10%.

10.B10 Complete Table 10.16 to show the conversion of pounds sterling to US dollars over the range £0 to £49. (Take £1 = 1.2 U.S. dollars.)

10.B11 Make out a table similar to Table 10.16 to show the conversion of pounds sterling to Yugoslav dinars over the range £0 to £29. (Take £1 = 200 dinars.)

10.B12 Calculate the length of the diagonal of:
(a) an office measuring 7.6 m by 10.18 m;
(b) a sheet of duplicating paper measuring 295 mm by 210 mm;
(c) a desk top measuring 1.76 m by 1.15 m;
(d) a notice board measuring 85 cm by 53 cm;
(e) a visual display screen measuring 32 cm by 28 cm;
(f) a credit card measuring 80 mm by 50 mm;
(g) a cash box measuring 28 cm by 19.5 cm;
(h) a wallchart measuring 68 cm by 35 cm.

TABLE 10.15

YEARS	ANNUAL RATE OF SIMPLE INTEREST							
	5% (£)	6% (£)	7% (£)	8% (£)	9% (£)	10% (£)	11% (£)	12% (£)
1	1.05	1.06	1.07	1.08	1.09	1.10	1.11	1.12
2	1.10	1.12	1.14	1.16	1.18	1.20	1.22	1.24
3	1.15	1.18	1.21	1.24	1.27	1.30	1.33	1.36
4	1.20	1.24	1.28	1.32	1.36	1.40	1.44	1.48
5	1.25	1.30	1.35	1.40	1.45	1.50	1.55	1.60
6	1.30	1.36	1.42	1.48	1.54	1.60	1.66	1.72
7	1.35	1.42	1.49	1.56	1.63	1.70	1.77	1.84
8	1.40	1.48	1.56	1.64	1.72	1.80	1.88	1.96
9	1.45	1.54	1.63	1.72	1.81	1.90	1.99	2.08
10	1.50	1.60	1.70	1.80	1.90	2.00	2.10	2.20
11	1.55	1.66	1.77	1.88	1.92	2.10	2.21	2.32
12	1.60	1.72	1.84	1.96	2.08	2.20	2.32	2.44
13	1.65	1.78	1.91	2.04	2.17	2.30	2.43	2.56
14	1.70	1.84	1.98	2.12	2.26	2.40	2.54	2.68
15	1.75	1.90	2.05	2.20	2.35	2.50	2.65	2.80

TABLE 10.16

	£									
	0	1	2	3	4	5	6	7	8	9
£	$									
0	0	1.2	2.4							
10	12									
20	24									
30										
40										

10.B13 Figure 10.33 shows the end view of a computer case. Determine the length L of the sloping face.

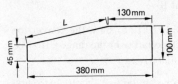

Fig. 10.33

10.B14 The table shows the bank balance held in the current account of a small trader at the end of seven consecutive weeks. Show this information on a graph. The horizontal scale is to be used for weeks.

Week number	1	2	3	4	5	6	7
Balance (£)	510	450	270	640	720	300	410

10.B15 A typing pool has an equal number of electric and manual typewriters. The number of letters produced on each type of machine over a period of six months is given in the following tables:

MANUAL TYPEWRITERS

Month	1	2	3	4	5	6
Number of letters	420	470	550	360	350	480

ELECTRIC TYPEWRITERS

Month	1	2	3	4	5	6
Number of letters	500	610	740	530	470	580

Display both sets of information on the same graph using the horizontal scale for months.

10.B16 Draw a conversion graph to give the value of each of the following currencies in pounds sterling up to £10. In each case the horizontal scale is to be used for pounds:
(a) Belgium, 80.5 francs = £1;
(b) Canada, 1.77 dollars = £1;
(c) Greece, 150 drachmas = £1;
(d) Irish Republic, 1.26 punts = £1;
(e) Sweden, 11.4 kronor = £1;
(f) Holland, 4.35 guilders = £1.

10.B17 The graph given in Fig. 10.34 shows the weekly wage of a typist when she works up to 8 h overtime. Use the graph to determine:
(a) the typist's basic weekly wage;
(b) her wage in a week when she works: (i) 4 h overtime; (ii) 7 h overtime; (iii) $3\frac{1}{2}$ h overtime and (iv) $5\frac{1}{2}$ h overtime;
(c) the hourly rate for overtime;
(d) the number of hours of overtime the typist must work to earn a weekly wage of £100.

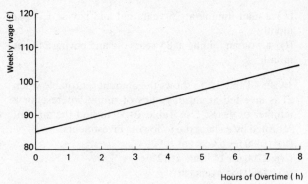

Fig. 10.34

10.B18 A bag of coins contains the following:
50p pieces 30
20p pieces 15
10p pieces 50
5p pieces 25
Show this information on (a) a horizontal bar chart and (b) a pie chart.

10.B19 A building society offers its members the following investment accounts:
A Ordinary share
B Extra interest
C Bonus
D Term share
E Regular savers
Use the pie chart shown in Fig. 10.35 to find the percentage of investors using each account.

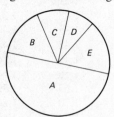

Fig. 10.35

10.B20 The bar chart given in Fig. 10.36 shows the amount of cash transactions made each day in a bank.

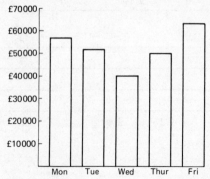

Fig. 10.36

176 Spotlight on Numeracy

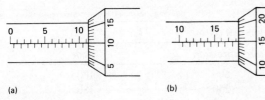

(a) (b) (c)

Fig. 10.37

(a) Use the bar chart to find the amount of each day's transactions.
(b) What were the total transactions for the week?
(c) What was the cash difference between money handled on Wednesday and on Friday?

Exercises 10.C Technical Services: Engineering and Construction

10.C1 State each of the metric micrometer readings shown in Fig. 10.37.

10.C2 State each of the metric vernier readings shown in Fig. 10.38.

(a)

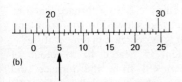
(b)

Fig. 10.38

10.C3 Figure 10.39 shows a dial gauge being used to measure the length of a block by comparison with a standard. Determine the length of the block L.

10.C4 Complete Table 10.17 to show the conversion of inches to millimetres over the range of 0 to 29 inches.

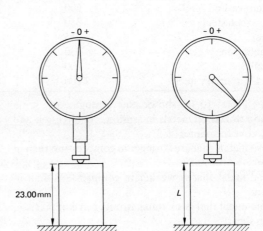

Fig. 10.39

10.C5 Use Table 10.17 to convert the dimensions of each component shown in Fig. 10.40 to millimetres.

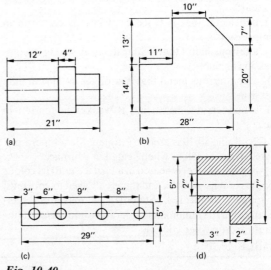

Fig. 10.40

TABLE 10.17

	INCHES									
	0	1	2	3	4	5	6	7	8	9
INCHES					MILLIMETRES					
0	0	25.4	50.8							
10	254									
20	508									

TABLE 10.18

MATERIAL	STRENGTH (N/mm^2)		
	Tension	Compression	Shear
Cast iron	150	600	100
Aluminium alloy	250	250	180
Medium-carbon steel	600	600	360
Low-carbon steel	350	350	210
Copper	215	300	240
Wrought iron	330	330	280
Zinc alloy	200	150	100
Nickel-chromium steel	1 200	1 200	720

10.C6 Table 10.18 shows the strength of some common workshop metals in tension, compression and shear. Give the names of:
(a) two metals that are stronger in compression than in tension;
(b) one metal that is weaker in compression than in tension;
(c) one metal that is six times stronger in compression than in shear;
(d) three metals that are weaker in tension than wrought iron;
(e) two metals that are stronger in tension than wrought iron;
(f) one metal that is twice as strong as aluminium alloy in shear;
(g) one metal that is twice as strong as cast iron in compression;
(h) a ferrous metal that is four times as strong in compression as zinc;
(i) a non-ferrous metal that is stronger in tension than cast iron;
(j) a non-ferrous metal that is weaker in tension in shear.

10.C7 Table 10.19 is used to estimate the temperature of steel from its colour when heated in a furnace.
(a) A piece of steel is heated in a furnace until its colour appears orange–yellow, and then is allowed to cool until it appears bright red. State (i) its maximum temperature; (ii) its minimum temperature and (iii) the temperature drop when cooling.

TABLE 10.19

COLOUR OF HEATED STEEL	APPROXIMATE TEMPERATURE (°C)
Faint dull red	600
Dull red	700
Bright red	800
Cherry red	900
Bright cherry red	1 000
Orange-red	1 100
Orange-yellow	1 200
Yellow-white	1 300
Bright white	1 400

(b) A steel bar at a room temperature of 20°C is heated in a furnace until its colour appears bright cherry red. What is the increase in its temperature?
(c) A steel block is removed from a furnace when its colour is yellow–white, and allowed to cool to a temperature of 80°C, what is its decrease in temperature?
(d) A steel component is heated to a colour of orange–red. It is removed from the furnace and quenched when its colour changes to dull red. State (i) its maximum temperature; (ii) the quenching temperature and (iii) the temperature drop before quenching.

10.C8 Table 10.20 shows recommended cutting speeds and rake angles for turning workshop metals.

TABLE 10.20

MATERIAL CUT	CUTTING SPEED		RAKE ANGLE (degrees)	
	m/min	m/seconds	Top rake	Side rake
Low-carbon steel	35	0.58	7	18
Medium-carbon steel	25	0.42	6	14
High-carbon steel	18	0.3	5	8
Cast iron	24	0.4	4	12
Copper	85	1.42	10	20
Brass	90	1.5	10	20
Aluminium	350	5.83	30	10

(a) State the recommended cutting speed in metres per second and top rake angle for each of the following metals: (i) medium-carbon steel; (ii) cast iron; (iii) brass; (iv) aluminium and (v) copper.
(b) (i) State the difference in the cutting speed in metres per minute required by high-carbon steel and aluminium.
(ii) What is the difference in side rake used for the two metals?

10.C9 Determine the centre distance D between the bored holes in the crank lever shown in Fig. 10.41.

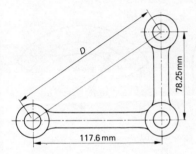

Fig. 10.41

10.C10 Determine the length L of the sloping face of the steel wedge shown in Fig. 10.42.

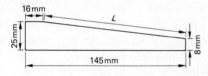

Fig. 10.42

10.C11 Calculate the centre distance C between the holes in the drilled plate shown in Fig. 10.43.

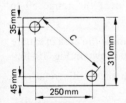

Fig. 10.43

10.C12 A builders' merchant supplies paint in different colours to a decorating firm in the amounts shown in the vertical bar chart (Fig. 10.44).
(a) Which is the more popular colour?
(b) Which colour of paint is least popular?
(c) What difference in quantity is shown between the green paint and the red paint?
(d) What is the total amount of paint supplied by the merchant?
(e) If the decorating firm uses all the paint in 4 weeks, how much paint would be used on an average day? (Take a working week as 5 days.)

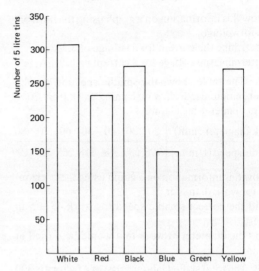

Fig. 10.44

10.C13 The table shows the number of finished components of the same type produced in a machine shop during a production run of 12 consecutive weeks. Show this information on a graph, use the horizontal axis for weeks.

Week no.	1	2	3	4	5	6
Components	435	680	874	910	886	730

Week no.	7	8	9	10	11	12
Components	515	485	612	683	779	940

10.C14 The table shows the number of parts rejected by inspection from eight batches of turned parts. Show this information on a graph using the horizontal axis for batch number.

Batch number	1	2	3	4	5	6	7	8
Number of rejects	9	6	5	7	7	8	4	3

10.C15 The table shows the electrical power P W taken by a circuit for increasing values of current I A. Show this data on a graph using the horizontal axis for current.

I(A)	0	1	2	3	4	5	6
P(W)	0	2	8	18	32	50	72

10.C16 The following results were obtained when testing a resistor in an electrical circuit.

Use of Tables, Graphs and Diagrams 179

Voltage (V)	0	1	2	3	4	5	6	7
Current (mA)	0	35	70	105	140	175	210	245

(a) Show this information on a graph using the horizontal axis for voltage.
(b) Determine the current for a voltage of 3.7 V.
(c) Determine the voltage for a current of 190 mA.

10.C17 The table shows the spindle speed in revolutions per minute used when machining work from 20 to 80 mm in diameter on a lathe.

Work diameter (mm)	20	30	40	50	60	70	80
Spindle speed (r/min.)	796	530	398	318	265	227	199

(a) Show this information on a graph using the horizontal axis for work diameter.
(b) Find the correct spindle speed for work 45 mm in diameter.
(c) Find the diameter of work that would be turned at 240 r/min.

10.C18 The total skilled labour force of a factory is 300 workers made up from the following trades:

- Machinists 130
- Mechanical fitters 60
- Electrical fitters 40
- Welders 30
- Sheet-metal workers 25
- Inspectors 15

Show this information on a pie chart.

10.C19 The table shows the percentage of accidents on a building site attributed to various causes.

- Moving machinery 14%
- Hand tools 26%
- Electrical 8%
- Falling objects 12%
- Falling from scaffolding 15%
- Other causes 25%

Show this information on (a) a bar chart and (b) a pie chart.

10.C20 The table shows the results of a series of tool-life tests carried out over a range of cutting speeds on a centre lathe.

CUTTING SPEED (m/min)	TOOL LIFE (min)
16–18	9.5
18–20	15.8
20–22	20.2
22–24	26.4
24–26	28.1
26–28	25.3
28–30	17.4

Show this information on a histogram.

Exercises 10.D General Manufacturing and Processing

10.D1 State the reading on each of the fuel level gauges shown in Fig. 10.45.

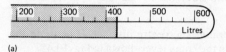
(a)

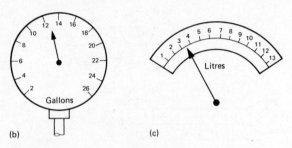
(b) (c)

Fig. 10.45

10.D2 Give **three** examples in each case of the use of linear or circular scales (a) in a factory, (b) on a farm and (c) on a ship.

10.D3 A farmer takes a loan of £3 500 from a credit company in order to replace the roof of his barn. He will repay the loan by 24 equal monthly payments. Use the information given in Table 10.10 to determine:
(a) the total amount to be repaid;
(b) the interest on the loan;
(c) the amount of the monthly payments.

10.D4 Complete Table 10.21 to show the conversion of square yards to square metres over the range 50 to 89 m². (Take 1 yd² = 0.84 m².)

10.D5 Use Table 10.21 to determine the area in square metres of each of the plots of land shown in Fig. 10.46.

10.D6 (a) Calculate the length of the diagonal of a rectangular field 85 m by 57 m.
(b) A sail is in the shape of a right-angled triangle. The two shortest sides are 3.6 m and 2.4 m, determine the length of the longest side.

10.D7 (a) Figure 10.47(a) shows the position of a trawler relative to two buoys, A and B. The range and bearings of the two buoys from the trawler are:
A 5.5 nautical miles, 0°
B 4.2 nautical miles, 90°
Calculate the distance between the two buoys.
(b) Figure 10.47(b) shows a piece of copper tubing to be used for connecting two vats in a processing plant.

TABLE 10.21

SQUARE YARDS	0	1	2	3	4	5	6	7	8	9
					SQUARE METRES					
50	42	42.84								
60	50.4									
70										
80										

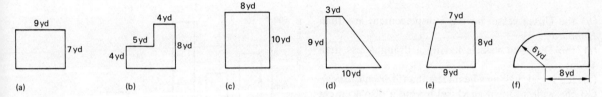

Fig. 10.46

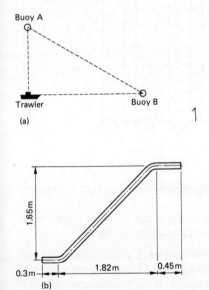

Fig. 10.47

Determine the length of tubing required to make the piece.

10.D8 Coal production over a six-month period is shown on the vertical bar chart (Fig. 10.48).
(a) Which month shows the best production rate?
(b) Which month had the lowest production?
(c) How much coal had been produced by the end of April?
(d) What was the average monthly production rate over the period shown?
(e) By what amount was the coal production in February less than that of May?
(f) How many tonnes of coal were mined over the whole of the period shown?

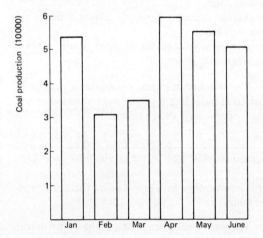

Fig. 10.48

10.D9 Table 10.22 gives details of six fishing vessels registered at two different ports. State the name of:
(a) the heaviest vessel;
(b) the longest Upsea vessel;
(c) the widest Downsea vessel;
(d) the shortest vessel;
(e) the Upsea vessel with the least draught;
(f) the vessel with the least displacement;
(g) the Downsea vessel having a greater displacement than *Ragwood Rose*;

TABLE 10.22

VESSEL	DISPLACEMENT (tonnes)	LENGTH (m)	BEAM (m)	DRAUGHT (m)	PORT
Lucky Lucy	199	31.4	6.7	3.5	Upsea
Hubbard	452	45.9	7.6	4.6	Downsea
Merry L	623	51.3	8.6	4.1	Downsea
Adrian Castle	394	41.6	7.9	3.6	Upsea
Ragwood Rose	467	45.3	8.2	3.9	Upsea
Foxy Lady	593	46.7	8.3	4.2	Downsea

(h) the Upsea vessel having a displacement less than *Hubbard*;
(i) the Downsea vessels having a draught less than 4.5 m;
(j) the vessels having a beam less than *Adrian Castle*;
(k) the widest Downsea vessel having a displacement less than 600 t.

10.D10 Use Table 10.22 to determine:
(a) the average displacement of all the vessels;
(b) the average length of Downsea vessels;
(c) the average beam of Upsea vessels;
(d) the range of displacements of vessels having a beam greater than 7.8 m;
(e) the average draught of vessels having a displacement less than 460 t;
(f) the average displacement of *Foxy Lady*, *Lucky Lucy* and *Ragwood Rose*.

10.D11 The number of plastic components produced by an injection moulding machine over a period of 8 h continuous operation is shown below:

Hours	1	2	3	4	5	6	7	8
Components	850	820	760	730	790	870	640	700

Show this information on a graph using the horizontal scale for hours.

10.D12 The total weekly wage of a night-shift worker was made up as follows:

Basic rate £74
Overtime payment £26
Productivity bonus £13
Meals allowance £ 7

Show this information on (a) a bar chart and (b) a pie chart.

10.D13 The bar chart shown in Fig. 10.49 shows the production of a particular component at a factory in one working week.
(a) State the number of components produced each working day.

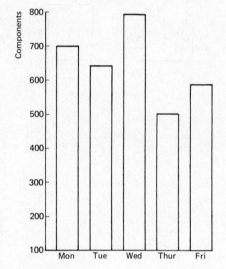

Fig. 10.49

(b) Find the average number of components produced in a working day.
(c) Give the difference between the highest and lowest number of components produced in one day.

10.D14 The pie chart given in Fig. 10.50 shows the cost elements which make up the wholesale price of an item of agricultural machinery. If the wholesale price is £720 determine the amount of each element of cost.

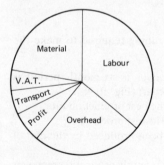

Fig. 10.50

10.D15 The fares charged on a rural bus service are shown in the Table 10.23.

TABLE 10.23

Lydiate Lane							
20	Doctors Lane						
30	20	The Green					
40	30	20	Withington Lane				
50	40	30	20	Mossy Road			
60	50	40	30	20	Chorley Road		
70	60	50	40	30	20	Cherry Tree	
80	70	60	50	40	30	20	Bus Station

(a) State the fares from:
(i) Lydiate Lane to Chorley Road;
(ii) The Green to the Bus Station;
(iii) Withington Lane to Doctors Lane;
(iv) Doctors Lane to Mossy Road;
(v) Lydiate Lane to Cherry Tree.
(b) A man living in Lydiate Lane was visiting a friend in Cherry Tree. He walked to Doctors Lane and took a bus to Chorley Road then walked the remaining distance. How much did he save by walking?

10.D16 A steel forging is taken from a furnace and left to cool in air. The measured temperature of the forging as it cools over a period of 2 h is shown below.

Time (min)	0	20	40	60	80	100	120
Temperature (°C)	850	560	340	220	150	95	45

(a) Show this information in a graph using the horizontal axis for time.
(b) Estimate the time taken for the forging to cool to 400°C.
(c) Estimate the temperature of the forging after 90 min.

10.D17 Draw a graph to convert temperature in degrees Celsius to degrees Fahrenheit for the range 0 to 300°C. Use the graph to convert process temperatures of (a) 145°C to °F and (b) 420°F to °C.

10.D18 The time taken to manufacture a component is made up as follows:
   Presswork   35 min
   Welding   45 min
   Riveting   40 min
   Fitting   60 min
Show this information on a pie chart.

10.D19 The table shows the results of a test measuring the corrosion resistance of a plated component. Each component was operated in a highly corrosive atmosphere and its operating life in hours before failure was recorded.

OPERATING LIFE (h)	NUMBER OF COMPONENTS
600– 700	4
700– 800	9
800– 900	16
900–1 000	21
1 000–1 100	13
1 100–1 200	5

Show this information on a histogram using the vertical axis for the number of components.

10.D20 (a) Figure 10.51 shows a graph of the number of rejected parts in eight batches of metal pressings. List the number of rejected parts in each batch.

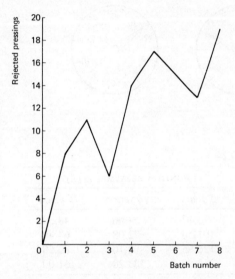

Fig. 10.51

(b) Electrical energy was consumed in a chemical laboratory at the following rates:

Time (h)	1	2	3	4	5	6	7	8	9
Electrical energy (kWh)	71	83	79	76	81	69	61	52	49

Plot a graph of these values using the horizontal axis for time.
(i) Which hour of the working day consumed the least amount of energy?
(ii) Which hour of the day had the peak load?
(iii) How many units were consumed in the whole 9-h working day?
(iv) What was the average consumption of units per hour?

Exercises 10.E Services to People I: Community Services

10.E1 State the readings on each of the scales shown in Fig. 10.52.

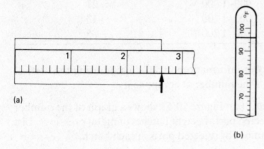

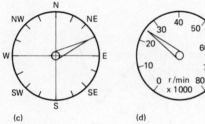

Fig. 10.52

TABLE 10.24

LOAN (£)	MONTHLY REPAYMENTS (£)		
	5 years	10 years	15 years
3 000	76.17	54.06	48.30
4 000	101.56	72.08	64.40
5 000	126.95	90.10	80.50
10 000	253.90	180.20	161.00

10.E2 Table 10.24 shows monthly payments to repay a loan over 5, 10 or 15 years. Determine the amount of monthly repayment to be made by:
(a) a youth club borrowing £7 000 for 10 years to convert a club house;
(b) an amateur football team borrowing £15 000 over 15 years to build changing rooms;
(c) a charity meals service borrowing £4 000 over 5 years to buy a van;
(d) a tennis club borrowing £6 000 over 10 years to provide new courts;
(e) a children's home borrowing £12 000 over 15 years to increase accommodation.

10.E3 Use the information given in Table 10.24 to draw a horizontal bar chart comparing the monthly repayments on a loan of £4 000 over 5 years, 10 years and 15 years.

10.E4 Table 10.25 shows the position of the five leading football teams in the Canon League First Division in February 1984. Complete the table to show the points earned by each team (win = 3 points, draw = 1 point).

10.E5 Update the league table (Table 10.25) to include the following results of matches played:
Luton 0 Liverpool 4
Sunderland 1 Notts F. 1
West Ham 2 Man. Utd 3
QPR 2 Birmingham 2

10.E6 (a) Use the information given in Table 10.25 to draw a bar chart showing the number of goals scored against each of the five teams in home matches.
(b) Draw another bar chart comparing the number of games drawn by each of the five teams both home and away.

10.E7 Calculate the length of the diagonal of each of the following:
(a) a playing field, 90 m by 65 m;
(b) a snooker table, 12 ft by 6 ft;
(c) a swimming pool, 25 m by 10 m;
(d) a bingo card, 15 cm by 7 cm;
(e) a boxing ring, 6.5 m square;
(f) a car park, 21 m by 16 m.

10.E8 A fitted carpet is cut to the shape shown in Fig. 10.53. Calculate the area of the carpet.

10.E9 Draw a graph to convert miles to kilometres for the range 0 to 8 miles, using the horizontal scale for miles. (Take 1 mile = 1.61 km.)

TABLE 10.25

			HOME				AWAY					
					Goals					Goals		
	PLAYED	W	D	L	F	A	W	D	L	F	A	POINTS
Liverpool	27	9	3	2	28	8	7	4	2	17	12	55
Notts Forest	27	9	3	2	31	13	7	1	5	22	18	
Man Utd	27	9	2	3	29	14	5	7	1	22	15	
West Ham	27	9	2	2	30	11	6	3	9	14	15	
QPR	26	8	3	3	23	9	5	1	6	20	15	

184 Spotlight on Numeracy

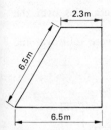

Fig. 10.53

10.E10 Use the graph drawn in Exercise 10.E9 to convert the following distances to kilometres:
(a) a running track, ½ mile long;
(b) a bus route, 6.7 miles long;
(c) a race track, 1.35 miles long;
(d) a sponsored walk, 5.4 miles long;
(e) a newspaper round, 1.7 miles long.

10.E11 The income of a household is £175 per week and is spent in the following way:
 (i) 15% rent and rates;
 (ii) 20% heating and lighting;
 (iii) 17% clothing;
 (iv) 26% food;
 (v) 12% entertainment;
 (vi) 6% savings;
 (vii) 4% insurance.

(a) Draw a pie chart to show this information.
(b) Determine how much is spent on each item.

10.E12 Table 10.26 shows the road distances between some main towns.

TABLE 10.26

Aberdeen									
445	Aberystwyth								
175	309	Ayr							
420	114	285	Birmingham						
318	169	200	108	Bradford					
493	125	362	81	188	Bristol				
458	214	346	100	152	159	Cambridge			
491	105	366	103	202	45	179	Cardiff		
221	224	90	196	110	277	264	289	Carlisle	
344	176	299	18	114	91	81	114	209	Coventry
384	138	266	40	74	127	96	142	176	40	Derby

(a) Give the distance between the following towns:
 (i) Aberystwyth and Cambridge;
 (ii) Aberdeen and Derby;
 (iii) Birmingham and Cardiff;
 (iv) Carlisle and Bristol;
 (v) Bradford and Ayr;
 (vi) Derby and Carlisle;
 (vii) Ayr and Coventry;
 (viii) Coventry and Birmingham.

(b) A man travelled by car from Aberystwyth to Coventry in 5 h. What was his average speed?

10.E13 In one day an ambulance service attends the following calls:

Under 3 miles	38
3 to 5 miles	23
5 to 8 miles	14
8 to 10 miles	10
Over 10 miles	5

Show this information on a pie chart.

10.E14 A van driver's journey is made up of 8 km town traffic, 14 km on country roads and 23 km on motorway. Display this information on a pie chart.

10.E15 An investigation into the major causes of road accidents on a trunk road give the following results:

Weather conditions	28%
Driver error	37%
Speeding	19%
Poor road surface	12%
Other causes	4%

Show these results in (a) a pie chart and (b) a bar chart.

10.E16 The consumption of central-heating fuel oil at a small hospital over a period of six months is listed below.

Month	Jan	Feb	Mar	Apr	May	Jun
Litres	9050	8760	7670	7830	7210	5860

(a) Show this data on a graph using the horizontal axis for months.
(b) Comment on the shape of the graph.

10.E17 A car-hire company is to purchase three saloon cars to the following specification: De Luxe 1.1L with cloth trim and radio fitted.
(a) Use Table 10.2 to determine the total cost of the three vehicles.
(b) To assist in buying the cars the company will raise a loan of £10000 over 5 years. Use Table 10.24 to determine the monthly repayments on the loan.

10.E18 The temperature of a patient in a hospital ward is taken at 1-h intervals over 6-h period, the results are shown below.

Hours	1	2	3	4	5	6
Temperature (°F)	103.5	103.2	101.6	102.2	100.8	99.4

Show these measurements on a graph using the horizontal scale for hours.

Use of Tables, Graphs and Diagrams 185

10.E19 The pie chart shown in Fig. 10.54 shows the percentage of electric lamps of different wattage used in a leisure centre. If the total number of lamps is 120, determine the number of (a) 100 W lamps, (b) 60 W lamps and (c) 40 W lamps.

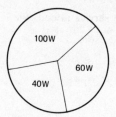

Fig. 10.54

10.E20 Table 10.27 shows the cost in pence of telephone calls.
(a) Determine the cost of each of the following calls:
 (i) 3 min local call at peak time;
 (ii) 2 min local call at cheap rate;
 (iii) call over 70 miles, at standard rate for the first 3 min, overlapping into peak rate for 2 min.
(b) How many (i) local and (ii) peak rate 4-min calls could be made for the price of a long distance standard rate 4-min call?
(c) A social worker made 27 phone calls in one day. Twelve calls were over 35 miles and lasted for 5 min at cheap rate. Nine calls were local at standard rate for 4 min, and the remaining calls were local at peak rate for 2 min. What was the total cost of the day's calls?

10.E21 Sales of metre lengths of carpet by a local firm are recorded on the vertical bar chart shown in Fig. 10.55. From the information given by the chart answer the following:
(a) Which month shows the best sales?
(b) How much carpet was sold in the best month?
(c) Which month shows the least amount sold?
(d) How much carpet was sold in May and February together?
(e) What were the average monthly sales over the period shown?
(f) How many metres of carpet were sold over the 6-month period?

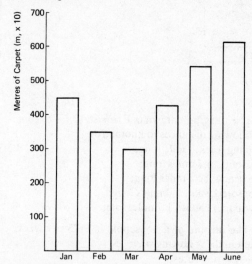

Fig. 10.55

10.E22 A road maintenance department employs the following categories of workers:

A = surface removers   B = machine drivers
C = surface layers   D = general workers
E = lorry drivers

The proportion of workers is shown in the pie chart (Fig. 10.56). What percentage of each category are employed?

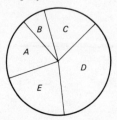

Fig. 10.56

TABLE 10.27

	1 MIN (P)	2 MIN (P)	3 MIN (P)	4 MIN (P)	5 MIN (P)
Local					
Cheap	5	6	7	8	9
Standard	5	7	8	9	10
Peak	5	7	9	10	11
Over 35 miles					
Cheap	10	20	25	29	33
Standard	15	30	43	50	62
Peak	20	40	49	58	65

Exercises 10.F Services to People II: Food and Clothing

10.F1 State the readings on each of the scales shown in Fig. 10.57:

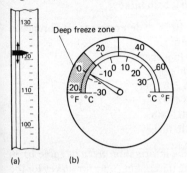

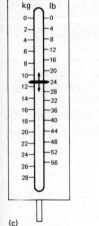

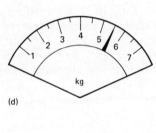

(a) an oven thermostat;
(b) a kitchen thermometer, (i) in degrees Fahrenheit and (ii) in degrees Celsius;
(c) a spring balance, (i) in kilogrammes and (ii) in pounds;
(d) a kitchen scale.

10.F2 Use Table 10.24 to determine the amount of monthly loan repayments to be made by:
(a) a restaurant owner borrowing £3 000 for 5 years for the purchase of dining-room furniture;
(b) a hotelier borrowing £15 000 for 15 years for building an extension;
(c) a café owner borrowing £6 000 for 10 years for the purchase of kitchen equipment;
(d) a boutique owner borrowing £2 000 for 5 years to purchase display stands.

10.F3 Complete Table 10.28 to show the conversion of yards to metres over the range 20 to 59 yd. (Take 1 yd = 0.91 m.)

10.F4 Use Table 10.28 to convert each of the following cloth lengths to metres:
(a) 27 yd (b) 55 yd (c) 36 yd (d) 22 yd (e) 41 yd

10.F5 Complete Table 10.29 to show the conversion of pounds to kilogrammes over the range 0 to 29 lb. (Take 1 lb = 0.454 kg.)

10.F6 Use Table 10.29 to convert the following quantities to kilogrammes:
(a) 12 lb carrots (b) 28 lb flour (c) 14 lb sugar
(d) 2 lb broad beans (e) 24 lb potatoes
(f) 18 lb swedes (g) $\frac{1}{4}$ lb tea (h) six 2-lb loaves
(i) $4\frac{1}{2}$ lb beef (j) $1\frac{1}{2}$ lb tomatoes

Fig. 10.57

TABLE 10.28

	YARDS									
	0	1	2	3	4	5	6	7	8	9
YARDS					METRES					
20	18.2									
30										
40										
50										

TABLE 10.29

	POUNDS									
	0	1	2	3	4	5	6	7	8	9
POUNDS					KILOGRAMMES					
0	0	0.454								
10	4.54									
20										

10.F7 Use the information given in Table 10.29 to draw a conversion graph of pounds to kilogrammes. The horizontal scale is to be used for pounds.

10.F8 Some conversion factors are listed below to convert kitchen quantities between imperial and metric weights and measures.

WEIGHTS	MEASURES
1 oz = 28.35 g	$\frac{1}{4}$ pint = 142 ml
2 oz = 56.7 g	$\frac{1}{2}$ pint = 284 ml
4 oz = 113.4 g	1 pint = 568 ml
8 oz = 226.8 g	$\frac{1}{2}$ litre = 0.88 pints
12 oz = 340.2 g	1 litre = 1.76 pints
1 lb = 454 g	
2.2 lb = 1 kg	

Use the conversion factors to convert:
(a) 8 oz butter to grammes;
(b) 3 pints milk to litres;
(c) 5 kg potatoes to pounds;
(d) $1\frac{1}{2}$ litres stock to pints;
(e) 3 oz cherries to grammes;
(f) $\frac{3}{4}$ pint water to millilitres;
(g) 14 oz sugar to grammes;
(h) 2 pints cooking oil to litres.

10.F9 Use the conversion factors given in Exercise 10.F8 to draw conversion graphs for:
(a) ounces to grammes for the range 0 to 16 oz;
(b) pints to millilitres for the range 0 to 2 pints;
(c) litres to pints for the range 0 to 4 litres.

10.F10 Table 10.30 gives the cooking times and electric oven settings for various weights of meat. State the oven setting and cooking time for each of the following:
(a) 3 lb beef (b) 5 lb chicken (c) 2 lb lamb
(d) 4 lb pork (e) 12 lb turkey (f) 5 lb beef
(g) 3 lb chicken (h) 16 lb turkey

10.F11 Use Table 10.30 to determine the difference in cooking time between:
(a) 5 lb beef and 14 lb turkey;
(b) 3 lb lamb and 2 lb pork;
(c) 5 lb chicken and 2 lb beef;
(d) 10 lb turkey and 4 lb chicken;
(e) 5 lb pork and 4 lb lamb;
(f) 3 lb beef and 18 lb turkey.

10.F12 (a) From the information given in Table 10.30 draw a graph showing the cooking time for turkey for the range 10 to 30 lb. Use the horizontal axis for pounds weight.
(b) Use the graph to find the cooking time for a turkey of weight 15 lb 8 oz.

10.F13 Table 10.31 shows a wine vintage chart.
(a) In what year was the vintage claret at its worst?
(b) What year was Rhone at its best?
(c) State two years when champagne was at its best.
(d) What year were claret and champagne at their joint best?
(e) Describe the vintage of port in 1977.

TABLE 10.30

	OVEN SETTING	WEIGHT OF MEAT			
		2 lb	3 lb	4 lb	5 lb
Beef	400/425	1 h	1 h 20 min	1 h 40 min	2 h
Lamb	400/425	1 h 15 min	1 h 40 min	2 h 5 min	2 h 30 min
Pork	400/425	1 h 15 min	1 h 40 min	2 h 5 min	2 h 30 min
Chicken	375/400	1 h 15 min	1 h 20 min	1 h 40 min	1 h 50 min
Turkey	375	15 min per lb + 15 mins			

TABLE 10.31

YEAR	CLARET	RHONE	LOIRE	RHINE	CHAMPAGNE	PORT
1970	7[a]	6	5	3	6	6
1971	6	5	6	7	7	4
1972	2	5	2	3	5	4
1973	4	5	4	4	6	3
1974	3	4	4	2	5	2
1975	7	4	5	5	7	6
1976	6	5	6	7	6	4
1977	3	4	3	3	5	7
1978	6	7	6	4	6	6
1979	5	6	5	5	6	5
1980	3	5	5	4	5	7

[a]Scale 0–7: 0 = worst, 7 = best.

10.F14 Table 10.32 shows the calories contained by various foods.

TABLE 10.32

FOOD	WEIGHT (g)	CALORIES
Cabbage	120	10
Carrots	120	10
Chicken	120	170
Coffee	cup	30
Egg	60	90
Grapefruit	120	30
Ham	120	140
Lettuce	60	10
Tomatoes	120	20

(a) A man dines in a café eating a chicken salad meal which consists of: 480 g chicken, 360 g tomatoes, 180 g egg, 120 g lettuce, 60 g grated carrots, 1 cup coffee. What is his total intake of calories?

(b) How many calories are contained in a chicken weighing 3.23 kg?

(c) A woman is on a diet which allows her to take 270 cal for breakfast. If she eats 420 g of grapefruit and 90 g of egg, can she have a cup of coffee and remain within her diet?

(d) How many calories has a boiled ham of 1.5 kg?

10.F15 Table 10.33 shows equivalent sizes in British, American and Continental clothes.

(a) Use the information given in Table 10.33(a) to draw the following conversion graphs; use the horizontal axis for the first-named quantity: (i) British to American sizes; (ii) American to Italian sizes and (iii) British to French sizes.

(b) Use the information given in Table 10.33(b) to draw the following conversion graphs, the horizontal axis to be used for the first-named quantity: (i) American to Continental and (ii) Continental to British.

10.F16 Figure 10.58 shows a conical hood for a kitchen fumes extractor fan. Calculate the height h.

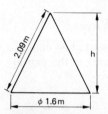

Fig. 10.58

10.F17 Figure 10.59 shows a vegetable chopping board.

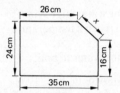

Fig. 10.59

(a) Calculate the length of the sloping side x.
(b) Find the area of the board.

10.F18 Table 10.34 shows sales and costs of a small café over a period of 9 weeks.

(a) Plot sales and costs in two different colours against the week number and use the graph to find:
(i) The week in which the most profit was made.
(ii) The week showing the greatest loss.

TABLE 10.33

(a) Women's coats, suits, dresses and blouses

British	10	12	14	16	18	20
American	8	10	12	14	16	18
Italian	44	46	48	50	52	54
French	40	42	44	46	48	50

(b) Shoe sizes

British	6	7	8	9	10	11
American	7	8	9	10	11	12
Continental	$39\frac{1}{2}$	$40\frac{1}{2}$	$41\frac{1}{2}$	$42\frac{1}{2}$	$43\frac{1}{2}$	$44\frac{1}{2}$

TABLE 10.34

Week	1	2	3	4	5	6	7	8	9
Sales (£)	2 100	2 350	1 890	1 760	2 150	2 460	2 010	2 320	1 830
Costs (£)	1 850	1 700	1 950	1 870	1 800	1 750	1 920	1 930	1 810

(iii) The difference between the highest sales and the least costs.
(iv) The average profit over the 9-week period.
(b) The costs in week 5 were made up of:
   Materials 45%
   Labour   25%
   Overheads 30%
(i) Show this information on a pie chart
(ii) What was the cost of materials in week 5?

10.F19 A pan of vegetables were set to boil on a gas cooker. The temperature of the water in the pan increased as shown in the table below.

Time (min)	0	1	2	3	4	5	6	7	8	9	10
Temperature (°C)	11	16	20	26	32	39	46	53	61	70	79

(a) Draw a graph of these values with the time on the horizontal axis.
(b) What was the temperature of the water after $2\frac{1}{2}$ min?
(c) How long did the water take to reach a temperature of 50°C?
(d) Estimate how long it took for the water to boil.

10.F20 The vertical bar chart shown in Fig. 10.60 shows the number of shoe sales made at a shop over 6 years.

(a) Which year had the highest sales?
(b) Which year had the lowest sales?
(c) What was the difference between the number of sales made in the second year and the sixth year?
(d) What were the total sales over the six years?
(e) What were the average sales per year over the six-year period?

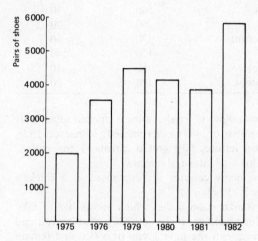

Fig. 10.60

11 Use of Calculators and Computers

The exercises contained in this chapter are intended for use by all readers and consequently are not divided into different vocational options.

11.1 Use of Calculators

Electronic calculators are practical and valuable aids to making job-related calculations in many occupations. Calculators are fast and simple to operate, and give accurate answers without the need for making rough estimates in order to position the decimal point. Calculators are a modern and sensible alternative to the use of logarithms for most routine calculations, and many calculators can directly evaluate squares and square roots, so making the use of tables unnecessary.

At this stage of studies it is advisable to purchase a relatively cheap and simple calculator, and become practised in the basic operations of addition, subtraction, multiplication and division of numbers. The method of operating a calculator keyboard is now fairly standard and will be described in the manufacturer's instructions. The following examples of calculator keyboard operation are shown for an algebraic logic model with an eight-figure display.

The worked examples are set out so that the reader may enter the values on his own calculator keyboard and check the **display** reading at each stage.

Example 11.1 $3.87 + 6.054 + 27.348$.

KEYBOARD ENTRY	DISPLAY READS
3.87	3.87
+	3.87
6.054	6.054
+	9.924
27.348	27.348
=	37.272

$3.87 + 6.054 + 27.348 = 37.272$ (*Ans.*)

Example 11.2 $453.69 - 128.91$.

KEYBOARD ENTRY	DISPLAY READS
453.69	453.69
−	453.69
128.91	128.91
=	324.78

$453.69 - 128.91 = 324.78$ (*Ans.*)

Example 11.3 57.36×14.85.

KEYBOARD ENTRY	DISPLAY READS
57.36	57.36
×	57.36
14.85	14.85
=	851.796

$57.36 \times 14.85 = 851.796$ (*Ans.*)

Example 11.4 $19.74 \div 3.58$.

KEYBOARD ENTRY	DISPLAY READS
19.74	19.74
÷	19.74
3.58	3.58
=	5.5139664

$19.74 \div 3.58 = 5.513\,966\,4$ (*Ans.*)

The answer displayed by the calculator in the last example has seven decimal places. For most work this figure would not be convenient and the answer would be rounded-off to a lesser number of decimal places, e.g. $19.74 \div 3.58 = 5.514$ correct to 3 d.p.

Example 11.5 $\dfrac{11.83 \times 7.09}{23.24}$.

KEYBOARD ENTRY	DISPLAY READS
11.83	11.83
×	11.83
7.09	7.09
÷	83.8747
23.24	23.24
=	3.6090662

$\dfrac{11.83 \times 7.09}{23.24} = 3.609$ correct to 3 d.p.   (*Ans.*)

Example 11.6 $\dfrac{283.61 - 97.8}{74.33}$.

KEYBOARD ENTRY	DISPLAY READS
283.61	283.61
−	283.61
97.8	97.8
÷	185.81
74.33	74.33
=	2.4997981

$\dfrac{283.61 - 97.8}{74.33} = 2.500$ correct to 3 d.p.   (*Ans.*)

On the calculator keyboard the key used for finding the **square** of a number is labelled X^2, and the key for finding the **square root** is labelled $\sqrt{X}$.

Example 11.7 Find the value of 2.78^2.

KEYBOARD ENTRY	DISPLAY READS
2.78	2.78
X^2	7.7284

$2.78^2 = 7.7284$ (*Ans.*)

Example 11.8 Find the square root of 8 390.56.

KEYBOARD ENTRY	DISPLAY READS
8390.56	8390.56
$\sqrt{X}$	91.6

$\sqrt{8390.56} = 91.6$ (*Ans.*)

Example 11.9 Given that $P = \dfrac{2.17LQ}{M}$, find the value of P when $L = 489$, $Q = 0.35$ and $M = 7.6$. Give the value of P correct to 3 decimal places.

Substituting the values of L, Q and M in the equation gives

$$P = \frac{2.17 \times 489 \times 0.35}{7.6}$$

KEYBOARD ENTRY	DISPLAY READS
2.17	2.17
×	2.17
489	489
×	1061.13
.35	0.35
÷	371.3955
7.6	7.6
=	48.867828

$P = 48.868$ correct to 3 d.p. (*Ans.*)

On some calculator keyboards a key is provided for π, otherwise the value of $\frac{22}{7}$ or 3.142 must be entered.

Example 11.10 Given that $r = \sqrt{\dfrac{V}{\pi l}}$, find the value of r when $V = 27.8$ and $l = 6.9$. Give the value of r correct to 4 d.p.

Substituting values

$$r = \sqrt{\frac{27.8}{\pi \times 6.9}}$$

KEYBOARD ENTRY	DISPLAY READS
27.8	27.8
÷	27.8
π	3.1415926
÷	8.8490149
6.9	6.9
=	1.2824659
$\sqrt{X}$	1.1324601

$r = 1.1325$ correct to 4 d.p. (*Ans.*)

Many calculators have a **memory** which is convenient to use when finding the sum of a number of simple calculations, such as a shopping list. The memory keys are usually labelled:

MC   clear memory
M+   add to memory
M−   subtract from memory
MR   recall memory

Example 11.11 Find the total cost of:
21 articles at 14p each
7 articles at £1.05 each
48 articles at 63p each
14 articles at £2.11 each
127 articles at 17p each

KEYBOARD ENTRY	DISPLAY READS	
MC	0	
21	21	
×	21	
.14	0.14	
M+	2.94	(cost of 21 at 14p)
7	7	
×	7	
1.05	1.05	
M+	7.35	(cost of 7 at £1.05)
48	48	
×	48	
.63	0.63	
M+	30.24	(cost of 48 at 63p)
14	14	
×	14	
2.11	2.11	
M+	29.54	(cost of 14 at £2.11)
127	127	
×	127	
.17	0.17	
M+	21.59	(cost of 127 at 17p)
MR	91.66	(total cost)

The total cost = £91.66 (*Ans.*)

Example 11.12 A customer purchases 17 articles at 32p each. If he pays with a £10 note how much change will he receive?

KEYBOARD ENTRY	DISPLAY READS	
MC	0	
10	10	
M+	10	
17	17	
×	17	
.32	0.32	
M−	5.44	(cost of 17 at 32p)
MR	4.56	(change)

Change received = £4.56 (*Ans.*)

11.2 Use of a Microcomputer

Inexpensive **microcomputers** are now widely used in small businesses and education. Calculators are usually adequate for most job-related calculations, but the use of a microcomputer is an advantage in circumstances such as:

    making a large number of similar calculations;
    keeping files and records which can be quickly assessed for reference and up-dating;
    producing graphic displays, bar charts and graphs;
    simulating processes to test the probable effects of changes;
    word handling and processing;
    communicating information over a network;
    controlling other equipment.

Microcomputers use a programming language called BASIC (Beginners' All-purpose Symbolic Instruction Code). This language resembles English and can be entered using the keyboard of the computer. Most popular microcomputers have a BASIC interpreter fitted as standard which converts the BASIC language into the **machine code** language used by the computer. Computer manufacturers use different versions or 'dialects' of BASIC, which means that programs written for one computer may have to be changed to some extent before they will run on a different computer.

The programs in this book have been run on the popular Sinclair ZX Spectrum, but may be very easily adapted to suit other popular models of microcomputer.

The worked programs are set out so that the reader may enter the program on his own computer keyboard and check the television display when the program is run. Explanatory notes are given where commands or routines are met for the first time.

PROGRAM 1

```
10 REM Sum of two numbers
20 LET a=5
30 LET b=7
40 LET c=a+b
50 PRINT c
60 STOP
```

Line 10 uses the BASIC word REM to tell the computer that this is a remark statement to inform the user of the name or the purpose of the program. The computer will ignore the rest of the line after REM.
Line 20 assigns the value of 5 to the variable a.
Line 30 assigns the value of 7 to the variable b.
Line 40 assigns to the variable c the value of the sum of the variables a and b.
Line 50 tells the computer to display the value of c on the TV screen.
Line 60 tells the computer that the program is ended.

The program is typed into the microcomputer using the keyboard. The BASIC words are labelled on the Spectrum keys for single-press entry and must not be typed in letter by letter. Each line must be numbered and is entered by pressing the ENTER key. To run the program the keys RUN and ENTER must be pressed. The display on the TV screen will then appear as:

12

9 STOP statement, 60:1

The program listing can be recovered by pressing ENTER or LIST and ENTER. The sum of two different numbers can be found by using the EDIT facility on the computer to change the values of a and b in lines 20 and 30.

PROGRAM 2

```
10 REM Multiplication
20 READ a
30 READ b
40 PRINT a*b
50 DATA 17,23
60 STOP
```

Line 20 instructs the computer to obtain the value of a from the first value given in the DATA statement shown in line 50.
Line 30 instructs the computer to obtain the value of b from the second value of the DATA statement.
Line 40 tells the computer to display the product of a and b, the symbol * is used to show multiplication.
Line 50 holds the value of a and b in consecutive order.

Press RUN, press ENTER.

391

9 STOP statement, 60:1

Other numbers can be multiplied by changing the DATA in line 50.

PROGRAM 3

```
10 REM Cubes of numbers
20 FOR n=1 TO 20 STEP 1
30 PRINT n,n↑3
40 NEXT n
50 STOP
```

Line 20 includes the commands FOR, TO and STEP which are used to inform the computer of the range of values of n (the number to be cubed) over which the program is to be run. Thus numbers from 1 to 20 increasing by steps of 1 will be cubed.

Use of Calculators and Computers 193

Line 30 tells the computer what calculation must be carried out on each value of *n*. The instruction $n \uparrow 3$ means raise *n* to the power of 3, i.e. *n* is cubed.

Line 40 instructs the computer to move to the next value of *n* when the cubing of the first value has been carried out. On reaching this instruction, the computer moves back to line 30 and repeats the calculation. This process continues until the state range of values of *n* has been covered.

Line 50 tells the computer that the program has ended when the last value of *n* (*n* = 20) has been cubed.

Press RUN, press ENTER.

```
 1        1
 2        8
 3       27
 4       64
 5      125
 6      216
 7      343
 8      512
 9      729
10     1000
11     1331
12     1728
13     2197
14     2744
15     3375
16     4096
17     4913
18     5832
19     6859
20     8000
```

A different range of cubes may be produced by changing the statement in line 20. Different powers, such as squares, may be produced by changing the statement in line 30.

PROGRAM 4

```
10 REM Metric Conversion
20 FOR I=0 TO 10 STEP 0.1
30 PRINT "inches","millimetres"
40 PRINT I,I*25.4
50 NEXT I
```

Line 30 instructs the computer to display the words in inverted commas on the TV screen. The comma between the words will cause them to be spaced apart on the screen.

Line 40 instructs the computer to display the dimension in inches (*I*) and its equivalent in millimetres ($I \times 25.4$).

Line 50 completes the FOR … NEXT loop which causes the computer to carry out lines 30 and 40 for each value of *I* defined by the range statement in line 20.

Press RUN, press ENTER.

```
inches            millimetres
0                 0
inches            millimetres
0.1               2.54
inches            millimetres
0.2               5.08
inches            millimetres
0.3               7.62
inches            millimetres
0.4              10.16
inches            millimetres
0.5              12.7
inches            millimetres
0.6              15.24
inches            millimetres
0.7              17.78
inches            millimetres
0.8              20.32
inches            millimetres
0.9              22.86
inches            millimetres
1                25.4
```

The TV displays a full screen of values and the query "scroll?" appears at the bottom of the screen. Pressing the ENTER key will cause the next full screen of values to be displayed. When the whole range has been covered the statement 0 OK, 50:1 will appear on the screen.

PROGRAM 5

```
10 REM Area of rectangle
20 INPUT "length,mm",l
30 INPUT "breadth,mm",b
40 PRINT "area,square mm",l*b
50 PAUSE 200: CLS
60 GO TO 20
```

Line 20 instructs the computer to display the words in inverted commas and then wait while the user types in the value of length.

Line 30 repeats the process for breadth.

Line 40 calculates the area as length × breadth and displays it on the screen.

Line 50 allows a short pause while the answer is read and then clears the screen (CLS). The reading time may be increased or decreased by changing the number after the PAUSE command.

Line 60 instructs the computer to go back to line 20 to receive dimensions of the next rectangle.

Press RUN, press ENTER

length, mm   (cursor flashing)

User enters 33.65.

   breadth, mm

User enters 12.28.

   area, square mm   413.222

After the pause the screen is cleared and the display for the next rectangle appears.

   length, mm

If the command STOP is entered the program can be halted and cleared from the memory by using the command NEW.

PROGRAM 6

```
10 REM Area of circle
20 INPUT "radius",r
25 IF r=999 THEN STOP
30 PRINT "area is",PI*r↑2
40 PAUSE 200: CLS
50 GO TO 20
```

Line 25 this is a stop device and uses an unlikely value of radius so that the program can be halted by entering 999.

Line 30 calculates the area of the circle using the formula: area = πr^2.

The television display is similar to Program 5.

PROGRAM 7

```
10 REM Fahrenheit to Celsius
20 PRINT "deg F","deg C"
30 PRINT
40 INPUT "enter deg F",F
50 PRINT F,(F-32)*5/9
60 GO TO 40
```

Line 20 displays the headings for Fahrenheit and Celsius.
Line 30 provides a blank line after the headings.
Line 50 converts degrees Fahrenheit to degrees Celsius using the formula $C° = \frac{5}{9}(F° - 32)$. Note that division is shown by the oblique line / in the program and brackets are used in the normal manner.

Press RUN, press ENTER.

deg F   deg C

enter deg F

The user can now type in any value of degrees Fahrenheit and press ENTER. The computer displays the equivalent value in degrees Celsius and asks for the next entry. A typical screen display is shown below.

```
deg F               deg C
40                  4.4444444
50                  10
60                  15.555556
250                 121.11111
3000                1648.8889
```

The following programs can now be entered and run. The user should try to relate the statements in the program lines to the display that appears on the screen. Where appropriate the program can be listed and edited to include different values or other data.

PROGRAM 8

This program finds the cost of tiling a floor when the cost per square metre and the dimensions of the floor are typed in. The program includes a counting routine in lines 20, 190 and 200 to allow calculations to be made for four floor areas before the program stops.

```
10 REM Cost of tiling
20 LET c=0
30 PRINT "Enter cost per square metre in £"
40 INPUT p
50 PRINT
60 PRINT "Enter length in metres"
70 INPUT l
80 PRINT
90 PRINT "Enter width in metres"
100 INPUT w
110 PRINT
120 LET t=p*l*w
130 PRINT "Tiling cost is","£";t
140 PRINT
150 PRINT
160 PRINT
170 PRINT
180 PAUSE 100
190 LET c=c+1
200 IF c<4 THEN GO TO 30
```

PROGRAM 9

```
10 REM Addition of three numbers
20 INPUT "Enter first number.";a
30 INPUT "Enter second number";b
40 INPUT "Enter third number";c
50 LET sum=a+b+c
60 PRINT "Sum of three numbers = ";sum
70 PAUSE 150: CLS
80 GO TO 20
```

Use of Calculators and Computers

PROGRAM 10

This program checks the films available for hire at a video film library. The name of the film requested must be entered in capital letters exactly as listed in the program. Use RUN and ENTER before each request. (Ask for a film which is not in the listing.)

```
10 REM VIDEO SHOP
20 INPUT "ENTER FILM REQUESTED"; F$
30 READ A$
40 IF A$="XXX" THEN GO TO 70
50 IF F$=A$ THEN GO TO 90
60 GO TO 30
70 PRINT F$,"OUT ON LOAN"
80 GO TO 999
90 PRINT F$,"AVAILABLE FOR HIRE"
100 GO TO 999
110 DATA "JAWS 2"
120 DATA "SILENT PARTNER"
130 DATA "SOPHIES STORY"
140 DATA "THE THING"
150 DATA "ALIEN"
160 DATA "E.T."
170 DATA "XXX"
999 STOP
```

PROGRAM 11

This program supplies the address when the name is entered in capital letters. Use RUN and ENTER before each request. (Ask for a name which is not in the listing.)

```
10 REM ADDRESS BOOK
20 INPUT "ADDRESS REQUIRED"; B$
30 READ A$,C$
40 IF A$="XXX" THEN GO TO 70
50 IF B$=A$ THEN GO TO 90
60 GO TO 30
70 PRINT B$;" ";"NOT IN BOOK"
80 GO TO 999
90 PRINT B$;" ";"ADDRESS IS";" ";C$
100 GO TO 999
110 DATA "JANE","12 HIGH ST."
120 DATA "JILL","8 LOW LANE"
130 DATA "TOM","45 CENTRAL DRIVE"
140 DATA "FRED","14 UPPER AVENUE"
150 DATA "SUE","FLAT 5,TOPPER HOUSE"
160 DATA "XXX","END"
170 STOP
```

PROGRAM 12

This program gives the lathe spindle speed in revolutions per minute to be used when machining different work diameters at a cutting speed of 25 m per minute. The program can be stopped by entering 999 for the value of d.

```
10 REM Lathe Speeds
20 PRINT "dia","rev/min"
30 PRINT
40 INPUT "Enter dia,mm",d
45 IF d=999 THEN STOP
50 PRINT d,1000*25/(PI*d)
60 GO TO 40
```

PROGRAM 13

This program uses Ohm's Law to find the current flowing in amperes when a resistance in ohms is connected across a 24-V supply. The current is calculated by using the formula:

$$\text{amperes} = \frac{\text{volts}}{\text{ohms}}$$

```
10 REM Ohms Law
20 PRINT "OHMS","AMPS"
30 PRINT
40 INPUT "Enter OHMS",R
50 PRINT R,24/R
60 GO TO 40
```

PROGRAM 14

This program draws a **bar chart** from listed data. The values of x and y give the plotted position of the bottom left-hand corner of each bar. The value of b decides the width of each bar and in this example is constant. The value of h gives the height of the bars.

```
10 REM Bar Chart
20 READ x,y,b,h
30 IF x=999 THEN STOP
40 PLOT x,y
50 DRAW b,0
60 DRAW 0,h
70 DRAW -b,0
80 DRAW 0,-h
90 GO TO 20
100 DATA 0,0,20,50
110 DATA 20,0,20,69
120 DATA 40,0,20,110
130 DATA 60,0,20,85
140 DATA 80,0,20,50
150 DATA 100,0,20,39
160 DATA 120,0,20,76
170 DATA 140,0,20,45
180 DATA 160,0,20,94
190 DATA 180,0,20,38
200 DATA 999,999,999,999
```

The height of the bars may be changed by altering the value of h in the DATA statements.

PROGRAM 15

This program calculates the **mean** or **average** of any number of values which are listed in the DATA statements. Further values can be added in additional DATA statements but they must precede line 200 which contains the value −999.

```
 10 REM Mean
 20 LET c=1
 30 LET s=0
 40 READ v
 50 IF v=-999 THEN GO TO 90
 60 LET s=s+v
 70 LET c=c+1
 80 GO TO 40
 90 LET n=c-1
100 PRINT "MEAN=",s/n
110 DATA 52,36,29,87,13,75
120 DATA 31,44,65,43,78,19
130 DATA 11,34,66,17,83,53
140 DATA 29,62,95,10,41,22
150 DATA -999
```

PROGRAM 16

This program uses the high-resolution graphics of the Spectrum to draw a holiday poster. REM statements are used to show how the different features are drawn. FLASH 1 (on) and FLASH 0 (off) are used in line 180 to draw attention to the slogan. Line 240 gives continuous running.

```
 10 REM Poster
 15 REM Hull
 20 PLOT 100,45
 30 DRAW 100,0
 40 DRAW 5,15
 50 DRAW -120,0
 60 DRAW 15,-15
 65 REM Mast
 70 PLOT 150,60
 80 DRAW 0,85
 85 REM Sails
 90 DRAW -75,-80
100 DRAW 140,0
110 DRAW -30,70
120 DRAW -35,-15
125 REM Horizon
130 PLOT 0,90
140 DRAW 95,0
150 PLOT 205,90
160 DRAW 50,0
165 REM Sun
170 CIRCLE 40,125,15
175 REM Slogan
180 FLASH 1: PRINT AT 19,4;"COME TO SUNNY SOUTHPORT": FLASH 0
185 REM Border
190 PLOT 0,0
200 DRAW 255,0
210 DRAW 0,175
220 DRAW -255,0
230 DRAW 0,-175
235 REM Continuous running
240 PAUSE 120: CLS : GO TO 20
```

PROGRAM 17

This program finds the complementary angle when two angles in a right angle are entered. Decimal fractions of degrees may be entered. Line 125 rejects input of angles whose sum exceeds 90°.

```
  5 REM Angles in a right-angle
 10 PLOT 100,170
 20 DRAW 0,-75
 30 DRAW 75,0
 40 PLOT 100,95
 50 DRAW 75,25
 60 PLOT 100,95
 70 DRAW 50,60
 80 PRINT AT 9,18;"A"
 90 PRINT AT 6,17;"B"
100 PRINT AT 4,14;"C"
110 INPUT "Enter angle A deg ";A
120 INPUT "Enter angle B deg ";B
125 IF A+B>90 THEN GO TO 110
130 PRINT AT 20,4;"Angle C deg=";90-A-B
140 PAUSE 150: CLS : GO TO 10
```

PROGRAM 18

This program is more difficult to enter and requires some keyboard experience. When run the program will ask the user to round-off decimal numbers to a stated number of decimal places. The program will not allow the user to proceed until the correct answer is entered.

```
  5 REM Rounding-off
 10 LET NUMBER=INT (RND*100000)/10000
 20 LET NOOFPLACES=INT (RND*4)
 25 CLS
 30 PRINT AT 10,8;"What is ";NUMBER;
 40 PRINT AT 12,3;"Rounded To ";NOOFPLACES;" Decimal Places"
 50 INPUT "ENTER CORRECT NUMBER ";ANSWER
 60 LET CORRECTANSWER=INT (NUMBER*10↑NOOFPLACES+.5)/10↑NOOFPLACES
 70 IF ABS (ANSWER-CORRECTANSWER)>.00001 THEN GO TO 500
 80 PRINT AT 14,1;"WELL DONE!!!! ";ANSWER;" is correct"
100 PRINT AT 21,0;"PRESS any key to continue": PAUSE 0: GO TO 10
500 PRINT AT 14,3; INK 2;"WRONG!!!! ";ANSWER;" is incorrect"
510 PRINT AT 21,0;"PRESS any key to try again"
520 PAUSE 0: GO TO 25
```

PROGRAM 19

This simple program shows the use of colour on the Spectrum; on black-and-white televisions the blocks will appear in shades of grey. Notice that line numbers can have any value providing they occur in ascending order.

```
  1 REM Logo
  4 CLS
  5 PAUSE 80
  6 PAPER 6
  7 PLOT 0,0: DRAW 255,0: DRAW
0,175: DRAW -255,0: DRAW 0,-175
  8 LET a$="▮▮▮▮▮▮▮▮"
 20 FOR y=1 TO 10
 25 INK 5: PRINT AT y,21;a$
 30 INK 4: PRINT AT y,1;a$
 35 INK 2: PRINT AT y+10,11;a$
 40 NEXT y
130 INK 0: PRINT AT 5,12;"NUMER
ACY"
140 PRINT AT 16,2;"COMPUTER"
150 PRINT AT 16,22;"PROGRAMS"
153 PRINT AT 5,3;"PITMAN"
156 PRINT AT 4,22;"H.Ogden"
157 PRINT AT 6,21;"R.Woodward"
158 STOP
```

PROGRAM 20

Try this short program to simulate movement across the screen, the simple graphic represents a car. List the program and experiment with different values of PAUSE to change the speed of the car. Try moving the car in the opposite direction and altering the graphics to produce different shapes.

```
10 REM Movement
20 FOR c=0 TO 28
30 PRINT AT 11,c;"  ▆ "
40 PAUSE 5
50 NEXT c
60 CLS
70 GO TO 20
```

To reverse direction

```
20 FOR c=28 TO 0 STEP -1
30 PRINT AT 11,c;"▆   "
```

Exercises 11

Exercises 11.1 to 11.35 are intended to provide practice in the use of the calculator keyboard. Give the full display answers.

11.1 $2.76 + 193.8 + 17.44$

11.2 $3\,192 + 167.38 + 44.69$

11.3 $1.381\,4 + 0.707\,2 + 0.069\,1$

11.4 $1.731\,24 + 869.317 + 187.043\,4$

11.5 $0.767\,67 + 1.003\,139 + 0.062\,184$

11.6 $325\,779.5 + 141.398\,2 + 5\,649.318$

11.7 $1\,947.39 - 806.414$

11.8 $13.383\,62 - 7.707\,07$

11.9 $201\,009 - 86\,434.8$

11.10 $0.064\,392 - 0.010\,523$

11.11 5.35×1.64

11.12 7038×4.196

11.13 17.717×8.143

11.14 $44.7 \times 13.8 \times 1.95$

11.15 $601.2 \times 0.984\,14 \times 33.645$

11.16 $81.209 \times 362.3 \times 0.449 \times 1.364$

11.17 $0.245 \times 0.318 \times 0.04$

11.18 $13.965 \div 24.76$

11.19 $7\,043 \div 168.2$

11.20 $0.319\,5 \div 0.786$

11.21 0.83^2

11.22 56.9^2

11.23 1.617^2

11.24 $0.93^2 \times 2.64$

11.25 $\sqrt{1.714}$

11.26 $\sqrt{3\,962}$

11.27 $\sqrt{484.4}$

11.28 $\dfrac{26.51 \times 17.83}{102.4}$

11.29 $\dfrac{6.55 \times 13.7 \times 0.964}{86.319}$

11.30 $\dfrac{7.364 + 12.085}{22.09}$

11.31 $\dfrac{0.043 + 0.209}{0.060\,6}$

11.32 $\dfrac{4.87 \times 12.19}{3.65 \times 14.71}$

11.33 $\dfrac{4\pi}{17 \times 9.6}$

11.34 $\dfrac{\sqrt{133.8}}{6.042}$

11.35 $\dfrac{41\,562 - 9\,875}{1.08 \times 207.3}$

11.36 Given that $T = \dfrac{4.9SL}{n}$, find the value of T when $S = 0.6$, $L = 410$ and $n = 1.5$.

11.37 Given that $p = \dfrac{\sqrt{m}+n}{q}$, find p when $m = 14$, $n = 0.155$ and $q = 20.7$. Give the value of p correct to 3 decimal places.

11.38 Given that $V = \dfrac{\pi d N}{1\,000}$, complete the following table.

d	N	V
34	281	
13	1 075	
75	127	
12.5	1 183	
62	203	
114		63.2

Give V correct to 1 d.p.

11.39 The volume of a rectangular block is given by
$V = L \times W \times H$
where V = volume (mm³), L = length (mm), W = width (mm) and H = height (mm). Use the formula to complete the following table.

LENGTH (mm)	WIDTH (mm)	HEIGHT (mm)	VOLUME (mm³)
64	43	17	
53	29	23	
90	61.5	32.4	
109	78.2	50.5	
163.7	44.7	31	
208	107.3	77.9	
324.5	125.4	86.3	

Give the volume to the nearest cubic millimetre.

11.40 The cross-sectional area of a tube is given by
$A = \pi(R^2 - r^2)$
where A = cross-sectional area (mm²), R = external radius (mm) and r = internal radius (mm). Use the formula to complete the following table.

EXTERNAL RADIUS (mm)	INTERNAL RADIUS (mm)	CROSS-SECTIONAL AREA (mm²)
20	15	
11	9	
25	22	
16	14	
32	27	

Give the answers correct to 1 d.p.

11.41 Use the formula given below to complete the table, give answer to 3 significant figures.

$F = \dfrac{g}{h}, \qquad N = Fg$

g	h	F	N
20	0.5		
7.5	35		
50	10.5		
12	7		
100	24		

11.42 Use the table of Useful Conversion Factors given at the front of this book to complete the following (answer to 4 significant figures):

(a) 3.746 in = millimetres
(b) 8 246 in² = square millimetres
(c) 41.6 m³ = litres
(d) 7.304 y³ = cubic metres
(e) 85.6 lb = kilogrammes
(f) 416 ft per minute = metres per second
(g) 1.097 ft = metres
(h) 54.39 in³ = cubic millimetres
(i) 2.108 yd = metres
(j) 25.7 miles = kilometres
(k) 356.3 yd² = square metres
(l) 844 ft² = square metres
(m) 216 gal = litres
(n) $3\frac{1}{2}$ horsepower = kilowatts

11.43 Find the total cost of:
17 reams of duplicating paper at £3.95 per ream
35 notepads at 63p each
28 ring files at 97p each
7 rolls of sealing tape at 71p per roll
13 boxes of carbon paper at £2.07 per box

11.44 Find the total cost of:
43 gal of petrol at £1.81 per gallon
3 fanbelts at £2.99 each
11 litres of antifreeze at 96p per litre
4 tyres at £29.85 each
2 windscreen wipers at £4.85 each

11.45 Find the total cost of:
14 packets of tea at 29p each
9 jars of coffee at 96p each
23 large loaves at 49p each
38 bottles of milk at 26p each
9 tins of salmon at £1.18 each
11 tins of ham at £1.07 each
8 kg of sugar at 44p per kilogramme
$4\frac{1}{2}$ lb of rump steak at £2.74 per pound
$2\frac{1}{4}$ lb of braising steak at £1.34 per pound
12 tins sardines at 53p each

11.46 A motorist drives a total distance of 575.4 km in one week. His journeys are made up of
183.9 km on motorways
247.2 km on 'A' roads
84.6 km on 'B' roads
remainder in the city centre.
How many kilometres does he travel in the city centre?

11.47 Calculate the area of:
(a) a rectangle, 14.93 mm by 27.64 mm;
(b) a square, 1.073 m side;
(c) a triangle, 84.17 cm base and 109.46 cm vertical height;
(d) a circle, 253.65 mm diameter.
Give the answers correct to 3 d.p.

11.48 Calculate the volume of:
(a) a rectangular block, 187.6 mm × 51.9 mm × 21.55 mm;
(b) a cylinder, 8.07 cm diameter and 33.23 cm long;
(c) a sphere, 52.75 mm diameter;
(d) a cone, 11.82 cm diameter and 14.96 cm vertical height.
Give the answers correct to 1 d.p.

11.49 Find the average of:
(a) 216 319, 152 064, 73 429, 586 427, 53 094;
(b) 7.106, 11.0295, 18.616, 29.309, 2.693, 101.75;
(c) 0.134 5, 0.097 16, 0.880 8, 0.041, 0.710 7, 0.0710 7.

11.50 Express as a percentage correct to 2 d.p.
(a) $\frac{2}{13}$ (b) $\frac{17}{19}$ (c) $\frac{2}{3}$
(d) $\frac{41}{111}$ (e) $\frac{12}{37}$ (f) $\frac{6}{29}$

Exercises 11.51 to 11.70 are intended to provide practice in writing and entering simple BASIC programs. For each exercise list the program and show the screen display.

11.51 The sum of three numbers using LET statements.

11.52 The product of two numbers using LET statements.

11.53 The sum of five numbers using READ and DATA.

11.54 The value of $\frac{3.749 \times 1.052}{15.647}$ using LET statements.

11.55 The value of $\frac{28.3^3 - 14.75}{1\,076}$ using LET statements.

11.56 The value of $5.6^2 \times 407$ using READ and DATA.

11.57 The product of two numbers using INPUT.

11.58 The squares of numbers from 1 to 20.

11.59 The squares roots of numbers from 30 to 60.

11.60 The cubes of numbers from 0.8 to 1.7 by steps of 0.1.

11.61 Convert millimetres to inches for the range 50 to 100 mm by steps of 2 mm.

11.62 Convert yards to metres for the range 0 to 25 yd.

11.63 Convert gallons to litres for the range 0 to 30 gal.

11.64 Convert pounds to kilogrammes using INPUT.

11.65 Convert feet per second to metres per minute using INPUT.

11.66 The area of a triangle.

11.67 The circumference of a circle.

11.68 The volume of a sphere.

11.69 Convert Celsius to Fahrenheit degrees.

11.70 Convert £ sterling to US dollars (take £1 = 1.2 dollars).

Exercises 11.71 to 11.90 are intended to provide practice in entering and editing existing programs. Each exercise uses a program which is listed in this chapter. The program should be entered and run, then listed and altered before running again. The answer should be given in the form of the new listing and a copy of the display.

11.71 Alter Program 1 to make $a = 14.95$ and $b = 7.392$.

11.72 Alter Program 1 to find half the sum of the numbers.

11.73 Alter Program 2 to find the value of $48.7 \times \pi$.

11.74 Alter Program 3 to find the squares of numbers.

11.75 Alter Program 4 to cover the range 0 to 1 in by steps of 0.01 in.

11.76 Alter Program 5 to suit dimensions in inches.

11.77 Alter Program 6 to take input of diameter.

11.78 Add a STOP device of $F = -999$ to Program 7.

11.79 Alter Program 8 to suit a standard cost of £4.62 per square metre.

11.80 Alter Program 9 to add five numbers and divide the answer by 2.

11.81 Modify Program 10 as follows,
removed from stock: JAWS 2, ALIEN
add to stock: TRON, FIREFOX, SEA RAIDERS

11.82 Change the names and addresses in Program 11 to those of your friends.

11.83 Alter Program 12 to suit a cutting speed of 63 m/min.

11.84 Alter Program 13 to suit a supply of 110 V and add a STOP device.

11.85 Modify Program 14 to draw a bar chart of the following values:
 17, 62, 39, 44, 90, 8, 25, 79, 81, 54, 40

11.86 Modify Program 15 to find the mean of,
 13.9 17.2 18.0 21.3 19.6
 24.7 11.3 39.7 13.2 41.3
 18.2 32.6 21.7   9.1 16.4

11.87 Show the effect of adding the following lines to Program 16:

```
81 PLOT 150,145
82 DRAW 0,20: DRAW 15,-8: DRAW -15,-8
83 PLOT 150,145
```

11.88 Alter Program 16 to:
(a) remove the FLASH from the slogan;
(b) increase the interval between CLS;
(c) change the slogan to Sailing on the Lakes.

11.89 Alter Program 19 to (a) change the colours, (b) change the wording and (c) make the wording flash.

11.90 Alter Program 20 to simulate the movement of (a) a van, (b) a ship and (c) a train.

Appendix 1 Project Work

Practical project work forms a useful component of pre-vocational studies. Projects can be designed to include many aspects of numeracy studies which can be readily integrated into the practical situation. Some examples of suggested projects are shown in this appendix.

Project 1 Furnishing a Bathroom

The bathroom of a semi-detached house measures 8 ft by 6 ft and is 8 ft high. There is a window 3 ft high and 4 ft wide in the middle of the end wall. Plan the layout of the bathroom which is to be furnished with:
 (i) a bath;
 (ii) a washbasin on a pedestal;
 (iii) a toilet;
 (iv) a hot water cistern and airing cupboard;
 (v) a radiator connected to the central-heating system.

Compare the prices of different suites of bathroom furniture from catalogues or inquiries.

Draw a bar chart to compare the prices of each item.

Select a particular suite and calculate the volume required to house it.

Find what percentage of the total volume of the room is occupied when the suite has been fitted.

Determine the size and area of floor space available for carpeting.

Compare the cost of (i) carpet tiles, (ii) foam-backed carpet, (iii) good-quality carpet and (iv) lino tiles.

The vertical walls are to be tiled with ceramic tiles, calculate the number of tiles required to cover each wall.

Compare the price and size of various tiles in both metric and Imperial measure.

The ceiling is to be papered with water-repellant paper, calculate the area to be covered and the number of rolls of paper required.

Estimate the cost of all the materials used.

Project 2 Carpeting a Hotel Ballroom

A hotel ballroom floor has a dance floor 7 m square and a raised dais for musicians as shown in Fig. P1.

Calculate the area of (i) the floor of the room; (ii) the dance floor and (iii) the dais used by musicians.

Determine the percentage of the floor space occupied by (i) the dance floor and (ii) the dais.

The remaining area of the floor is to be carpeted or carpet tiled. Calculate the area to be carpeted and compare the prices of covering it with (a) good-quality carpet, (b) cheap carpet and (c) carpet tiles.

Draw a bar chart to compare the costs of each covering.

The musicians' dais is 0.5 m high, calculate the volume of the room occupied by the dais.

If the room is 6 m high, estimate what percentage of the volume of the room is occupied by the dais.

What is the ratio of carpeted area to the rest of the room?

Sketch the plan of the ballroom and give its dimensions in feet and inches.

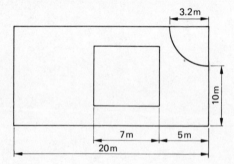

Fig. P1

Project 3 Design of a Cold Frame

A cold frame is used for propagating seeds before planting out. It can be regarded as a smaller version of a greenhouse and usually has a wooden surround with glass lid which can be opened.

Examine a cold frame and measure, in centimetres, the following:
 (i) the overall size;
 (ii) the shape and height of the framework;
 (iii) the size of the opening area;
 (iv) the size of the panes of glass in the opening area;
 (v) the size of the frames holding the glass;
 (vi) any other measurements you find that you need to carry out this project.

Determine:
(a) the length of the perimeter;
(b) the surface area of the ground covered by the cold frame;
(c) the total surface area of the four sides;
(d) the surface area of the opening lid;
(e) the surface area of the glass in the opening area;
(f) the volume of glass in the construction;
(g) the percentage area covered by glass;
(h) the fraction of the whole area covered by glass;
(i) the area of soil available for plant life;
(j) the number of seed boxes, 45 cm × 30 cm, which the cold frame will hold on one level.

If seeds are grown at the rate of 250 per seed box and 35% do not germinate, how many seedlings will germinate?

Assume all the germinated seeds are transplanted in open ground after propagation in the cold frame, how much space would be required if each seed needs a circle of radius 9″ in which to grow?

List the costings of all the materials for the manufacture of a cold frame and all the equipment needed to produce lettuce plants by this method.

Draw graphs to illustrate:
(k) the initial cost of setting up the cold frame;
(l) the numbers of lettuces produced;
(m) the cost of producing each lettuce;
(n) the break-even point for equipment alone;
(o) the break-even point for labour and equipment.

Project 4 Building a Garage

This project considers the building of a garage attached to a house. It is assumed that plans have been passed for the work and the foundations have been laid.

Measure up the site (in metric) and compare it with the sizes on the plans.

Calculate the area of the brick wall required.

Establish the dimensions of a brick and find the number of bricks required to build 1 m² of wall.

Estimate the total number of bricks required for the garage.

Estimate the volume of mortar required in cubic metres.

Estimate the quantities of sand and cement required using a ratio of 5:1.

Measure the sizes of the doors and windows.

Calculate the areas of the doors and windows.

Calculate the area of wall in the end walls of the garage.

Calculate the area of roof and find the lengths of spar and board material, and the area of roofing felt required.

Determine what percentage of the wall is brick and what percentage of the wall is occupied by windows and doors.

Assume an up-and-over door is used and find the area of wall it will cover when closed.

Find the angle the door will rotate through from the open to the closed position.

Calculate the distance the door will protrude outside and inside the garage when open.

Determine the length of guttering required.

Draw a pie chart comparing the area of roof, wall, doors and windows.

Estimate the material costs of the garage.

Project 5 Concrete Base for a Garage

The design of the concrete base is shown in Fig. P2.
Determine:
(a) the length of the perimeter of the base in metres;
(b) the thickness of the concrete;
(c) the area of the base;
(d) the volume of the base;
(e) the size, area and volume of the base in imperial units.

The ratio of aggregate to cement is 6:1. Determine:
(f) the mass of (i) aggregate and (ii) cement;
(g) the percentage of the volume of the base which is (i) aggregate and (ii) cement;
(h) the cost of a cubic metre of concrete;
(i) the estimated cost of laying the base including labour charges.

Obtain, from local firms, estimates for (i) the complete job, (ii) the laying of the concrete only, (iii) the base preparation before the concrete is laid and (iv) the cost of the concrete only. Draw a pie chart to compare the cost of labour, material and base preparation.

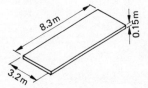

Fig. P2

Project 6 Comparison of Duplicating Costs

Compare the different methods of duplicating used in office work, e.g. Gestetner, Banda, Offset Litho, skin cutters, etc.

(a) List the current prices of the various models.
(b) Measure the main dimensions of each machine in millimetres. Convert these measurements to inches. Use a calculator or conversion table for this work.

(c) Calculate the volume of space in cubic centimetres required for each machine. List this information and compare each method.

(d) List the size of prints obtainable from each machine in ascending order of size. Compare the lists and state the range for each machine.

(e) List the software requirements for each machine. Price each item and calculate the cost of producing 100 copies. Compare the results as percentages.

(f) List the automatic speed ranges of each machine and compare these ranges in terms of average production of copies per minute.

(g) Estimate the time required to prepare each machine to produce 100 copies and to produce 500 copies. Draw a bar chart to compare times for each machine.

(h) Calculate the average production rate from each machine per working day and compare these rates on a bar chart.

(i) Draw a graph of the cash layout for each process for a given period of time to include the initial cost and running costs and compare these results.

(j) Choose a number of copies required from each machine and calculate the cost per copy.

Appendix 2 Computer Program

The microcomputer offers a useful means of enabling students to familiarise themselves with many job-related situations. By the use of a computer program which is designed to simulate a set of working conditions, the student is able to experiment with and investigate many of the factors concerned in a practical task through the medium of the computer keyboard.

The following program is written for the Sinclair ZX Spectrum and simulates a shopping expedition to a supermarket.

```
   2 REM SUPERMARKET
   5 RANDOMIZE
  10 LET V$="......................
........."
  50 GO SUB 5000
  60 CLS : BORDER 5: INK 1
  70 REM CALCULATE NUMBER OF
        ITEMS TO BE BOUGHT
 100 LET num=2+INT (RND*10)
 105 DIM M(NUM)
 110 LET TOTAL=NUM
 300 CLS : GO SUB 3200
 310 RESTORE 6000
 315 PRINT AT 5,10;"DEPARTMENTS"
;TAB 9;"-------------"
 316 REM PRINT NUMBER OF
             DEPARTMENTS
 320 READ N
 330 FOR Q=1 TO N
 340 READ T$
 350 PRINT AT 7+Q,5;CHR$ (Q+64);
"/    ";T$
 360 NEXT Q
 365 PRINT AT 18,1;"YOU HAVE TO
PURCHASE ";NUM;" ITEMS
 FROM ANY DEPARTMENT"
 370 INPUT AT 0,0;"WHICH DEPARTM
ENT DO YOU WISH TO BUY  FROM? ";
LINE a$
 380 LET pos=CODE a$-96
 390 IF pos<1 OR pos>n THEN GO T
O 370
 395 REM PRINT LIST OF ITEMS IN
        SELECTED DEPARTMENT
 500 CLS : GO SUB 3200
 505 PRINT AT 4,4;"ITEM";TAB 22;
"PRICE";TAB 22;"£"
 510 RESTORE 6900+POS*100
 520 READ NO
 530 FOR Q=1 TO NO
 540 READ S$,PRICE
 545 LET v=price: GO SUB 4000
 550 PRINT AT 5+Q,2;CHR$ (64+Q);
"/   ";S$;
 560 LET L=LEN S$
 570 PRINT V$( TO 18-L);
 580 PRINT V$
 590 NEXT Q
 600 GO SUB 3000
 610 PRINT AT 19,0;"ENTER Z TO D
ISPLAY INITIAL MENU"
 615 REM INPUT ITEM SELECTED
 620 INPUT "ENTER LETTER OF SELE
CTED ITEM ";LINE A$
 625 IF A$="Z" THEN GO TO 300
 630 IF A$<"a" OR A$>CHR$ (96+NO
) THEN GO TO 620
 640 LET M(total+1-NUM)=POS*100+
CODE A$-96
 650 LET NUM=NUM-1: IF NUM>0 THE
N GO TO 600
 900 REM PRINT BILL
1000 CLS : GO SUB 3200
1005 LET sum=0
1010 PRINT AT 3,13;"BILL"
1015 PRINT AT 4,22;"£"
1017 REM PRINT LIST OF ITEMS
            BOUGHT
1020 FOR Q=1 TO total
1030 LET C=INT (M(Q)/100)*100
1040 RESTORE 6900+C
1045 READ n
1050 FOR W=1 TO M(Q)-C
1060 READ T$,PRICE
1070 NEXT W
1072 LET v=price: GO SUB 4000
1075 LET sum=sum+price
1080 PRINT AT 4+Q,3;T$;TAB 24-f;
v$
1085 BEEP .2,20
1090 NEXT Q
1095 LET v=sum: GO SUB 4000
1097 REM PRINT TOTAL COST
1100 PRINT AT 6+total,3; INK 2;"
TOTAL";TAB 24-f;v$
1102 BEEP .4,40
1105 PLOT 169,130-total*8: DRAW
40,0
1107 REM INPUT AMOUNT TENDERED
1110 INPUT AT 0,0;"ENTER AMOUNT
TENDERED IN         PAYMENT £";pa
y
1115 PRINT AT 21,0;"
1120 IF pay<sum THEN LET v=pay:
GO SUB 4000: PRINT AT 21,0; INK
2;"SORRY £";v$;" IS INSUFFICIENT
": GO TO 1110
```

```
1126 LET u=pay: GO SUB 4000
1130 PRINT AT 8+total,3;"AMOUNT
 RECEIVED";TAB 24-f;u$
1135 BEEP .3,50
1137 LET u=INT ((pay-sum)*100+.5
)/100: GO SUB 4000
1140 PRINT AT 10+total,3; INK 4;
"CHANGE";TAB 24-f;u$
1150 BEEP .5,30
1200 INPUT AT 0,0;"ENTER S TO ST
OP OR C TO CONTINUE"; LINE a$
1210 IF a$="S" THEN STOP
1220 IF a$="C" THEN GO TO 60
1230 GO TO 1200
1300 REM END OF MAIN PROGRAMM
2999 REM THIS ROUTINE PRINTS THE
 NUMBER OF ITEMS TO BE SELECTED
3000 FLASH 1: INK 3
3005 PRINT AT 20,2;"SELECT ";NUM
;
3010 IF TOTAL<>NUM THEN PRINT "
MORE"
3020 PRINT " ITEMS FROM THIS"; F
LASH 0;TAB 7; FLASH 1;"OR ANY DE
PARTMENT"
3025 FLASH 0: INK 1
3030 RETURN
3100 REM TITLE
3200 PRINT TAB 10; INK 2;"SUPERM
ARKET";TAB 9;"-------------"
3210 RETURN
3999 REM THIS ROUTINE ADDS EXTRA
 .00'S AND 0.'S TO ALL NUMBERS
BEFORE THEY ARE PRINTED
4000 LET u$=STR$ u
4005 LET f=0
4010 FOR i=1 TO LEN u$
4020 IF u$(i)="." THEN LET f=i
4030 NEXT i
4040 IF f=0 THEN LET f=LEN u$+1:
 LET u$=u$+".00": GO TO 4100
4045 IF f=1 THEN LET u$="0"+u$:
LET f=2
4050 LET ll=LEN u$-f
4060 IF ll=1 THEN LET u$=u$+"0"
4100 RETURN
4999 REM INSTRUCTIONS
5000 CLS : GO SUB 3200
5010 BORDER 6
5020 INK 1
5030 PRINT AT 6,0;"THE PROGRAM W
ILL TELL YOU HOW   MANY ITEMS YO
U HAVE TO BUY FROM A SUPERMARKET
. YOU MAY THEN    SELECT THE IT
EMS FROM SEVERAL   DEPARTMENTS.T
HE COMPUTER WILL   THEN PRODUCE
A BILL FOR YOU TO  PAY AND CALCU
LATE ANY CHANGE."
5040 PRINT : PRINT "N.B.THE PRIC
ES OF THE ITEMS MAY BE CHANGED.S
EE PROGRAM LINE 6500FOR DETAILS.
"
5100 INPUT "PRESS ENTER TO CONTI
NUE "; LINE a$
5110 RETURN
5500 REM DATA
6000 DATA 6,"MEAT","VEGETABLES +
 FRUIT","TINS","CEREALS","FROZEN
 FOODS","GENERAL"
6500 REM CHANGING PRICES
---------------
THE DATA STATEMENTS BELOW
CONTAIN THE LIST OF ITEMS
AVAILABLE IN THE DEPARTMENTS.
THE FIRST NUMBER IN EACH LINE I
THE NUMBER OF ITEMS IN THAT
DEPARTMENT, FOLLOWED BY A LIST
OF ITEMS, AND PRICES.
6999 REM MEAT DEPARTMENT
7000 DATA 8,"BACON",1.09,"BRAI
NG STEAK",2.04,"LOIN CHOPS",1.9
,"RIB CHOPS",1.35,"PIG LIVER",.
6,"OX. TAILS",1.04,"LEG OF LAMB
,8.2,"SAUSAGES",.51
7099 REM VEG. DEPARTMENT
7100 DATA 11,"POTATOES",.2,"ORA
GES",.42,"TOMATOES",.68,"CUCUMB
R",.46,"LETTUCE",.21,"APPLES",
6,"BANANAS",.42,"CABBAGE",.28,"
AULIFLOWER",.58,"CARROTS",.34,"
NIONS",.35
7199 REM CANNED DEPARTMENT
7200 DATA 11,"BAKED BEANS",.23,
SOUP",.29,"SPAGHETTI",.19,"RAV
LI",.35,"PROCESSED PEAS",.2,"M
ED VEG.",.29,"SWEETCORN",.29,"
AM",.79,"CORNED BEEF",.83,"PAS
 SPREAD",.23,"SARDINES",.21
7299 REM cereal department
7300 DATA 7,"CORN FLAKES",.64,
ICE CRISPIES",.6,"WEETABIX",.5
"SHREDDED WHEAT",.73,"ALL-BRAN
.87,"ALPEN",.89,"PORAGE OATS",.
4
7399 REM frozen department
7400 DATA 10,"BEEFBURGERS",.84
MEATBALLS",.75,"PIES",1.02,"GA
AU",1.89,"ICE CREAM",1.88,"OVE
 CHIPS",.45,"FISH FINGERS",1.09
PLAICE FILLETS",1.99,"PRAWNS",.
59,"CHICKEN",1.92
7499 REM general department
7500 DATA 13,"SUGAR",.45,"JAM",
42,"FLOUR",.37,"KETCHUP",.47,"
STANT COFFEE",1.11,"TEA",.61,"
EAD",.29,"CHEESE",.8,"MILK",.2
YOGURT",.22,"EGGS",.47,"BUTTER
.54,"WASHING POWDER",.79
7999 REM ALTER THE SOUND EFFE
 IN LINE 1102 OR 1135 OR 1150
```

When the program is RUN the shopper is told a random number of items which are to be purchased from the supermarket. The program then displays a list of the departments within the supermarket. Selecting a department causes a list of items available to be displayed together with their cost. The shopper may buy one or more items from any number of departments. When the full complement of items has been selected the program proceeds to the check-out stage and simulates the action of a till. The shopper is presented with a bill listing the selected items with their individual and total costs. To pay the bill, the shopper may tender a round number of pounds and the till will make change.

The advantage of this simulation program is that it can be easily modified, using the editing techniques described in Chapter 11, to update prices, add or delete items and even change the names of departments within the supermarket. Thus the whole process of supermarket shopping can be experienced through the computer.

Screen 1: Selecting a Department

```
       SUPERMARKET
       -----------


       DEPARTMENTS
       -----------

   A/       MEAT
   B/       VEGETABLES + FRUIT
   C/       TINS
   D/       CEREALS
   E/       FROZEN FOODS
   F/       GENERAL


   YOU HAVE TO PURCHASE 2 ITEMS
        FROM ANY DEPARTMENT
```

Screen 2: Selecting Items

```
          SUPERMARKET
          -----------

    ITEM                   PRICE
                             £
   A/ BEEFBURGERS........0.84
   B/ MEATBALLS..........1.79
   C/ PIES...............1.02
   D/ GATEAU.............1.89
   E/ ICE CREAM..........1.86
   F/ OVEN CHIPS.........0.45
   G/ FISH FINGERS.......1.09
   H/ PLAICE FILLETS.....1.99
   I/ PRAWNS.............1.59
   J/ CHICKEN............1.92


   ENTER Z TO DISPLAY INITIAL MENU
     SELECT 9 ITEMS FROM THIS
         OR ANY DEPARTMENT
```

Screen 3: At the Check-out

```
          SUPERMARKET
          -----------

             BILL
                            £
   BACON                   1.09
   SAUSAGES                0.51
   PROCESSED PEAS          0.20
   SARDINES                0.21
   POTATOES                0.20
   CARROTS                 0.34
   CABBAGE                 0.28
   BEEFBURGERS             0.84
   FISH FINGERS            1.09
   CHICKEN                 1.92
                          -----
   TOTAL                   6.68

   AMOUNT RECEIVED        10.00

   CHANGE                  3.32
```

Answers

Exercises 1.A

1.A1 (a) 1514 (b) 2294 (c) 446 (d) 14966 (e) 1242 (f) 2570

1.A2 (a) 108 (b) 81 (c) 129 (d) 105 (e) 75

1.A3 (a) £16.42 (b) £2.95 (c) £124.05 (d) £54.43 (e) £2.76 (f) £70.91

1.A4 (a) 91 mm (b) £14 609 (c) 270 m (d) 2 087 kg (e) 4 h 38 min (f) 12 weeks (g) 2461 (h) 3 580 km

1.A5 13.00, 14.15, 16.02, 19.25, 22.23, 01.40

1.A6 £16 793.52

1.A7 (a) 449 (b) 1248 (c) 1413 (d) 22 324 (e) 2 085 (f) 67 772 (g) 7 814 (h) 33 343

1.A8 (a) 116 m (b) 1 114 kg (c) 2791 (d) 5 075 km (e) 489 g (f) 730 mm (g) 6 h 41 min (h) 10 h 39 min

1.A9 (a) £59.55 (b) 54p (c) £6 356.50 (d) £101.42 (e) £655.84 (f) £225.34 (g) £3.43 (h) £27 083.66

1.A10 (a) 50 mm (b) 18 mm (c) 33 mm (d) 53 mm (e) 18 mm (f) 16 mm

1.A11 $x = 19$ mm, $y = 47$ mm

1.A12 (a) 99, 1 118 (b) 73, 1 176 (c) 182, 7 440 (d) 56, 703 (e) 32, 240 (f) 46, 480 (g) 32, 192 (h) 67, 850

1.A13 (a) 1971 (b) 2 160 (c) 5 871 (d) 14 784 (e) 49 152 (f) 27 447 (g) 378 840 (h) 433 912 (i) 672 (j) 702 (k) 4 200 (l) 35 100 (m) 13 376 (n) 52 896

1.A14 360, 1 404, 4 644, 910, 896, 54 400, 6 111

1.A15 (a) £1 100 (b) £121.50 (c) £15.36 (d) £138.45 (e) £1 814.40 (f) £252 (g) £18 703.65 (h) £2 142

1.A16 (a) 5850 m (b) 5850 g (c) 2 184 km (d) 22 h 45 min (e) 5 625 mm (f) 10 906 kg (g) 261 h 36 min (h) 25 h 54 min

1.A17 £32.63

1.A18 (a) 56 (b) 25 (c) 865 (d) 89 (e) 38 (f) 235 (g) 1 894 (h) 209

1.A19 (a) £3.56 (b) 5p (c) £9.32 (d) 356 m (e) 43 mm (f) £18.30 (g) £108.56 (h) 248 kg

1.A20 (a) 504 (b) 10 (c) 98 (d) 53 (e) 5 040 (f) 134 (g) 2 511 (h) 41

Exercises 1.B

1.B1 (a) 61p (b) 50p (c) £1.06 (d) 90p (e) £1.85

1.B2 £2.50

1.B3 £2.01

1.B4 (a) £2.80 (b) £145.60

1.B5 (a) £256 (b)(i) £79 (ii) £316

1.B6 £819

1.B7 (a) £3.02 (b) 64p, 66p, 32p, 81p, 59p

1.B8 (a) 2 min 5 seconds (b) 30 min (c) 6 h 40 min (d) 1 h 40 min

1.B9 (a) £29 (b) £20

1.B10 (a) £2.92 (b) £1.66 (c) £2.25 (d) £1.42 (e) £1.93 (f) £2.30 (g) £1.72 (h) £1.66

1.B11 Across: 12 713, 13 119, 12 389, 38 301 Down: 18 535, 19 508, 258, 38 301

1.B12 Across: £1 199.12, £877.09, £965.35, £970.84, £1 483.05, £674.47, £6 169.92 Down: £4 381.62, £335.03, £659.80, £793.47, £6 169.92

1.B13 (a) £615.68 (b) £86.64

1.B14 (a) £62.60 (b) £178.30 (c) £40.77 (d) £53.78 (e) £40.61 (f) £543.84

1.B15 (a) 562 (b) 274 (c) 288 (d) Thursday (e) £35.62

1.B16 (a) 896 (b) £1 280

1.B17 (a) £78.50

1.B18 (a) £6.25 (b) £24.75

1.B19 (a) £4.50 (b) £5.00 (c) £10.50 (d) £11.80 (e) £4.30

1.B20 (a) £285.14 (b) £112.95 (c) £144.75

Exercises 1.C

1.C1 (a) 86 mm (b) 449 mm (c) 510 mm (d) 142 cm

1.C2 (a) 514 mm (b) 151 mm (c) 90 mm (d) 458 mm

1.C3 46 min 40 seconds

1.C4 31 min 15 seconds

1.C5 (a) 82 mm (b) 24 mm (c) 14 mm (d) 48 mm (e) 127 mm

1.C6 4 h 32 min

1.C7 (a) 1 097 mm (b) 7 679 mm

1.C8 647 mm

1.C9 (a) 680 mm (b) 15 kg (c) 12 kg

1.C10 9 880

1.C11 (a) 877, 718, 803, 853, 818, 983, 956, 939, 926, 767 (b) 8 640

1.C12 (a) £1 666.30, £1 364.20, £1 525.70, £1 620.70, £1 554.20, £1 867.70, £1 816.40, £1 784.10, £1 759.40, £1 457.30 (b) £16 416

1.C13 (a) 11 mm (b) 4 675 mm^2 (c) 3 mm

1.C14 (a) 89 mm (b)(i) 2 mm (ii) 270 mm (iii) 416 mm
1.C15 (a) 28 mm, 37 mm (b) 14 mm, 17 mm (c) 32 mm, 29 mm
1.C16 (a) 4 mm (b) 76 500 mm³
1.C17 (a) 27 kW (b) 14 kW (c) 7
1.C18 4 A, 120 ohms, 5 A, 12 V, 3 A
1.C19 33 ohms, 66 ohms, 720 ohms, 25 ohms, 450 ohms
1.C20 (a) 283 (b) 127 (c) 194 (d) 38

Exercises 1.D

1.D1 £6.44
1.D2 (a) 106 (b) 35 m
1.D3 (a) 14 712 (b) 11 868
1.D4 (a) 7 140 (b) £22.56
1.D5 (a) 102 000 (b) 653
1.D6 (a) £129.10, £126.45, £123.94, £130.37, £130.78, £128.25 (b) £768.89
1.D7 (a) £1.20 (b) £1.02 (c) £1.93 (d) £1.06 (e) £1.55
1.D8 (a) 425 (b) £8.50
1.D9 (a) 25 t (b) 152 t
1.D10 (a) 5 280 litres (b) 56 min
1.D11 (a) 93 (b) 3 999 km
1.D12 (a)(i) 2 020 t (ii) 87 m (iii) 116 t (b) 505 t (c)(i) 44 m (ii) 8 m (iii) 4 m
1.D13 (a) 162 m (b) £202.50
1.D14 £1 032.59
1.D15 (a) £28.56 (b) £2.89
1.D16 (a) £6 (b) £15
1.D17 (a) 197 g (b) 6 480 g, 11 760 g, 3 600 g, 1 800 g
1.D18 (a) 6 650 litres (b) 33 250 litres (c) 19 950 litres
1.D19 (a) 33 258 litres (b) 4 117 litres
1.D20 (a)(i) 25 m (ii) 9 m (b)(i) 56 m² (ii) 102 m² (iii) 204 m² (iv) 63 m² (v) 325 m² (c) 269 m²

Exercises 1.E

1.E1 (a) 56 m, 66 m, 51 m, 101 m (b) 274 m (c) 216
1.E2 (a) 14 730 m (b) 982
1.E3 £24.59
1.E4 (a) £2.46 (b) 6p
1.E5 (a) £5 (b)(i) 520 (ii) 125 km
1.E6 1 089 m
1.E7 2 508 km
1.E8 (a) 515 (b) 3 575 m²
1.E9 24 km
1.E10 205
1.E11 (a) £9 (b) £468
1.E12 £300, £1 050, £1 800, £600
1.E13 (a) £1 479 651 (b) £715 522
1.E14 (a) £6 589.15 (b) £1 180.29 (c) £21
1.E15 (a) 11.30, 14.08, 16.46, 21.00, 02.12 (b) £165.96 (c) 1 h 52 min (d) £72.41
1.E16 £1 702
1.E17 (a) 48 m² (b) 64 m² (c) 64 m² (d) 80 m²
1.E18 (a) £800 (b)(i) £1 280 (ii) £768 (iii) £512
1.E19 £1 980
1.E20 (a) £75 782 (b) £22 697

Exercises 1.F

1.F1 (a) £1.86 (b) £2.59 (c) £4.33 (d) £3.45 (e) £3.69 (f) £4.05 (g) £4.47 (h) £6.69
1.F2 (a) £3.14 (b) £2.41 (c) 67p (d) £1.55 (e) £1.31 (f) 95p (g) 53p (h) £3.31
1.F3 (a) £4.66 (b) £4.91 (c) £6.43 (d) £3.81 (e) £7.38 (f) £5.11
1.F4 (a) £15.34 (b) £15.09 (c) £12.37 (d) £16.19 (e) £12.62 (f) £14.89
1.F5 (a) 424 g (b) £3 (c) 855
1.F6 (a) 15 (b) 33 (c) £1.68 (d) 22p
1.F7 438
1.F8 (a) 368 kg (b) 251 kg
1.F9 £8.48
1.F10 (a) £19.19 (b) £43.94 (c) £30.94 (d) £4.41 (e) £60.10 (f) £18.40
1.F11 (a)(i) £14 (ii) £713 (b) 723
1.F12 Across: 1 031, 974, 1 415, 680, 1 187, 1 867, 1 879, 9 033 Down: 3 686, 4 005, 1 342, 9 033
1.F13 (a) £425, £295, £126, £84, £34 (b) 964
1.F14 (a) 243, 96, 73, 51, 44 (b) £778 (c) Thursday (d) Thursday, Friday, Saturday
1.F15 (a) 1 740 (b)(i) £16.75 (ii) £19.52 (iii) £34.10 (iv) £77.72
1.F16 (a) 85 (b) 72 m
1.F17 (a) 1 128 (b) Tuesday (c) Saturday (d) 282
1.F18 £24.75
1.F19 (a) £14 (b) £113.62
1.F20 (a) 1 650, 550, 180, 5, 5, 600, $2\frac{1}{2}$, 1 100 (b) $2\frac{1}{2}$, $2\frac{1}{2}$, 60, 1 100, $2\frac{1}{2}$, 550, $2\frac{1}{2}$, 550 (c) 1 100, 160, 5, $2\frac{1}{2}$, 10, 10, 10

Exercises 2.A

2.A1 (a) 12 (b) 6 (c) 8 (d) 6 (e) 18
2.A2 (a) $\frac{5}{7}$ (b) $\frac{6}{19}$ (c) $\frac{8}{13}$ (d) 1 (e) $\frac{19}{21}$ (f) $1\frac{2}{3}$ (g) $2\frac{1}{4}$ (h) $1\frac{5}{8}$
2.A3 (a) $\frac{1}{7}$ (b) $\frac{1}{2}$ (c) $\frac{1}{2}$ (d) $\frac{2}{5}$ (e) $\frac{1}{2}$ (f) $\frac{2}{5}$ (g) $\frac{2}{3}$ (h) $1\frac{1}{2}$
2.A4 (a) $\frac{11}{7}$ (b) $\frac{12}{5}$ (c) $\frac{23}{4}$ (d) $\frac{25}{2}$ (e) $\frac{34}{9}$ (f) $\frac{59}{11}$ (g) $\frac{68}{55}$ (h) $\frac{169}{16}$
2.A5 (a) $1\frac{3}{8}$ (b) $1\frac{3}{5}$ (c) $2\frac{3}{5}$ (d) $2\frac{5}{6}$ (e) $2\frac{5}{12}$ (f) $3\frac{7}{8}$ (g) $2\frac{16}{53}$ (h) $14\frac{5}{9}$
2.A6 (a) $\frac{1}{4}$ (b) $\frac{3}{4}$ (c) $\frac{3}{8}$ (d) $\frac{2}{15}$ (e) $\frac{3}{25}$ (f) $\frac{49}{60}$ (g) $\frac{3}{11}$ (h) $\frac{3}{10}$
2.A8 (a) $\frac{7}{8}$ (b) $\frac{15}{16}$ (c) $\frac{9}{10}$ (d) $\frac{3}{4}$ (e) $1\frac{1}{4}$ (f) $1\frac{5}{12}$ (g) $2\frac{1}{24}$ (h) $1\frac{17}{18}$
2.A9 (a) $4\frac{1}{4}$ (b) $3\frac{1}{16}$ (c) $5\frac{11}{15}$ (d) $5\frac{11}{16}$ (e) $10\frac{3}{8}$ (f) $8\frac{7}{18}$ (g) $18\frac{7}{12}$ (h) $19\frac{7}{12}$
2.A10 (a) $\frac{13}{24}$ (b) $\frac{11}{20}$ (c) $\frac{7}{15}$ (d) $\frac{5}{24}$ (e) $\frac{14}{15}$ (f) $1\frac{7}{16}$ (g) $4\frac{7}{12}$ (h) $1\frac{17}{30}$
2.A11 (a) 2 (b) $2\frac{13}{24}$ (c) 8 (d) $9\frac{1}{12}$ (e) $3\frac{5}{18}$ (f) $6\frac{2}{15}$ (g) $5\frac{7}{32}$ (h) $2\frac{8}{21}$
2.A12 (a) $\frac{3}{8}$ (b) $\frac{8}{15}$ (c) $\frac{5}{24}$ (d) $\frac{4}{17}$ (e) $\frac{6}{25}$ (f) $\frac{63}{104}$ (g) $\frac{1}{4}$ (h) $\frac{5}{28}$ (i) $\frac{1}{30}$ (j) $\frac{1}{9}$ (k) $\frac{1}{8}$ (l) $\frac{3}{25}$
2.A13 (a) $1\frac{1}{2}$ (b) $\frac{4}{7}$ (c) $\frac{2}{3}$ (d) $\frac{2}{3}$ (e) $1\frac{1}{6}$ (f) $1\frac{1}{4}$ (g) $4\frac{1}{2}$ (h) $2\frac{9}{28}$ (i) $3\frac{1}{5}$ (j) 10 (k) $1\frac{13}{20}$ (l) $\frac{3}{4}$
2.A14 (a) 170 (b) 60 (c) 350 (d) 84 (e) 360 (f) 1 320 (g) 1 875 (h) 135 (i) 3 500 (j) 520

2.A15 (a) 120 m (b) 40 cm (c) 120 kg (d) 850 litres
(e) 224 g (f) £66 (g) 36 mm (h) 390 km
(i) 540 m (j) 72 kg
2.A16 (a) $\frac{3}{7}$ (b) $\frac{2}{7}$ (c) $\frac{1}{8}$ (d) $\frac{5}{16}$ (e) $\frac{1}{5}$ (f) $\frac{1}{6}$ (g) $\frac{1}{50}$
(h) $\frac{3}{8}$
2.A17 (a) $8\frac{3}{4}$ m² (b) $11\frac{11}{16}$ m² (c) $5\frac{5}{8}$ m² (d) 26 m²
(e) 28 m² (f) $28\frac{1}{8}$ m² (g) $56\frac{3}{8}$ m² (h) $28\frac{11}{16}$
2.A18 (a)(i) $\frac{2}{9}$ (ii) $\frac{1}{3}$ (b) B.S. £60 000, J & Sons £40 000,
N.E. £20 000
2.A19 (a) 46p (b) 60p (c) £1.35 (d) £2.13
(e) £10.50 (f) £8.40 (g) £6.25 (h) £60.40
(i) 84p (j) £15.625 (k) £2.38 (l) £6.69
(m) 15p (n) £32 (o) £7.50 (p) £1.6 (q) £3.50
(r) £6.60 (s) £14.16 (t) £19.14
2.A20 (a) 9 in (b) 10 oz (c) 7 in (d) 6 lb
(e) 22 pints (f) 3 quarts (g) 80 lb (h) 660 yd
(i) 1792 lb (j) $1\frac{1}{2}$ yd (k) 3960 ft (l) 112 oz
(m) 7 pints (n) 36 cwt (o) 62 stone

2.C4 $3\frac{3}{4}$ in
2.C5 (a) 160 (b) $1\frac{7}{8}$
2.C6 (a) $\frac{1}{4}$ (b) $9\frac{1}{2}$ gal
2.C7 $\frac{7}{11}$
2.C8 (a) $\frac{5}{9}$ (b) $\frac{1}{6}$ (c) $\frac{1}{12}$ (d) $\frac{7}{36}$
2.C9 $\frac{43}{45}$
2.C10 (a) $\frac{15}{26}$ (b) 4 in
2.C11 (a) $3\frac{9}{32}$ m³
2.C12 40 cm, 36 cm, 13 cm, 8 cm, 3 cm
2.C13 (a) $\frac{45}{48}$ (b) 6 litres
2.C14 (a) $\frac{157}{200}$ (b)(i) 168 (ii) 90 (iii) 942
2.C15 (a) $\frac{5}{24}$ (b)(i) 27 000 mm³ (ii) 30 000 mm³
(iii) 15 000 mm³
2.C16 (a) 10 mm (b) 13 mm (c) 51 mm
2.C17 (a) (i) 36 (ii) 564 (iii) $\frac{47}{50}$ (b)(i) 160 (ii) 120
(iii) 120
2.C18 $\frac{3}{10}, \frac{1}{2}, \frac{1}{9}, \frac{1}{18}, \frac{1}{30}$
2.C19 $\frac{1}{6}, \frac{1}{4}, \frac{1}{2}, \frac{1}{12}$
2.C20 $8\frac{1}{2}, 18\frac{3}{4}, 5\frac{3}{5}, 4\frac{1}{4}, 12\frac{1}{10}$

Exercises 2.B

2.B1 (a) $\frac{5}{19}$ (b) $\frac{9}{19}$ (c) $\frac{16}{95}$ (d) $\frac{9}{95}$
2.B2 (a) $\frac{1}{4}$ (b) $\frac{1}{15}$ (c) $\frac{1}{30}$
2.B3 (a) $\frac{1}{5}$ (b) $\frac{3}{10}$ (c) $\frac{2}{5}$ (d) $\frac{1}{10}$
2.B4 (a) $\frac{41}{115}$ (b) $\frac{6}{23}$ (c) $\frac{7}{46}$ (d) $\frac{53}{230}$
2.B5 (a) £16.64 (b) $\frac{3}{8}$
2.B6 (a) 300 (b)(i) 30 (ii) 240 (iii) $\frac{1}{8}$
2.B7 (a) £3 300 (b) £1 980
2.B8 (a) £22.40 (b) £595
2.B9 (a) 120 (b) 300 (c) 60
2.B10 (a) 1 h 30 min (b) 3 h (c) 3 h
2.B11 (a) $\frac{7}{18}$ (b) $\frac{1}{4}$ (c) $\frac{13}{36}$
2.B12 £141.80
2.B13 $41\frac{11}{16}$
2.B14 (a) 66p (b) 80p (c) £8.40 (d) £2.66
(e) £1.00 (f) £260 (g) 42p (h) £5.08
(i) £7.28 (j) £11.44 (k) £16.20

2.B15

	MATERIAL COST (£)	LABOUR COST (£)	OVERHEADS (£)
(a)	24.93	41.55	8.31
(b)	32.82	54.70	10.94
(c)	54.66	91.10	18.22
(d)	71.70	119.50	23.9

2.B16 (a) 844 items (b) £53.83
2.B17 (a) £10.62 (b) £16.14 (c) £22.98 (d) £55.41
(e) £19.98 (f) £14.40 (g) £26.19 (h) £28.23
(i) £11.25 (j) £19.92
2.B18 (a) $\frac{3}{5}$ (b)(i) 75 (ii) 135 (iii) 15
2.B19 (a) $\frac{7}{24}$ (b) 45 (c) 40
2.B20 (a) 900 (b) 600 (c) 500

Exercises 2.C

2.C1 (a) $\frac{2}{3}$ (b) 3 h 40 min
2.C2 (a) $19\frac{1}{2}$ lb/in² (b) $9\frac{3}{4}$ lb/in² (c) $6\frac{1}{2}$ lb/in²
2.C3 (a) $\frac{13}{27}$ (b) $\frac{16}{27}$ (c) $\frac{7}{9}$

Exercises 2.D

2.D1 (a) $\frac{3}{4}$ (b) $\frac{7}{8}$ (c) $\frac{1}{4}$ (d) $\frac{3}{8}$ (e) $\frac{1}{8}$ (f) $\frac{12}{25}$ (g) $\frac{33}{50}$
(h) $\frac{17}{50}$
2.D2 $\frac{60}{73}$
2.D3 $\frac{51}{60}$
2.D4 (a) $3\frac{1}{6}$ h (b) $\frac{1}{16}$
2.D5 (a) $\frac{5}{9}$ (b) $1\frac{1}{8}$
2.D6 (a) $\frac{50}{143}$ (b) $\frac{24}{143}$ (c) $\frac{7}{143}$ (d) $\frac{62}{143}$
2.D7 (a) $\frac{48}{187}$ (b) $\frac{84}{187}$ (c) $\frac{34}{187}$ (d) $\frac{21}{187}$ (e) $\frac{23}{64}$
2.D8 $\frac{13}{60}$
2.D9 (a) $\frac{5}{18}$ (b) $\frac{9}{40}$
2.D10 (a) $\frac{3}{5}$ (b) $\frac{2}{5}$
2.D11 £12 105, £8 070, £2 690, £9 415
2.D12 68
2.D13 £879.30
2.D14 7 200, 6 000, 3 000, 1 800
2.D15 (a) $\frac{1}{24}$ (b) $\frac{5}{12}$
2.D16 (a) 160 (b) 140
2.D17 £130.68
2.D18 (a) $\frac{1}{4}$ (b) $\frac{5}{24}$ (c) $\frac{1}{8}$ (d) $\frac{7}{24}$ (e) $\frac{1}{8}$
2.D19 (a) £1.60 (b) £4 (c) 80p
2.D20 £3 431.25

Exercises 2.E

2.E1 $\frac{12}{37}$
2.E2 (a) $\frac{1}{165}$ (b) $\frac{1}{7}$
2.E3 $48\frac{3}{4}$ m²
2.E4 (a) $\frac{23}{52}$ (b) $\frac{17}{52}$ (c) $\frac{3}{13}$
2.E5 $\frac{3}{40}$
2.E6 (a) $\frac{11}{18}$ (b) $\frac{48}{73}$
2.E7 $\frac{1}{3}$
2.E8 $\frac{3}{25}$
2.E9 19
2.E10 (a) $\frac{9}{16}$ (b) 2 880
2.E11 (a) $8\frac{1}{16}$ m (b) $29\frac{1}{32}$ m
2.E12 (a) $\frac{3}{8}$ (b) $\frac{3}{5}$ (c) $\frac{1}{40}$

2.E13 (a) 63 (b) 140 (c) 126
2.E14 (a) $\frac{3}{10}$ (b) $\frac{2}{5}$ (c) $\frac{1}{5}$ (d) $\frac{1}{10}$
2.E15 £305
2.E16 (a) $4\frac{1}{2}$ h (b) $\frac{16}{45}$
2.E17 $\frac{3}{4}$
2.E18 (a) 48 passengers (b) $\frac{25}{67}$
2.E19 Surgery = $35\frac{3}{4}$ m², reception = $12\frac{3}{4}$ m², waiting room = $14\frac{7}{8}$ m²

2.E20

TIME	NO.	FRACTION
09.00	640	$\frac{8}{13}$
11.00	880	$\frac{11}{13}$
13.00	960	$\frac{12}{13}$
15.00	720	$\frac{9}{13}$
17.00	480	$\frac{6}{13}$
19.00	800	$\frac{10}{13}$

Exercises 2.F

2.F1 $2\frac{1}{8}$ m²
2.F2 $\frac{10}{91}$
2.F3 (a) $\frac{52}{147}$ (b) $\frac{5}{147}$ (c) $\frac{90}{147}$
2.F4 (a) £530 (b) £742
2.F5 Food store = 15 m², Kitchen = $33\frac{3}{4}$ m², Dining room = 104 m²
2.F6 (a) $\frac{11}{35}$ (b) $\frac{3}{7}$ (c) $\frac{9}{35}$
2.F7 (a) $\frac{1}{16}$ (b) $\frac{1}{4}$
2.F8 (a) $\frac{1}{4}$ (b)(i) £2.10 (ii) £7 (iii) 70p
2.F9 $\frac{2}{15}$
2.F10 (a) $\frac{1}{2}$ (b) $\frac{3}{10}$ (c) $\frac{1}{5}$
2.F11 (a) $\frac{3}{13}$ (b) $\frac{20}{39}$ (c) $\frac{4}{39}$
2.F12 $\frac{1}{4}$ beef, $\frac{13}{60}$ ham, $\frac{7}{60}$ cheese, $\frac{7}{30}$ chicken, $\frac{11}{60}$ salad
2.F13 (a) $\frac{343}{600}$ (b)(i) $\frac{7}{16}$ (ii) 1 gal 6 pints (iii) 9
2.F14 (a) 20p (b) $\frac{99}{204}$ (c) 60p
2.F15 (a) Mon $\frac{13}{100}$, Tue $\frac{81}{500}$, Wed $\frac{21}{100}$, Thur $\frac{12}{125}$, Fri $\frac{9}{50}$, Sat $\frac{111}{500}$. (b) Mon £32.50, Tue £40.50, Wed £52.50, Thur £24, Fri £45, Sat £55.50
  (c) Total profit = £250, Total receipts = £1 000
2.F16 (a) raisins $\frac{1}{14}$, Cherries $\frac{3}{56}$, peel $\frac{5}{56}$, currants $\frac{17}{56}$, sultanas $\frac{1}{4}$, butter $\frac{9}{56}$, sugar $\frac{5}{28}$, almonds $\frac{1}{14}$
  (b) rhubarb $\frac{12}{21}$, plain flour $\frac{4}{21}$, self-raising flour $\frac{2}{21}$, margarine $\frac{2}{21}$, lard $\frac{1}{21}$
2.F17 Skirt = $\frac{1}{4}$, top = $\frac{3}{16}$, coat = $\frac{3}{8}$, scarf = $\frac{1}{32}$, shoes = $\frac{5}{32}$
2.F18 (a) £15 (b) £11.25 (c) £7.50
2.F19 (a) Mon $\frac{23}{120}$, Tue $\frac{103}{480}$, Wed $\frac{19}{96}$, Thur $\frac{1}{6}$, Fri $\frac{11}{48}$
2.F20 Suit £48, trousers £18, overcoat £51, hat £6, jeans £9, shoes £12, shirt £6.60, tie £1.80

Exercises 3.A

3.A1 (a) 4.6 (b) 3.6 (c) 16.1 (d) 16.6 (e) 80.7 (f) 39.2 (g) 57 (h) 129.6 (i) 39.9 (j) 26.4 (k) 9.6 (l) 49.9 (m) 186.1 (n) 169.9
3.A2 (a) 4.94 (b) 27.31 (c) 3.83 (d) 4.84 (e) 6.99 (f) 8.82 (g) 18.942 (h) 7.254 (i) 27.973 (j) 33.489 (k) 8.3009 (l) 12.0715 (m) 58.106 (n) 141.851
3.A3 (a) 1.423 (b) 1.373 (c) 4.59 (d) 20.57 (e) 4.838 (f) 0.3417 (g) 26.86 (h) 211.637 (i) 2.54 (j) 7.265 (k) 11.47 (l) 7.12 (m) 41.875 (n) 19.001
3.A4 (a) 125.57 (b) 224.33 (c) 71.39 (d) 268.66
3.A5 (a) 23.19 (b) 52.72 (c) 17.3 (d) 17.75
3.A6 (a) $x = 11.2, y = 9.23$ (b) $x = 20.86, y = 12.8$ (c) $x = 35.62, y = 15.9$
3.A7 (a) 9 (b) 11 (c) 41.04 (d) 9.35 (e) 46.8 (f) 39.6 (g) 175.68 (h) 251.12 (i) 538.06 (j) 1.316 (k) 78.764 (l) 3.4 (m) 236.643 (n) 333.914
3.A8 (a) 1.56 (b) 1.05 (c) 1.74 (d) 14.98 (e) 3.586 (f) 2.616 (g) 6.768 (h) 28.602 5 (i) 45.832 8 (j) 5.255 32 (k) 51.000 27 (l) 592.41 (m) 8.96 (n) 12.258 477
3.A9 (a) 14.2 (b) 8.5 (c) 15.7 (d) 26.9 (e) 4.53 (f) 1.85 (g) 31.6 (h) 0.84 (i) 0.037 (j) 1.987 (k) 0.178 1 (l) 23.47 (m) 6.24 (n) 18.62
3.A10 (a) 32 (b) 12 (c) 123 (d) 562 (e) 36.728 57 (f) 80 (g) 23 (h) 18 (i) 3.83 (j) 94.3 (k) 3.57 (l) 19.5 (m) 1.33 (n) 1.69
3.A11 23.4, 21.6, 44.2, 2.52, 5.68, 20.4, 80.04
3.A12 120, 15.84, 22.68, 98, 196
3.A13 (a) 1.045 (b) 3.798 (c) 12.611 (d) 8.040 (e) 1.7
3.A14 (a) 28.72 (b) 1.74 (c) 5.9 (d) 0.9432 (e) 7.65 (f) 1.400 (g) 18.1 (h) 11.10
3.A15 (a) 3 160 (b) 2.08 (c) 12.9 (d) 20 600 (e) 0.0843
3.A16 (a) 24.29 (b) 3.33 (c) 4.46
3.A17 46.910
3.A18 (a) 8.433 (b) 8.43
3.A19 (a) 2.794 (b) 0.05 (c) 8.1672 (d) 39.78 (e) 0.0455
3.A20 (a) 254 (b) 38.1 (c) 248.92 (d) 66.04 (e) 22.86

Exercises 3.B

3.B1 (a) £13.28 (b) £13.31 (c) £60.76 (d) £7.77
3.B2 (a)(i) 480 (ii) 1.44 (iii) 1 200 (b)(i) £7.80 (ii) £13 (iii) £41.66
3.B3 £376.73, £456.68, £144.4, £84.4, £516.07, £412.21
3.B4 £63.65
3.B5 £394.23
3.B6 (a) £8.1, £1.92, £1.74, £1.32, £0.34 = £13.42
  (b) £4.08, £7.26, £11.04, £9.82, £9.36 = £41.56
  (c) £0.84, £0.50, £1.72, £2.57, £2.54 = £8.17
3.B7 £91.17
3.B8 £16 045.92
3.B9 1.23
3.B10 (a) 245 (b) 168
3.B11 (a) £133.8 (b) £453.23
3.B12 (a) £76.08 (b) £3.43
3.B13 £1 440.75
3.B14 (a) £3 297.22 (b) £932.12
3.B15 (a) $960 (b) £640 (c) £40

3.B16 (a) £24 700 (b) £16 200 (c) £40 850
3.B17 (a) £825.50 (b) £596.90 (c) £419.10
3.B18 20.32 cm × 15.24 cm
3.B19 £61.77
3.B20 Mon 3 100, Tue 2 700, Wed 4 100, Thur 2 500, Fri 5 200

Exercises 3.C

3.C1 (a) 90.48 (b) 81.03 mm
3.C2 (a) £2.174 (b) 2.73 m^2
3.C3 £22.45
3.C4 (a) 258.29 cm (b) 478.4 m^2
3.C5 $x = 12.56$, $y = 45.17$
3.C6 (a) 111.04 (b) 47 and 0.15 m
3.C7 (a) 22.4 kg (b) 89.6 kg
3.C8 20.8
3.C9 (a) 104.375 (b) 62.625
3.C10 (a) 15.625 mm (b) 23.33
3.C11 1 182.43 mm
3.C12 1.2, 3.2, 0.5, 4.8, 2.25, 0.3, 2.75, 1.2
3.C13 (a) 31.39 litres (b) 26.17 litres (c) 17.48
3.C14 (a) 3 in (b) 4.375 litres (c) 2.52 litres
3.C15 (a) 23.77 mm (b) 22.86 mm (c) 92.09 mm (d) 47.25
3.C16 (a) $x = 15.355$ mm, $y = 14.84$ mm
      (b) $x = 18.18$ mm, $y = 103.39$ mm
      (c) $x = 26.75$ mm, $y = 13.46$ mm
3.C17 (a) 26.25 mm (b) 15 mm (c) 80.82 mm
3.C18 76.2, 38.1, 50.8, 114.3
3.C19 (a) 92.48 mm (b) 166.464
3.C20 68

Exercises 3.D

3.D1 (a) 1 643 (b) 2 393
3.D2 95
3.D3 2.268 g
3.D4 (a) 331.2 m^3 (b) 897 m^3
3.D5 12.391
3.D6 43.44 t
3.D7 403.77 t
3.D8 104.1 min
3.D9 (a) 7 932 litres (b) 2 644 litres
3.D10 (a) 0.554 (b) 5 202.05 yd^2
3.D11 383.38 min
3.D12 12.738 t
3.D13 (a) £1.43 (b) £1.43 (c) £2.33 (d) £6.73
3.D14 (a) 3.8 litres (b) 4.4 litres (c) 1.8 litres (d) 1.9 litres
3.D15 £11.20
3.D16 (a) 3.945 h (b)(i) 20 (ii) 42 (iii) 87
3.D17 (a) 2 100, 1 800, 2 000, 1 600, 2 100, 1 700
       (b) 1 260, 1 080, 1 200, 960, 1 260, 1 020
3.D18 (a) 7.12 kg (b) 183.1 kg
3.D19 (a) 58.2 g (b) 698.9 g (c) 2 096.6 g
3.D20 (a) 17.3 kW (b) 3.1 kW

Exercises 3.E

3.E1 28.96 m
3.E2 (a) 1 876 (b) 392 (c) 532
3.E3 (a) £40.12 (b) £6.54
3.E4 (a) 48.75 m^2 (b) 123.8 m^2
3.E5 £19.29 million
3.E6 (a) £515.61 (b) £177.55
3.E7 (a) £1.67 (b)(i) £207.6 (ii) £12.97$\frac{1}{2}$
3.E8 (a) £503.23 (b) £4.90
3.E9 (a) 0.91 min (b) 12.74 min
3.E10 (a) 840, 720, 720, 530, 420, 400
      (b) 655.2, 561.6, 561.6, 413.4, 327.6, 312 = 2 831.4 h
3.E11 (a) 18.6 kW (b) 2.28 kW
3.E12 (a) 0.254 litres (b) 0.096 litres (c) 5.76 litres
3.E13 0.037 litres
3.E14 (a) 17.3 min (b) 9.65 miles (c) 20.8 min
3.E15 (a) 644 (b) 308 (c) 252 (d) 196
3.E16 31.5, 20.3, 5.8, 16.8, 8.3, 19.3, 4.75
3.E17 (a) £58.03 (b) £15.73 (c) £29.94
3.E18 11.5 miles per gallon
3.E19 (a) 322.1 miles (b) 518.58 km
3.E20 53.1, 268.9, 318.8, 141.7, 58

Exercises 3.F

3.F1 £36.50
3.F2 £9.93
3.F3 (a) 2.75 m (b) £5.11$\frac{1}{2}$
3.F4 £8.09$\frac{1}{2}$
3.F5 (a) 6.5 kg (b)(i) £3.11 (ii) £8.45$\frac{1}{2}$
3.F6 (a) 207.12 (b) 83
3.F7 (a) £1 249.50 (b) £1 556.10
3.F8 £9.40 + £2.40 + £0.58$\frac{1}{2}$ = £12.37$\frac{1}{2}$
3.F9 (a) 24.62 m (b) £4.24 (c) 0.76
3.F10 (a) 6$\frac{1}{2}$p (b) £1.40 (c) £15.20
3.F11 (a) 4p (b) £7.02 (c) 8.45 kg
3.F12 3.105 kg
3.F13 (a) 1 500, 1 400, 1 500, 1 700, 1 300, 1 300
      (b) £1 626.90
3.F14 86.36 cm
3.F15 (a) 0.225 kg (b) 10.8 kg
3.F16 2.025 kg
3.F17 £516.66
3.F18 11, 4.4, 3.3, 5.28, 1.1
3.F19 (a) 26 (b) 3.52 gal
3.F20 (a) 14 (b) 25, 19, 13, 32, 38, 6

Exercises 4.A

4.A1 (a) 0.5 (b) 0.3 (c) 0.55 (d) 0.25 (e) 0.8
     (f) 0.625 (g) 0.375 (h) 0.35 (i) 0.38
     (j) 0.175 (k) 0.9 (l) 0.47 (m) 0.731
     (n) 0.42 (o) 0.39 (p) 0.425
4.A2 (a) 0.667 (b) 0.833 (c) 0.571 (d) 0.454
     (e) 0.461 (f) 0.380 (g) 0.889 (h) 0.714
4.A3 (a) 1.75 (b) 2.4 (c) 7.7 (d) 8.45 (e) 14.6
     (f) 12.175 (g) 6.18 (h) 2.31

4.A4 (a) $\frac{3}{10}$ (b) $\frac{1}{4}$ (c) $\frac{22}{25}$ (d) $\frac{1}{20}$ (e) $\frac{1}{8}$ (f) $\frac{7}{8}$ (g) $\frac{19}{20}$
      (h) $\frac{16}{25}$ (i) $\frac{41}{125}$ (j) $\frac{3}{8}$ (k) $\frac{61}{125}$ (l) $\frac{199}{200}$
4.A5 (a) $1\frac{1}{2}$ (b) $1\frac{3}{8}$ (c) $2\frac{1}{4}$ (d) $4\frac{7}{20}$ (e) $\frac{12}{25}$ (f) $7\frac{3}{200}$
      (g) $3\frac{91}{200}$ (h) $8\frac{21}{200}$
4.A6 (a) 30% (b) 55% (c) 48% (d) 75%
      (e) $62\frac{1}{2}$% (f) 31.9% (g) $80\frac{3}{5}$% (h) $25\frac{2}{5}$%
4.A7 (a) 75% (b) $62\frac{1}{2}$% (c) 85% (d) 70%
      (e) $22\frac{1}{2}$% (f) 52% (g) 79% (h) $68\frac{3}{4}$%
4.A8 (a) $\frac{7}{10}$ (b) $\frac{6}{25}$ (c) $\frac{11}{20}$ (d) $\frac{24}{25}$ (e) $\frac{11}{50}$ (f) $12\frac{1}{2}$
      (g) $37\frac{1}{4}$ (h) $\frac{17}{20}$
4.A9 (a) $1\frac{1}{2}$ (b) $2\frac{1}{4}$ (c) $1\frac{3}{4}$ (d) $3\frac{1}{4}$ (e) $1\frac{2}{5}$ (f) $1\frac{1}{4}$
      (g) $4\frac{11}{50}$ (h) $6\frac{7}{20}$
4.A10 (a) 65% (b) 28% (c) $31\frac{7}{10}$% (d) $40\frac{1}{2}$%
      (e) $12\frac{3}{10}$% (f) $45\frac{2}{5}$% (g) $7\frac{1}{5}$% (h) $\frac{1}{2}$%
4.A11 (a) 0.2 (b) 0.65 (c) 0.32 (d) 0.195 (e) 0.287
      (f) 0.864 (g) 0.939 (h) 0.0475
4.A12 (a) 10 (b) 33 (c) 102 (d) 280 (e) 165
      (f) 42 (g) 252 (h) 396
4.A13 (a) 150 m (b) £72 (c) 261 cm (d) 15 g
      (e) 8p (f) 54 kg (g) 180 g (h) 42.6 mm
4.A14 0.6%
4.A15 12 litres
4.A16 (a) 30% of 600 (b) 15% of 70 g
      (c) 60% of £130 (d) 35% of 30 kg
4.A17 (a) 60% (b) 16.66% (c) 12.5% (d) 22.5%
      (e) 4.75% (f) 25% (g) 40% (h) 5%
4.A18 (a) 6% (b) 4% (c) 90%
4.A19 (a) 34 kg copper, 6 kg tin
4.A20 Press operator £105, roll operator £126,
      plant fitter £147, fork-lift driver £115.5

Exercises 4.B

4.B1 (a) £196.50 (b) £24.56 (c) £3.93
4.B2 £9 094.75
4.B3 (a) £31 (b) £37.20 (c) £46.5 (d) £18.60
      (e) £52.70
4.B4 (a) 1 100 (b) 25% (c) 0.2 (d) 1 310
4.B5

SERVICE	FRACTION	DECIMAL	PERCENTAGE
Deposit A/c	$\frac{53}{100}$	0.53	53%
Cheque card	$\frac{4}{5}$	0.80	80%
Cash card	$\frac{11}{20}$	0.55	55%
Personal loan	$\frac{3}{20}$	0.15	15%
Mortgages	$\frac{3}{50}$	0.06	6%

4.B6 (a) Typing 5 h, phone 1.5 h, shorthand 0.75 h,
      photocopy/letters 0.75 h (b) Typing 62.5%, phone
      18.75%, shorthand 9.38%, photocopy/letters 9.38%
4.B7 (a) 70.20 (b) 5.09%
4.B8 (a) £95 (b) $\frac{19}{200}$
4.B9 (1) £5.40 (2) £7.20 (3) £6.75 (4) £9.45
      (5) £4.20
4.B10 Paper = 10.7%, tapes = 17.4%, files = 8.7%,
      typewriter supplies = 28.5%,
      duplicating supplies = 30.9%, sundry items = 12.5%
4.B11 Wages 56.45%, rent 13.71%, rates 5.65%,
      heating 7.26%, supplies 16.94%
4.B12 (a) £1 107 (b) £399.60 (c) £11.88
4.B13 (i) £150 (ii) £330 (iii) £384 (iv) £420 (v) £495
4.B14 42.67%
4.B15 13.33%
4.B16 £6 080
4.B17 (i) 21% less than £1 (ii) 65% £1 to £3
      (iii) 14% over £3
4.B18 12.5%
4.B19 (a) £5 400 (b) £5 940
4.B20 (a) £6 (b) £46.50 (c) £26.80 (d) £1 845
      (e) £55.68 (f) £55.50 (g) £10.08

Exercises 4.C

4.C1 (a) 3.75 gal (b) 62.5%
4.C2 (a) 13.83 mph (b)(i) 12.41 litres (ii) 33.32%
4.C3 (a) 1.625, 0.75, 1.25, 0.875, 2.19
      (b) 0.94, 2.375, 0.625, 1.375, 4.375, 1.25
      (c) 4.625, 0.563, 0.688, 3.25, 5.75, 0.875, 2.375
4.C4 4.1:1, 4 971.4 r/min
4.C5 2.53 mm
4.C6 (a) $\frac{1}{16} = 0.063$, $\frac{1}{8} = 0.125$, $\frac{3}{16} = 0.188$, $\frac{1}{4} = 0.25$,
      $\frac{5}{16} = 0.313$, $\frac{3}{8} = 0.375$, $\frac{7}{16} = 0.438$, $\frac{1}{2} = 0.5$ (b) 30%
4.C7 (a) 2.5 mm (b) 13 (c) 66.7%
4.C8 (a) 7.2 min (b) 7.7 mm/min, 12 min
4.C9 (a) 17.1 t (b) 11 cm
4.C10 1.8 m^3 sand, 0.48 m^3 cement, 1.72 m^3 aggregate
4.C11 (a) £324 (b) £372.60
4.C12 (a) £21.60 (b) £24.84
4.C13 5.26%
4.C14 £1 320
4.C15 (a) 20% (b) 1 200 km
4.C16 (a) 6.6 cm (b) 22%
4.C17 8.92%
4.C18 28%
4.C19 (a) 127.5 kg copper, 19.5 kg tin, 3 kg lead
      (b)(i) 6% (ii) 7% (iii) 87%
4.C20 (a) 70% (b) 96 kg (c) 2%

Exercises 4.D

4.D1 28.03 m
4.D2 134.38 km
4.D3 (a) $\frac{5}{8}$ (b) 281.25 m
4.D4 C120, 2 070; C90, 8 510; C60, 9 315; C30, 1 472
4.D5 £248.40
4.D6 (a) A = 1 472, B = 920, C = 1 288 (b) 35%
4.D7 Common Market countries 48%, U.S.A./Canada
      22%, Far East 12%, Middle East 18%
4.D8 Milking, etc., 2.25 h, field work 11.25 h, m/c repair
      3.38 h, resting 18 − 16.88 = 1.12 h
4.D9 (a) £65.83 (b) £75.71
4.D10 (a) £6.50 (b) £3.92 (c) £5.15 (d) £9.18
      (e) £5.94 (f) £4.93
4.D11 (a) 146 litres (b) 70
4.D12 £3 960.60
4.D13 (a) 6.5 m (b) 2.16 m
4.D14 (a) 1.9875 m (b) 292 (c) 79.95 m^3
4.D15 (a) £99.27 (b) 532

4.D16 (a) £85.33 (b) 26.7%
4.D17 (a) 77 kg (b) 175.2 kg carbon, 57.6 kg oxygen
4.D18 2nd = £42 660, 3rd = £46 072.80
4.D19 (a) £116.60 (b) £114.48 (c) £127.20
       (d) £118.72 (e) £143.10 (f) £148.40
4.D20 (a) 45–55 ohms (b) 0.109 in

Exercises 4.E

4.E1   (a) 50% (b) £49.98
4.E2   (a) 6.6% (b) 10%
4.E3   (a) £301.94 (b) £25.44
4.E4   (a) £225.76 (b) £23.10
4.E5   (a) 31 160 (b) 27 360 (c) 17 480
4.E6   (a) £141.12 (b) £952.56
4.E7   (a) 93.36% (b)(i) 37.5% (ii) 37.5% (iii) 375%
4.E8   42
4.E9   (a) 56.25% (b)(i) 57.14% (ii) 33.33%
       (iii) 9.5%
4.E10 9.9%
4.E11 0.63 h
4.E12 (a) 151.5 km (b) £18.94
4.E13 (a) £1.32 (b) 61.45 gal
4.E14 (a) £31.54 (b) £2.10
4.E15 (a) 47 (b)(i) 40% (ii) 40%
4.E16 (a) $87\frac{1}{2}$% (b)(i) $\frac{3}{4}$ (ii) 25%
4.E17 (a) (i) £2.88 (i) £50.88 (i) 16.2p
       (ii) £1.96 (c) (i) £3.18 (ii) £29.68
       (d) (i) £20.35 (ii) £390.35 (e) (i) £1.57
       (ii) £10.27
4.E18 (a) £4.27$\frac{1}{2}$ (b) 88$\frac{1}{2}$ (c) £4.84$\frac{1}{2}$ (d) 43$\frac{1}{2}$p
       (e) £62.55
4.E19 1 613.13 m^2
4.E20 (a) 1.6 h (b) 60% (c) 0.4

Exercises 4.F

4.F1   (a) 37.5 litres (b) 108 eggs (c) 100 (d) 2.7 kg
       (e) 3.2 kg
4.F2   (a) 89.47% (b) 94 g
4.F3   6.44 litres
4.F4   (a) 2.75 litres (b) 11%
4.F5   £5.04
4.F6   (a) £500.82$\frac{1}{2}$ (b) 1.6 kg (c) 474.15
4.F7   (a) £32.86 (b) £37.79
4.F8   (a) (i) £8.40 (ii) £16.90 (iii) £20.73
       (iv) £38.08 (v) £49.08 (vi) £139.65 (b) (i) £9.40
       (ii) £19.43 (iii) £23.84 (iv) £43.79 (v) £55.21
       (vi) £157.11
4.F9   (a) 22.99% (b) 25%
4.F10 0.4%
4.F11 60%
4.F12 (a) (i) £7.56 (ii) £1.21 (iii) £17.30 (iv) £2.73
       (b) £4.32 (c) £0.98 (d) £23.50
4.F13 (a) £0.69 (b) £1.02 (c) £1.53 (d) £3.27
       (e) £1.11
4.F14 (a) Food prep. = 10 h, ordering = 1$\frac{1}{2}$ h,
       kitchen work = 1$\frac{1}{2}$ h (b) Food prep. = 5 h,
       ordering = 3 h, kitchen work = 3 h

4.F15 (a) £0.86 (b) £3.0 (c) £2.60 (d) £9.20
       (e) £18.90 (f) £34.56
4.F16 22$\frac{1}{2}$%
4.F17 (a) £3 960 (b) £3.13 (c) 37$\frac{1}{2}$%
4.F18 (a) 87$\frac{1}{2}$% (b) 46.4% (c) 36.6%
4.F19 37.125 kg
4.F20 (a) £38.40 (b) £34.50 (c) £15.31 (d) £56.23
       (e) £40.91

Exercises 5.A

Estimates should be near the following answers which are correct.

5.A1   (a) 55.875 (b) 6.13 (c) 2 364.73 (d) 4.884
       (e) 0.271 (f) 18.76 (g) 92.24 (h) 17 171.82
5.A2   (a) 73.28 (b) 43.21 (c) 2 330.37 (d) 179.64
5.A3   (a) 38.578 g (b) £152.05 (c) 1 263.16 m
       (d) 761.9 kg (e) £343.64 (f) £48.74
5.A4   (a) to (j) Discussion in groups
5.A5   (a) 152.4 × 228.6 (b) 381 × 584.2 (c) 482.6
       (d) 50.8 × 101.6 × 762 (e) 1 014.4 × 1 727.2
       (f) 76.2 × 609.6 (g) 9 169.4 (h) 44.45
       (i) 152.4 (j) 215.9
5.A6   (a) 1 kg (b) 51 kg (c) 23 kg (d) 1 018 kg
       (e) 1.3 kg (f) 27 kg
5.A7   (a) 12 (b) 9 (c) 108 (d) 42 (e) 115
       (f) 0.41 (g) 15.14 (h) 10.11 (i) 50.16
5.A8   16.22
5.A9   392 mm
5.A10 50.4
5.A11 (a) £7.82 (b) 46 (c) 151 kg (d) 1 176 mm
5.A12 (a) 1:3 (b) 1:4 (c) 1:5 (d) 1.875:1
       (e) 1:4.28 (f) 1:18 (g) 1:2.33 (h) 1:8
5.A13 (a) 2:3 (b) 18
5.A14 (a) 540 (b) 7:6
5.A15 30 min
5.A16 (a) 52 kg (b) 6 kg
5.A17 (a) A = 10, B = 25, C = 40
       (b) 220 mm, 352 mm, 308 mm
5.A18 (a) 56p (b) £2.25 (c) £4.50 (d) £1.68
       (e) £4.56 (f) £16.49 (g) £33 (h) £132.75
5.A19 (a) 12 h (b) 70 litres
5.A20

	A	B	C
White	1.50	2.25	3.75
Blue	1.64	2.46	4.1
Green	1.32	1.98	3.3
Luxury finish	2.10	3.15	5.25
Document grey	2.66	3.99	6.65

Exercises 5.B

Estimates should be near the following answers which are correct.

5.B1   (a) 23 765
5.B2   8 750
5.B3   (a) £143.73$\frac{1}{2}$ (b) £608.58 (c) £368.55
       (d) £9.76$\frac{1}{2}$ (e) £1 838.82 (f) 153.23$\frac{1}{2}$ (g) £3.99
       (h) £123.72

5.B4   (a) £39.55 (b) £110.71 (c) £48.18 (d) £10.67
       (e) £171.65
5.B5   43
5.B6   £1 135.81
5.B7   26 years 11 months
5.B8   249
5.B9   (a) 42.28 (b) 9 h 42 min
5.B10 (a) 7 hr 29 min (b) Mon, Wed
5.B11 £91.40½
5.B12 (a) 120 words per minute (b) 3 min
       (c) 0.25 seconds
5.B13 £40.57½
5.B14 £1.05
5.B15 (a) 352 (b) 96
5.B16 (a) 6:1 (b) 144
5.B17 £1 800, £1 200, £3 000
5.B18 (a) A = £60 000, B = £80 000, C = £100 000
       (b) 6:3:1
5.B19 (a) 3 h 20 min typing, 4 h 10 min duplicating (b) 6
5.B20 (a) 35 kg wood, 15 kg plastics (b) 4:1 (c) 6

Exercises 5.C

Estimates should be near the following answers which are correct.

5.C1   5 mm = 7 968, 8 mm = 4 512, 10 mm = 6 528,
       12 mm = 2 784
5.C2   (a) 19.197 m² (b) 94.365 2 m² (c) 140.355 6 m²
       (d) 99.304 2 (e) 101.734 4 m²
5.C3   (a) £118.06 (b) £580.34 (c) £86 319
       (d) £610.72 (e) £625.67
5.C4   7 900.1 min or 1 316 h 41 min
5.C5   (a) £91.58 (b) £120.44 (c) £755.60
5.C6   (a) 1 036 mm (b) 731 mm (c) 816 mm
       (d) 779 mm (e) 1 207.5 (f) 5 229
       (g) 1 314 mm (h) 810 mm
5.C7   65.496
5.C8   (a) 780 r/mm (b) 22.04
5.C9   (a) 96.75 mm (b) 7 seconds
5.C10 (a) 2.73 gal (b) 1.4 min (c) 182.83 mm
5.C11 (a) 146.2 (b) 119.7 (c) 133.8 (d) 142.2
       (e) 136.3 (f) 163.8 (g) 159.3 (h) 156.5
       (i) 154.3 (j) 127.8
5.C12 357.5
5.C13 51.1 ohms
5.C14 (a) 27.7 mins (b) 3 h 18 mins
5.C15 (a) 9:5 (b)(i) 1:0.09 (ii) 1:0.57 (iii) 2:5:7
5.C16 (a) 3:1 (b) 3:4 (c) 4:9
5.C17 (a) 1.5 litres (b) 160 litres
5.C18 (a)(i) 2:5:7 (ii) 40 parts cement, 100 parts sand,
       140 parts stone (b) 1 h (c) 10 h (d) 30 m
5.C19 A = 8, B = 12, C = 28 (b) 264
5.C20 (a) 42 kg (b) 15 kg (c) 4.65 kg
5.C21 (a) Copper 135 kg, tin 15 kg, zinc 3 kg
5.C22 52.2 kg
5.C23 (a) 1.5 kg (b) 2.1 kg (c) 2.7 kg
5.C24 140
5.C25 (a) 96 kg, 40 kg, 8 kg (b) 12.5:5:1

Exercises 5.D

5.D1   (a) London 1 567, Birmingham 1 206, Manchester
       1 421, Bristol 949, Glasgow 1 517, Cardiff 1 254
       (b) Manual 6 033, technical 1 298, office 234,
       sales/management 349 (c) 7 914
5.D2   London £181 231.71, Birmingham £145 708.61,
       Manchester £171 075.65, Bristol £115 261.60,
       Glasgow £615 248.33, Cardiff £149 701.50
5.D3   (a) 712 800 (b) £4.20
5.D4   (a) 1 870.5 (b) 10 144.75 (c) 9 772.75
       (d) 33 306 (e) 187 960.5
5.D5   (a) 2 270.84 m² (b) 3 047.76 m² (c) 3 120 m²
5.D6   (a) 1 006 (b) 216 (c) 39 (d) 58
5.D7   95
5.D8   4.697
5.D9   455
5.D10 222°C
5.D11 2 h 52 min
5.D12 (a) 32 kg of A, 52 kg of B (b) 80 kg (c) 260 kg
5.D13 (a) 180 men, 80 women (b) 16
5.D14 (a)(i) 4 m (ii) 10 m (b) 5:1
5.D15 6:8:5
5.D16 Dismantle 2 h 9 min, produce parts 2 h 52 min,
       re-assemble 1 h 26 min
5.D17 (a) 7:5 (b) 21
5.D18 (a) 2 days (b) £1 080
5.D19 (a) £74.65
       (b) fruit £2 253, plants £751, veg. £5 257
5.D20 (a) Wood 21 kg, metal 6 kg, plastics 9 kg
       (b) 80 kg, 600 kg, 320 kg
5.D21 400 litres
5.D22 (a) 1st man £200, 2nd man £160 (b) 10:5:1
5.D23 (a) £1 800, £3 000, £4 200 (b) £120, £200, £280
5.D24 2 340
5.D25 (a) 6 days (b) 1 500 tonnes

Exercises 5.E

5.E1   Southbend 621, Eastlee 616, Weston 1 331,
       Northly 1 111
5.E2   £53 325.81
5.E3   £833.42
5.E4   (a) York St. 2 511, Parkside 3 682, Elm Sq. 4 609,
       Avon Rd. 1 776, Devon Dr. 1 878 (b) 14 456
5.E5   1 492
5.E6   (a) 1 118 (b) 1 463 (c) 311
5.E7   (a) £3.35 (b) £409.54
5.E8   25 145 miles
5.E9   1 206
5.E10 3 h 52 min
5.E11 (a) 19 320 (b) 9 982 (c) 51 842
5.E12 (a) 9:16 (b) 1:2.07
5.E13 (a) £4.84½ (b) £481.05
5.E14 (a) 1:1.1 (b) 21:2
5.E15 (a) 14 m × 22 m (b) 26.5 miles
5.E16 (a) 82.5 km (b) 115.5 km (c) 148.5 km
5.E17 216 houses, 360 bungalows
5.E18 25 girls, 20 boys

5.E19 7:6 or 1.17:1
5.E20 (a) 121.6 Australian (b) 78

Exercises 5.F

5.F1 (a) £31.46 (b) £29.42 (c) £54.21 (d) £90.91
5.F2 (a) 924 kg (b) 35 kg (c) 31.5 kg (d) 103.75 kg
5.F3 (a) £11.31 (b) £8.16 (c) £13.53 (d) £32.13
      (e) £10.78
5.F4 (a) £6.16 (b) £3.38 (c) £2.42 (d) £2.89
      (e) £11.61
5.F5 626
5.F6 (a) 13 920 g (b) Rhubarb 45 lb, flour 19.29 lb,
      S. R. flour 11.57 lb, marg. 9.64 lb, lard 5.79 lb
5.F7 (a) 0.333 kg (b) 0.11 litres
5.F8 (a) 391.2p (b) £18.05
5.F9 (a) £12.53 (b)(i) 59.33 (ii) £20.85 (iii)(1) 38p
      (2) 33p (3) 35p
5.F10 (a) Mon 4:9, Tue 8:9, Wed 7:9, Thur 2:3,
      Fri 3:5, Sat 14:15, Sun 2:5 (b) 1:1.49
5.F11 (a) 1.27:1 (b) 1.12 m
5.F12 (a) 4 (b) £1.56
5.F13 (a) £63 (b) £120 (c) £185
5.F14 37.5 kg
5.F15 (a) 9.62 (b) 5
5.F16 2.39:1
5.F17 300 g Fish, $1\frac{1}{2}$ egg, 375 ml bechamel,
      75 g mushrooms, 300 g potatoes

5.F18

	(a)	(b)	(c)
Chicken stock	2.25 litres	3.75 litres	8 litres
Flour	112.5 g	187.5 g	400 g
Onion, etc	225 g	375 g	800 g
Milk	562.5 g	937.5 ml	2 l
Butter	112.5 g	187.5 g	400 g
Chicken	56.25 g	93.75 g	200 g

5.F19 (a) 35 men:25 women (b) 4 h
5.F20 (a) 1:35 (b) 29.16 g

Exercises 6.A

6.A1 (a) 13 mm (b) 2500 mm (c) 800 mm
      (d) 4650 mm (e) 1238 mm (f) 47 mm
      (g) 236 mm (h) 70 mm
6.A2 (a) 2.1 m (b) 0.65 m (c) 0.75 m (d) 0.125 m
      (e) 0.486 m (f) 1.035 m (g) 1.49 m
      (h) 1.732 m
6.A3 (a) 2700 m (b) 3.844 km (c) 1.965 m
      (d) 6.75 m (e) 38.5 cm (f) 950 mm
      (g) 2.675 m (h) 104.7 cm
6.A4 (a) 101.6 mm (b) 304.8 mm (c) 914.4 mm
      (d) 38.1 mm (e) 182.88 mm (f) 22.86 mm
      (g) 914.4 mm (h) 396.24 mm
6.A5 (a) 4 in (b) 11 in (c) 3.6 in (d) 14 in
      (e) 0.3 in (f) 0.25 in (g) 4.95 in (h) 1.5 in
6.A6 (a) 1.83 m (b) 18.28 m (c) 6.44 km (d) 38.1

(e) 1.27 mm (f) 120.65 mm (g) 76.25 cm
(h) 50.29 m
6.A7 1.52 = 38.61 mm, 1.28 = 32.51 mm,
      2.75 = 69.85 mm, 4.36 = 110.74 mm
6.A8 27.4 m × 11 m
6.A9 (a) 200 mm² (b) 7.5 m² (c) 120 mm²
      (d) 104 cm² (e) 12 cm² (f) 1000 mm²
6.A10 (a) 12 in² (b) 475.8 mm² (c) 3.36 m²
      (d) 1583.4 mm² (e) 2.7 cm² (f) 7245 mm²
6.A11 7.2 cm², 150 mm², 38.7 cm², 5.25 m², 936 mm²,
      11 475 mm²
6.A12 (a) 78.3 cm² (b) 64 cm² (c) 5 cm²
      (d) 169 mm² (e) 1.04 m² (f) 50.75 cm²
      (g) 1.5625 m²
6.A13 (a) 1171 cm² (b) 2685 mm² (c) 1760 m²
      (d) 3765 mm² (e) 1108 m² (f) 5526 cm²
6.A14 (a) 7.5 cm² (b) 15.96 cm² (c) 2000 mm²
      (d) 270 mm² (e) 350 mm² (f) 2646 cm²
6.A15 (a) 1106 cm² (b) 2163 mm² (c) 6210 mm²
      (d) 1932 cm² (e) 120 m² (f) 10 150 m²
6.A16 (a) 94.5 cm² (b) 43.5 mm² (c) 360 mm²
      (d) 32.5 cm² (e) 990 mm² (f) 12.6 cm²
6.A17 (a) 97.76 cm² (b) 58.27 cm² (c) 15.75 cm²
      (d) 103.08 cm² (e) 44.5 cm² (f) 53.22 cm²
6.A18 (a) 52 mm² (b) 5.06 cm² (c) 4305 mm²
      (d) 0.018 m² (e) 98.8 cm² (f) 0.14 mm²
      (g) 48 720 mm² (h) 3.04 cm²
6.A19 (a) 264 mm (b) 308 mm (c) 462 mm
      (d) 88 cm (e) 11 m (f) 33 mm (g) 528 cm
      (h) 495 mm
6.A20 (a) 91.75 mm (b) 79.81 mm (c) 6.85 m
      (d) 10.02 cm (e) 154.90 mm
6.A21 (a) 154 mm² (b) 1386 mm² (c) 616 cm²
      (d) 67 914 mm² (e) 18 634 mm²
6.A22 (a) 169 100 mm² (b) 1.674 m² (c) 211 600 mm²
      (d) 7015 mm² (e) 5411 cm²
6.A23 (a) 648 cm² (b) 4219 cm²
6.A24 12 250 mm²
6.A25 6738 mm²
6.A26 (a) 1067 mm (b) 341 ft (c) $31 \times 1\frac{3}{4}$ in
      (d) 10 034 m² (e) 4.8 mile² (f) 1452 mm²

Exercises 6.B

6.B1 (a) mm (b) m (c) mm (d) m (e) mm
      (f) cm (g) m (h) km (i) cm (j) cm
      (k) mm (l) cm
6.B2 (a)(i) 240 (ii) 300 (iii) 1500 (iv) 175 (v) 320
      (b)(i) 90 (ii) 16.5 (iii) 15 (iv) 180 (v) 80
      (c)(i) 0.6 (ii) 0.72 (iii) 1.8 (iv) 2.3 (v) 1.5
6.B3 (a) 300 × 210 (b) 11.81 × 8.27
6.B4 (a) 12 in (b) 100 in (c) 48 in (d) 12 in
      (e) 64 in (f) 16 in
6.B6 (a)(i) 6.8 (ii) 7.44 (b) 45.72
6.B7 (a)(i) 2.12 mm (ii) 0.08 in (b) 0.007 mm
6.B8 7.69 yd
6.B9 (a) 229 × 101 (b) 270 × 216 (c) 305 × 254
      (d) 381 × 254
6.B10 (a) 721.78 yd (b) 2.357 miles

6.B11 (a) 2.6 m (b) 12.6 cm
6.B12 (a)(i) 6 300 (ii) 630 (b)(i) 94 300 (ii) 943
 (c)(i) 45 000 (ii) 450 (d)(i) 24 000 (ii) 240
 (e)(i) 6 300 (ii) 630 (f)(i) 1 800 (ii) 180
6.B13 (a) Office manager = $31.2\,m^2$, audio typing office = $42.25\,m^2$, word processor room = $40.3\,m^2$, records store = 48.75, stationery store = $42.25\,m^2$, general office = $140\,m^2$, duplicating room = $112\,m^2$
 (b) $546.28\,yd^2$
6.B14 (a) 250.14 m (b) $8.75\,m^2$
6.B15 (a) $38.5\,mm^2$ (b) $38\,500\,mm^2$
6.B16 (a)(i) $368\,mm^2$ (ii) 78 mm
 (b)(i) $23\,000\,mm^2$ (ii) 660 mm (c) $22\,632\,mm^2$
6.B17 (a)(i) $5.76\,cm^2$ (ii) $7.7\,cm^2$ (iii) $3.46\,cm^2$
 (b) 243.25 cm
6.B18 (a) 101.4 cm (b) $623.7\,cm^2$ (c) $215\,cm^2$
6.B19 (a)(i) 100 ft (ii) $548\,ft^2$ (b) $516.54\,ft^2$
6.B20 (a) 37.7 cm (b) 62.83 cm (c) $527\,cm^2$
6.B21 (a) $2\,889\,m^2$ (b) $6362\,m^2$ (c) $119\,m^2$
 (d) $1\,357\,m^2$ (e) $408\,m^2$ (f) $221\,m^2$ (g) $486\,m^2$
 (h) $248\,m^2$

Exercises 6.C

6.C1 (a) m (b) cm (c) mm (d) mm (e) m
 (f) cm (g) m (h) cm (i) mm (j) mm
 (k) mm (l) m
6.C2 (a) kg (b) g (c) kg or g (d) g (e) kg
 (f) kg
6.C4 0.4 = 10.16, 0.5 = 12.7, 0.6 = 15.24, 1.2 = 30.48, 1.3 = 33.02, 1.75 = 44.45, 1.8 = 45.72, 2.25 = 57.15, 2.3 = 58.42, 0.125 = 3.18
6.C5 16 = 0.63 in, 20 = 0.787 in, 22 = 0.866 in, 23 = 0.906 in, 29 = 1.142 in, 30 = 1.181 in, 32 = 1.26 in, 36 = 1.417 in
6.C6 (a) $9141\,mm^2$ (b) 600 mm (c) 200 mm
 (d) 287.5 mm (e) $15\,000\,mm^2$
6.C7 (a) 2.7 m (b) 9 m (c) 16.2 m (d) 11.25 m
 (e) 2.1 m
6.C8 35.03 mm
6.C9 69.8 mm
6.C10 $9.3\,mm^2$
6.C11 (a)(i) 45.3 mm (ii) 44.7 mm
 (b)(i) 20.2 mm (ii) 9.8 mm
6.C12 (a) 7.48 yd (b) 909
6.C13 (a) $669\,mm^2$ (b) $2065\,mm^2$ (c) $5177\,mm^2$
 (d) $838\,mm^2$ (e) $3716\,mm^2$ (f) $1297\,mm^2$
6.C14 1 000
6.C15 494
6.C16 (a) 92 cm (b) $358\,cm^2$
6.C17 150 m
6.C18 £243.20
6.C19 (a) $352.75\,cm^2$ (b) $625\,cm^2$ (c) $110.5\,m^2$
 (d) $1.3\,m^2$ (e) $32.5\,m^2$ (f) $87.5\,m^2$
 (g) $28.27\,cm^2$ (h) $3.68\,m^2$
6.C20 (a) $1041.89\,cm^2$ (b) $1500\,cm^2$ (c) $947\,cm^2$
 (d) $1490\,cm^2$ (e) $2883.59\,cm^2$ (f) $2555\,cm^2$
6.C21 $16\,304\,mm^2$
6.C22 $16\,250\,mm^2$
6.C23 $3236.25\,mm^2$
6.C24 (a) $246.5\,mm^2$ (b) $1075.65\,mm^2$ (c) $693\,mm^2$
 (d) $782\,mm^2$ (e) $2135\,mm^2$ (f) $787\,mm^2$

Exercises 6.D

6.D1 (a) km or m (b) km or m (c) cm (d) m
 (e) mm (f) km (g) m (h) cm (i) cm
 (j) m (k) mm (l) mm
6.D2 (a) kg (b) kg (c) g (d) g (e) kg (f) kg
6.D3 (a) 7 cm (b) 4.02 m
6.D4 (a) 57.87 miles (b) 92.59 km
6.D5 (a) 914.4 mm (b) 609.6 mm (c) 457.2 mm
 (d) 381 mm
6.D6 (a) 1.49 (b) 2 622.4 yd
6.D7 (a) Storage room 38.22 × 36.4, small store 20.02 × 27.3, loading bay 27.3 × 16.38
 (b) S.R. = $1391.21\,m^2$, S.S. = $546.55\,m^2$, L.B. = $447.17\,m^2$
6.D8 (a) $600\,cm^2$ (b) $1200\,cm^2$ (c) $800\,cm^2$
 (d) $706\,cm^2$
6.D9 (a) 24.2 m (b) 34 m (c) 39.2 m (d) 56.2 m
 (e) 24.45 m
6.D10 (a) $303.75\,m^2$ (b) $677.97\,m^2$ (c) $3579.39\,m^2$
 (d) $1632.96\,m^2$ (e) $2633.77\,m^2$ (f) $1616.99\,m^2$
6.D11 (a) $3.125\,m^2$ (b) £969.26
6.D12 (a) 206 (b) 15.2 cm
6.D13 (a) 72 (b) 5
6.D14 (a) 18.25 m (b) 15.9 m (c) 24.3 m
 (d) 56.03 m (e) 45.34 (f) 9 m (g) 43.67 m
6.D15 (a) 54 (b) £126.90
6.D16 $4.78\,ft^2$
6.D17 (a) 33 (b) £159.09
6.D18 (a) 51.3 litre (b) $89.89\,m^2$
6.D19 3.43 km
6.D20 28.44 m
6.D21 (a) $962.5\,m^2$ (b) $1682.54\,m^2$ (c) $109.97\,m$
6.D22 (a) 13.48 m (b) $1.04\,m^2$ (c) $5.04\,m^2$ (d) $1.04\,m^2$

Exercises 6.E

6.E1 (a) km (b) m (c) m (d) m (e) km (f) cm
 (g) m (h) cm (i) mm (j) m (k) km (l) m
6.E2 (a) g (b) kg (c) kg (d) kg (e) g (f) g
6.E4 9 = 2.7, 10 = 3.1, 28 = 8.5, 38 = 11.6, 125 = 38.1
6.E5 (a) 158.8 (b) 1473.2 (c) 31.8
 (d) 139.7 long, 22.2 dia. (e) 419.1
6.E6 (a) $2734.4\,mm^2$ (b) 900 mm (c) 2.3 m
 (d) $7421.9\,mm^2$
6.E7 (a) 10.08 (b) 11.02
6.E8 (a) 1648 (b) 20.125
6.E9 (a) 126 m, $962\,m^2$ (b) 125 m, $970\,m^2$
 (c) 216 m, $2592\,m^2$ (d) 142 m, $1310\,m^2$
 (e) 242 m, $3252\,m^2$ (f) 266 m, $2719\,m^2$
 (g) 260 m, $3612\,m^2$ (h) 204 m, $1578\,m^2$
6.E10 (a) 2370 m (b) 8.24 m (c) 7.39 m
6.E11 (a) 1.144 5 yd (b) £3 907.33

6.E12	(a) 85.7 km	(b) 67.8 km	(c) 33.23 km
	(d) 61.82 km	(e) 21.85 km	(f) 12.6 km
6.E13	(a) 8.16 m²	(b) 11.25 m²	(c) 12.2 m
6.E14	531.96 m
6.E15	(a) 176.3 ft²	(b) 59.112 ft
6.E16	(a) 3 637.5 cm²	(b) 1 417.5 cm²	(c) 148.55 m²
	(d) 113.1 m²	(e) 242 m²	(f) 5 980 cm²
	(g) 2 520 cm²	(h) 730 m²
6.E17	(a) 40.06 in²	(b) 6 089.2 mm²	(c) 103.69 cm²
	(d) 13 857 mm²
6.E18	(a) 1.68 m²	(b) £2.22
6.E19	(a) 28	(b) 352.9 m²
6.E20	(a) S.B. = 37.5 m², O. = 25 m², Rec. = 21 m², W. = 150 m², P.S. = 50.4 m² (b) S.B. = 25 m, O. = 20 m, Rec. = 18.4 m, W. = 49 m, P.S. = 32.4 m (c) 283.9 m²

Exercises 6.F

6.F1	(a) m (b) cm (c) km (d) mm (e) cm (f) cm (g) cm (h) mm (i) mm (j) m (k) mm (l) cm
6.F2	(a) g (b) kg (c) kg (d) g (e) g (f) kg
6.F3	(a) 13 = 33.02, 15 = 38.1, 17 = 43.18, 38 = 96.52, 47 = 119.38 (b) 9 = 22.86, 12 = 30.48, 14 = 35.56, 15 = 38.1, 16 = 40.64, 21 = 53.34 (c) 8 = 20.32, 10 = 25.4, 11 = 27.94, 19 = 48.26, 22 = 55.88, 30 = 76.2, 58 = 147.32 (d) 3 = 7.62, 4 = 10.16, 5 = 12.7, 9 = 22.86, 13 = 33.02, 18 = 45.72, 23 = 58.42, 28 = 71.12
6.F4	(a) 5.16 m (b) 1.45 m² (c) 1 720 (d) 1.45 m × 0.53 m
6.F5	14.14 m
6.F6	(a) 2.513 m² (b) £3.42
6.F7	2 110 cm²
6.F8	(a) 23 cm² per slice (b) 110.25 cm²
6.F9	287.18 m²
6.F10	48 m
6.F11	(a)(i) 975 cm (ii) 310 cm (iii) 7.55 m² (b) 0.322 m²
6.F12	(a) 113.76 m² (b)(i) 625 cm² (ii) 18 201.6 (c) 45.25 cm (d)(i) 3.2 m² (ii) 110.56 m²
6.F13	(a) 706.95 cm² (b) 1 696.68 cm²
6.F14	(a) 158 cm (b) 1 504 cm²
6.F15	(a)(i) £5.26 (ii) £8.04 (b) 7.8 m²
6.F16	(a) 490.9 mm² (b) 297 mm² (c) 216 mm² (d) 666 mm² (e) 532 mm² (f) 432 mm²
6.F17	(a) 362 m² (b) 1 023 m² (c) 1 404 m² (d) 455 m² (e) 678 m² (f) 1 127 m²
6.F18	(a) 114 m, 758.96 m² (b) 53 m, 159.96 m² (c) 55.8 m, 172.98 m² (d) 31 m, 59.5 m² (e) 32.6 m, 65.1 m² (f) 39.4 m, 88.09 m²
6.F19	(a) 26.65 m² (b) 9.6 m² (c) 47.25 m² (d) 12.5 m²
6.F20	(a) 49 (b)(i) 7.2 m² (ii) 66.98 ft²

Exercises 7.A

7.A1	(a) 1 000 (b) 1 728 (c) 1 000 000 000 (d) 27 (e) 16 400 (f) 283 392 (g) 35.335 7
7.A2	(a) 3 500 000 (b) 56.537 (c) 1.2 (d) 1 200 000 000 (e) 2 800 (f) 1 650 (g) 750 (h) 45.4 (i) 653
7.A3	(a) 3 600 mm³ (b) 1.44 m³ (c) 230.4 cm³ (d) 51 000 mm³ (e) 0.031 5 m³
7.A4	(a) 14 112 mm³ (b) 124 950 mm³ (c) 29 835 mm³ (d) 26 130 mm³
7.A5	(a) 0.462 m³ (b) 200 000 mm³ (c) 2 400 cm³ (d) 37.8 m³
7.A6	(a) 4 563 cm³ (b) 7 560 cm³ (c) 13 568 cm³ (d) 3 718 cm³ (e) 5 130 cm³ (f) 22 274 cm³
7.A7	21 264 mm³
7.A8	(a) 19 899 mm³ (b) 77 448 mm³ (c) 36 295 mm³ (d) 8 946 mm³
7.A9	(a) 18 923.8 in³ (b) 11 216.5 in³ (c) 211.6 in³ (d) 479.1 in
7.A10	(a) 145 800 mm³ (b) 92 400 mm³ (c) 90 000 mm³ (d) 462 300 mm³ (e) 327 600 mm³ (f) 326 700 mm³
7.A11	2 988 000 mm³
7.A12	(a) 6 158 mm³ (b) 26 707 mm³ (c) 125 680 mm³ (d) 245 469 mm³ (e) 141 390 mm³
7.A13	(a) 8 640 mm³ (b) 31 110 mm³ (c) 49 725 mm³
7.A14	56 163.25 mm³
7.A15	27 200 mm³
7.A16	37 152 mm³
7.A17	(a) 1 436.941 2 cm³ (b) 6 000 cm³ (c) 14 143 cm³ (d) 38 797.4 cm³
7.A18	603 litres
7.A19	9 828 litres
7.A20	(a) 8 100 g (b) 2.1 t (c) 4 800 g (d) 612.69 g (e) 22.5 t (f) 2 860 g

Exercises 7.B

7.B1	10 500 cm³
7.B2	245 760 cm³
7.B3	(a) 2 100 000 mm³ (b) 20 250 mm³ (c) 8 000 mm³ (d) 698 750 mm³
7.B4	74 250 cm³
7.B5	(a) Manager's office = 107.016 m³ (b) audio = 129.654 m³ (c) corridor = 111.09 m³ (d) general office = 458.64 m³ (e) accounts office = 280.14 m³ (f) 1 086.54 m³
7.B6	(a) 960 ft³ (b) 27.2 m³
7.B7	(a) 3 300 cm³ (b) 1 650 000 cm³
7.B8	(a) 800 cm³ (b) 6 600 000 mm³ (c) 2.052 m³
7.B9	(a) 2.856 m³ (b) 0.39 m³ (c) 9.675 m³ (d) 0.36 m³ (e) 0.69 m³
7.B10	108 000 mm³
7.B11	28 734.64 mm³
7.B12	(a) 2.24 m³ (b) 0.12 m³ (c) 0.336 m³ (d) 0.283 5 m³
7.B13	(a) Desk = 79.06 ft³, safe = 4.24 ft³, filing cab. = 11.86 ft³, photocop. = 10 ft³ (b) desk = 2.93 yd³, safe = 0.157 yd³, filing cab. = 0.44 yd³, photocop. = 0.37 yd³
7.B14	2 328 litres
7.B15	(a) 2.376 litres (b) 10.054 litres (c) 21.239 litres
7.B16	315 kg

7.B17 34.65 litres
7.B18 7.673 litres
7.B19 (a) 70.695 cm³ (b) 91 200 mm³ (c) 9 049 mm³
       (d) 1 232 mm³ (e) 10 800 mm³
7.B20 (a) 95 438.25 mm³ (b) 527.856 cm³
       (c) 4 021.76 mm³

Exercises 7.C

7.C1 (a) 7 271.81 mm³ (b) 11 193 mm³
      (c) 5 422.56 mm³ (d) 22 455.3 mm³
      (e) 1 160.3 mm³ (f) 2 033.2 mm³
7.C2 (a) 146 250 mm³ (b) 112 500 mm³
      (c) 33 750 mm³
7.C3 (a) litres (b) litres (c) litres (d) cm³ (e) ml
      (f) litres
7.C4 (a) 27 000 mm³ (b) 381.75 cm³ (c) 3 600 cm³
7.C5 55 ft³
7.C6 (a) 2 771.244 cm³ (b) 2.771 litres
7.C7 (a) 2 025 cm³ (b) 2.025 litres
7.C8 (a) 27 000 cm³ (b) 5 008 cm³ (c) 1 767.37 mm³
      (d) 22 034.782 cm³
7.C9 1.3 m³
7.C11 5.236 cm
7.C12 (a) 17 667 mm³ (b) 34 798 mm³ (c) 25 995 mm³
      (d) 12 874 mm³ (e) 21 856 mm³ (f) 133 254 mm³
7.C13 (a) 199 124.25 mm³ (b) 188 944.17 mm³
7.C14 31 158 mm³
7.C15 (a) 43 807.335 mm³ (b) 61 480.299 mm³
      (c) 47 657.856 mm³
7.C16 169 857 mm³
7.C17 (a) 87.96 m³ (b) 5 302 mm³
7.C18 13 m³
7.C19 (a) 62.84 litres (b) 3.817 53 litres (c) 180 litres
      (d) 0.147 3 litres
7.C20 (a) 25 200 000 mm³ (b) 2 337 500 mm³
      (c) 15 312 500 mm³
7.C21 214 200 mm³
7.C22 11 250 mm³
7.C23 (a) 4.212 kg (b) 0.84 kg (c) 1.011 kg
      (d) 196.1 g
7.C24 (a) 60.606 g (b) 100.24 g (c) 96.174 g
      (d) 124.4 g (e) 622 g
7.C25 (a) 0.11 kg (b) 688.8 mm (c) 4.27 kg

Exercises 7.D

7.D1 (a) 1.792 m³ (b) 0.76 m³ (c) 0.115 5 m³
      (d) 5.832 m³ (e) 0.405 m³
7.D2 (a) 512 cm³ (b) 7 000 cm³ (c) 2 052 cm³
      (d) 12 600 cm³ (e) 20 000 cm³
7.D3 27.552 m³
7.D4 (a) 429 m³ (b) 468 m³ (c) 357.5 m³ (d) 195 m³
      (e) 195 m³ (f) 299 m³
7.D5 (a) 19.114 m³ (b) 1.976 m³ (c) 40.619 8 m³
7.D6 (a) 157 500 mm³ (b) 115 200 mm³
      (c) 256 622.85 mm³ (d) 148 200 mm³
      (e) 92 374.8 mm³ (f) 216 000 mm³
      (g) 116 550 mm³ (h) 162 000 mm³
7.D7 (a) 6 539.28 cm³ (b) 45 000 cm³
7.D8 (a) 9.204 m³ (b) 348.75 m³
7.D9 (a) litres (b) ml (c) cm³ (d) cm³ (e) litres
7.D10 (a) 4 m³ (b)(i) 1 608.7 cm³ (ii) 1.608 7 litres
7.D11 (a) 5 891.25 cm³ (b) 32.4 m³
7.D12 (a) 4 m (b) 201 600 cm³
7.D13 (a) 4.2 cm (b) 54.32 cm
7.D14 (a) 6 600 litres (b) 8.4 m
7.D15 (a)(i) 3 000 m³ (ii) 72 000 m³ (b) 3.96 m³
7.D16 (a) 22 200 cm³ (b) 57.8 litres
7.D17 (a) 0.248 m³ (b) 248.5 litres
7.D18 21 t
7.D19 (a) 3 920 lb (b) 1 779.68 kg (c) 1.75 tons
      (d) 1.779 7 t
7.D20 (a) 435.515 g (b) 1 198.512 g (c) 548.1 g
      (d) 409.1 g
7.D21 3 456 kg

Exercises 7.E

7.E1 (a) litres (b) litres (c) cm³ (d) litres (e) cm³
      (f) litres
7.E2 (a) 179.62 cm³ (b) 18.75 m³
7.E3 (a) 1 680 (b) 750 yd³
7.E4 (a) 24 (b) 50.4 t
7.E5 (a) 42 (b) 67.2 t
7.E6 (a) 21 160 cm³ (b) 15 456 000 mm³
      (c) 1 296 000 cm³ (d) 39 375 cm³
7.E7 (a) 3 808 m³ (b) 1.944 m³ (c) 186.984
      (d) 105.84 (e) 72.864
7.E8 (a) 22.2 m³ (b) 22 200 litres
7.E9 35.52 t
7.E10 (a) 144 litres (b) 150 litres (c) 286 litres
      (d) 194.4 litres (e) 157.5 litres
7.E11 (a) 13.33 ml (b) 22p
7.E12 26 cm
7.E13 15 m
7.E14 2.55 m
7.E15 (a) 102.336 m³ (b) 114.816 m³ (c) 97.344 m³
      (d) 242.42 m³ (e) 556.92 m³
7.E16 7.31 m³
7.E17 (a) 234.4 m³ (b) 312.5 m³
7.E18 1 508.16 m³, 804. 35 m³
7.E19 196 875 gal
7.E20 (a) 74.775 in³ (b)(i) 1.178 m³ (ii) 1 178 litres
      (c) 203.125 gal

Exercises 7.F

7.F1 (a) litres (b) litres (c) litres (d) cm³/litre
      (e) litres (f) ml
7.F2 (a) 165 000 cm³ (b) 12 725 cm³ (c) 3 142 cm³
      (d) 2 356.5 cm³
7.F3 (a) 24 ft³ (b) 2 ft³
7.F4 63 625 cm³
7.F5 10 752 cm³
7.F6 (a) 669 375 cm³ (b)(i) 223 125 cm³ (ii) 0.669 m³
7.F7 486 litres

7.F8 18 102 cm³
7.F9 38 313 cm³
7.F10 (a)(i) 9 048.96 cm³ (ii) 9.05 litres
(b)(i) 2 651.06 cm³ (ii) 2.65 litres
(c)(i) 16 259.85 cm³ (ii) 16.26 litres
(d)(i) 15 876 cm³ (ii) 15.88 litres
(e)(i) 4 032.266 (ii) 4.03 litres
7.F11 (a) 415 m³ (b) 345 m³ (c) 380 m³
7.F12 377.664 m³
7.F13 (a) 14 139 cm³ (b) 8.208 m³ (c) 4.332 m³
(d) 6 270 000 mm³ (e) 4 332 000 cm³
(f) 294 562 cm³
7.F14 (a) 660 cm³ (b) 0.495 kg
7.F15 (a) 148 460 cm³ (b) 375 100 cm³ (c) 27 147 cm³
(d) 232 510 cm³ (e) 11 453 cm³
7.F16 10.648 ml
7.F17 (a) 24.25 kg (b) 330 g Barbados sugar, 120 g butter, 300 g treacle
7.F18 3.09 kg
7.F19 0.79 litres
7.F20 (a) 5.7 cm³ (b) 7.07 cm³ (c) 17.6 cm³
(d) 8.4 cm³ (e) 20.48 cm³ (f) 16.087 cm³

Exercises 8.A

8.A1 (a) 34° (b) 123° (c) 70° (d) 260°
8.A2 (a) 300′ (b) 192′ (c) 872′ (d) 30′ (e) 150′
(f) 527′ (g) 1 639′
8.A3 (a) 1°30′ (b) 1°50′ (c) 4°35′ (d) 11°40′
(e) 2°14′
8.A4 (a) 240″ (b) 133″ (c) 945″ (d) 4 102″
(e) 37 936″
8.A5 (a) 3′20″ (b) 1°30′30″ (c) 4°26′40″ (d) 5°50′
8.A6 (a) 6°39′ (b) 88°39′ (c) 22°20′13″ (d) 14°56′26″
8.A7 (a) 14°30′ (b) 9°18′ (c) 121°54′ (d) 7°17′
(e) 38°10′
8.A8 (a) 2.6° (b) 43.25° (c) 19.7° (d) 2.5°
(e) 13.5°
8.A10 (a)(i) 57° (ii) 12° (iii) 46°25′ (iv) 68°48′
(b)(i) 76° (ii) 115° (iii) 139°40′ (iv) 40°44′
8.A11 (a) 37° (b) 49° (c) 62° (d) 35° (e) 46°29′
(f) 12°52′
8.A12 (a) 85° (b) 109°49′ (c) 96° (d) 110°49′
(e) 126° (f) 56°11′
8.A13 (a) 51° (b) 48° (c) 85° (d) 21°53′ (e) 50°25′
(f) 127°12′
8.A14 (a) $x = 57°$, $y = 123°$, $z = 123°$ (b) $x = 86°30′$, $z = 86°30′$, $y = 93°30′$ (c) $z = 49°$, $x = 131°$, $y = 131°$ (d) $z = 158°44′$, $x = 21°16′$, $y = 21°16′$ (e) $y = 13°$, $x = 167°$, $z = 167°$ (f) $y = 17°29′$, $x = 162°31′$, $z = 162°31′$
8.A15 (a) $b, e, g = 135°$, $a, c, d, f = 45°$
(b) $b, e, g = 52°$, $a, c, d, f = 128°$
(c) $b, e, g = 76°$, $c, a, f, d = 104°$
8.A16 (a) $x, y = 87°$ (b) $x = 85°30′$, $y = 94°30′$
(c) $x = 112°$, $y = 68°$
8.A17 (a) 41° (b) 19° (c) 38° (d) 61° (e) 48°
(f) 57°43′
8.A18 (a) $y = 39°$, $x = 47°$, $z = 133°$ (b) $x = 27°$, $y = 78°$, $z = 105°$ (c) $x = 57°25′$, $y = 44°35′$, $z = 135°25′$ (d) $y = 119°$, $z = 76°$, $x = 104°$
8.A19 (a) 42° (b) 43° (c) 78° (d) 37°
8.A20 (a) $x = 90°$, $y = 49°28′$ (b) $y = 90°$, $x = 21°40′$
(c) $x = 90°$, $y = 12°44′$ (d) $y = 90°$, $x = 7°$
8.A21 (a) 139.6 mm² (b) 102.67 cm² (c) 96.03 cm²
(d) 4.19 cm²
8.A22 1 047.3 mm²
8.A23 (a) $x = 110°$, $y = 105°$, $z = 75°$ (b) $x = 105°$, $y = 85°$, $z = 95°$ (c) $x = 86°$, $y = 65°$, $z = 115°$
(d) $z = 63°$, $x = 117°$, $y = 63°$ (e) $y = 98°$, $z = 82°$, $x = 84°$ (f) $z = 77°$, $y = 73°$, $x = 90°$
8.A24 (a)(i) 225 cm² (ii) 60 cm (b)(i) $x = 90°$
(ii) $y = 45°$ (c) $\frac{1}{4}$
8.A25 (a) 1 225 mm² (b) 75 cm² (c) 1 385.62 cm²
(d) 18 750 mm²
8.A26 (a) Hexagon (b) pentagon (c) quadrilateral
(d) triangle (e) octagon (f) heptagon
8.A27 (a) Acute (b) isosceles (c) equilateral
(d) right angled (e) obtuse
8.A28 (a) $x = 83°$, $y = 97°$ (b) $x = 16°$, $y = 164°$
(c) 79°30′ (d) 38°30′ (e) $y = 65°$, $x = 50°$
(f) $x = 72°$, $y = 108°$

Exercises 8.B

8.B1 24°
8.B2 120°
8.B3 19°
8.B4 65°
8.B5 (a) 180° (b) 90° (c) 90° (d) 90° (e) 30°
(f) 30° (g) 60° (h) 90° (i) 150° (j) 60°
(k) 150° (l) 180°
8.B6 (a) 35 cm × 35 cm (b) 140 cm (c) 1 225 cm²
(d) ✗ (e) 90° (f) 45°
8.B8 (a)(i) 36 mm (ii) 113.11 mm (iii) 1 018 mm²
(b) 4.37 cm²
8.B9 (a) Parallelogram (b) 12.5 cm² (c) 125°
8.B10 (a) 51°26′ (b) 51.43°
8.B11 (a) $2\frac{1}{2}$ (b) $\frac{4}{9}$ (c) 32° (d) 64°
8.B12 $z = 75°$, $x = 68°$, $y = 88°$
8.B13 (a) 7 (b) heptagon
8.B14 $z = 73°$, $x = 153°$, $y = 102°$
8.B15 (a) $x = 77°07′$ (b)(i) 160°56′ (ii) 195°12′
(iii) 146°52′

Exercises 8.C

8.C1 (a) 57° (b) 77° (c) 56° (d) 44°44′ (e) 89°
8.C2 (a) $x = 17°48′$, $y = 51°34′$ (b) $x = 43°$, $y = 137°$
(c) $x = 30°$, $y = 20°$ (d) $x = 28°35′$, $y = 35°10′$
8.C3 (a) 90° (b) 45° (c) 30° (d) 60°
8.C4 (a) 120° (b) 84°47′
8.C5 30°04′
8.C6 (a) Wedge angle = 65° (b) wedge angle = 61°
(c) wedge angle = 55° (d) rake angle = 15°
(e) clearance angle = 10° (f) clearance angle = 9°

8.C7	(a) 20° (b) 12° (c) 50°	
8.C8	a = rake angle = 10°, c = clearance angle = 20°	
8.C9	(a) 180° (b) 6° (c) 60° (d) 90° (e) 198°	
8.C10	19 divisions	
8.C11	(a) $x = 80°$, $y = 100°$ (b) $y = 100°$, $z = 135°$, $x = 60°$ (c) $z = 95°$, $x = 80°$, $y = 85°$ (d) $x = 55°40'$, $z = 55°40'$, $y = 124°20'$ (e) $y = 62°$, $x = 105°$ (f) $x = 112°$, $y = 107°$, $z = 35°$	
8.C12	(a) 6.06 m (b) 2.37 m^2	
8.C13	132.45 cm^2	
8.C14	(a) 1031 mm^2 (b) 57736 mm^3 (c) 67°	
8.C15	(a) 16°22' (b) 72° (c) 275.5 mm^2	

Exercises 8.D

8.D1	(a) 39° (b) 61°30' (c) 135° (d) 53°45'
8.D2	$x = 30°22'$, $y = 103°23'$
8.D3	(a) 90° (b) 60° (c) 45° (d) 30° (e) 20°
8.D4	(a) $x = 116°$ (b) $y = 116°$ (c) $z = 64°$
8.D5	(a) 50° (b) 0.9 m^2 (c) 27 m^3
8.D6	(a) 211.7 m^2 (b) 585.18 m^2 (c) 357 m^2 (d) 764.25 m^2 (e) 102.06 m^2
8.D7	(a) 37° (b)(i) 151° (ii) 200° (iii) 209° (iv) 266°
8.D8	(a) 73°7' (b)(i) 203°37' (ii) 187°8' (iii) 238°36' (iv) 260°15'
8.D9	(a) $x = 26°$, $y = 64°$ (b) 46.545 m^2
8.D10	(a) 18.79 m^2 (b)(i) 5.0396 m^2 (ii) 22.317 m^2
8.D11	9.4 m
8.D12	(a) $x = 50°$, $y = 65$ (b)(i) 1963.75 cm^2 (ii) 490.94 cm^2
8.D13	(a) 106°36' (b) 106°36' (c) 73°24'
8.D14	(a) 40° (b) 60° (c) 72°
8.D15	(a) $x = 59°$, $y = 38°$, $z = 142°$ (b) $x = 77°$, $y = 113°$, $z = 10°$ (c) $x = 148°$, $y = 70°$, $z = 102°$ (d) $x = 44°24'$, $y = 45°36'$, $z = 134°24'$ (e) $x = 90°$, $y = 71°$, $z = 109'$ (f) $y = 38°8'$, $x = 50°36'$, $z = 129°24'$

Exercises 8.E

8.E1	(a) 81° (b) (i) 202° (ii) 136° (iii) 120° (iv) 163°
8.E2	(a) 60° (b) 180° (c) 60° (d) 30° (e) 30° (f) 120° (g) 180° (h) 180°
8.E3	88°50'
8.E4	(a) 73° (b) 107° (c) 159°
8.E5	(a) 11.55 m^2 (b) 120°
8.E6	4311.5 mm^2
8.E7	1 m^2
8.E8	$y = 106°$, $x = 77°$, $z = 103°$
8.E9	(a) 1.44 m^2 (b) 1.15 m^2
8.E10	$a = 28°$, $b = 45°$, $c = 20°$, $d = 37°$, $e = 90°$, $f = 73°$, $g = 70°$
8.E11	$x = 120°$, $y = 60°$
8.E12	$x = 76°31'$, $y = 103°29'$, $z = 76°31'$
8.E13	(a) 1455.83 cm^3 (b) 10%
8.E14	(a) 90° (b) 400 cm^2 (c) 1200 cm^2
8.E15	(a)(i) 90°54' (ii) 136°41' (iii) 320°12' (b)(i) 1200 cm^2 (ii) 1420 cm^2 (iii) 1442.9 cm^2 (iv) 1400 cm^2

Exercises 8.F

8.F1	(a) 120° (b) 90° (c) 72° (d) 60°
8.F2	(a) $x = 15°$, $y = 63°$ (b) 254 cm^2 (c) 132.75 cm^2
8.F3	(a) $x = 26°$ (b) 2802 cm^2
8.F4	(a) 131.96 m (b) 1385.62 m^2 (c) 52°
8.F5	(a) 51°25' (b) 154°17' (c) 102°52'
8.F6	(a) 68° (b) 176.74 cm^2 (c) 78.55 cm^2
8.F7	(a) 180° (b) 120° (c) 150° (d) 90° (e) 60° (f) 30°
8.F8	(a) 36° (b) 45.25 cm^2 (c) 226.2 cm^2
8.F9	(a) 72° (b) 4524.48 cm^2
8.F10	(a) 24° (b) 141.39 cm^3
8.F11	$x = 77°$, $y = 80°$, $z = 180°$
8.F12	(a) Circle (b) equilateral triangle (c) hexagon (d) heptagon (e) isosceles triangle (f) square (g) pentagon (h) rectangle (i) octagon (j) parallelogram (k) quadrilateral
8.F13	(a) 188.52 mm (b) 12° (c) 10
8.F14	(a) 58° (b) 67° (c) 22° (d) 114°
8.F15	(a) $x = 135$, $y = 130°$ (b) 640 cm^2

Exercises 9.A

9.A1	(a) $x + y$ (b) $x - y$ (c) $y - x$ (d) $x + y + z$ (e) xy (f) yz (g) $\dfrac{x}{y}$ (h) $\dfrac{xy}{z}$ (i) $2x + 3y$ (j) $3x - 3z$ (k) $4x + 2y + 5z$ (l) $x^2 + \sqrt{z}$ (m) $y^3 - z^2$
9.A2	(a) $\frac{1}{2}$ base × perpendicular height (b) $2l + 2b$ (c) $l \times b \times h$ (d) base × perpendicular height = $b \times h$
9.A3	(a) $142° - b = a$ (b) $142° - a = b$
9.A4	(a) $m \times n$ (b) $2x \times y$ (c) $\dfrac{x}{2} \times y$ (d) $3a \times 4b$ (e) $(2y \times x) + (x \times y)$ (f) $(x \times y) + (x \times y) + (x \times y)$ (g) $(3y \times x) + (2x \times y)$ (h) $b \times h$
9.A5	(a) $\pi r^2 L = V$ (b) $\dfrac{V}{\pi r^2} = L$
9.A6	$x = 2b + 2.5p + 3e$
9.A7	$x = £140$
9.A8	Trail angle = 180° − (approach angle + plan angle)
9.A9	(a) km × 1000 = m (b) $\dfrac{mm}{1000}$ = m (c) m × 100 = cm (d) $\dfrac{g}{1000}$ = kg (e) $\dfrac{mm}{25.4}$ = inches
9.A10	(a) $13x$ (b) $21y$ (c) $5x + 4y$ (d) $15x$ (e) $16x$

(f) $11x \times 3y$ (g) $8a + 4b$ (h) $2a + 3b + c$
(i) $21x + 4y$ (j) $6x + 9y$

9.A11 (a) $6b$ (b) $5x^2$ (c) $8x^2$ (d) $8x + 5x^2$
(e) $7a^2 + 8a$ (f) $3a^2$ (g) $3b^2 + 9b$ (h) $6b^2$

9.A12 (a) $4x$ (b) $15b$ (c) $8x$ (d) $5xy$ (e) $2a$

9.A13 (a) $3a + 3b$ (b) $7x + 7y$ (c) $2x - 2y$
(d) $6x + 8$ (e) $12k - 6$ (f) $20x + 30y$
(g) $15x^2 + 10x$ (h) $36b^2 + 27b + 36$

9.A14 (a) 9 (b) 40 (c) 73 (d) 27 (e) 45 (f) 50
(g) 52 (h) 41

9.A15 (a) 38. (b) -5 (c) 18 (d) 18 (e) 4 (f) 7
(g) 7 (h) 7

9.A16 (a) 4 (b) 6 (c) 6 (d) -16 (e) -24 (f) 4
(g) -1 (h) -32

9.A17 (a) 13 (b) 48 (c) 2 (d) 22 (e) 12 (f) 16
(g) 50 (h) 24 (i) 25 (j) 99 (k) 4 (l) 228

9.A18 (a) 0.12 (b) 41.2 (c) 100.48 (d) 0.412
(e) 34 (f) 103.2 (g) 69 (h) 0.18

9.A19 $15^2 = 225$, $16^2 = 256$, $17^2 = 289$, $18^2 = 324$, $19^2 = 361$,
$20^2 = 400$, $21^2 = 441$, $22^2 = 484$, $23^2 = 529$, $24^2 = 576$,
$25^2 = 625$, $26^2 = 676$, $27^2 = 729$, $28^2 = 784$, $29^2 = 841$,
$30^2 = 900$

9.A20 (a) 175 (b) 1040 (c) 875 (d) 153 (e) 480
(f) 24

9.A21 $7^3 = 343$, $8^3 = 512$, $9^3 = 729$, $10^3 = 1\,000$,
$11^3 = 1\,331$, $12^3 = 1\,728$, $13^3 = 2\,197$, $14^3 = 2\,744$,
$15^3 = 3\,375$

9.A22 (a) 599 (b) 1468 (c) 446 (d) 3014 (e) 313

9.A23 (a) 7 (b) 20 (c) 9 (d) 11 (e) 13

9.A24 (a) 14 (b) 35 (c) 349 (d) 77 (e) 1370

9.A25 (a) a^3 (b) b^2 (c) $a^2 + b^3$ (d) $2b^2$ (e) $6a^3$
(f) $K^3 - T^2$ (g) $\sqrt{a^2}$ (h) a

9.A26 (a) $x = 14$ (b) $x = 6$ (c) $x = 10$ (d) $x = 20$
(e) $x = 1\frac{1}{2}$ (f) $x = \frac{3}{4}$ (g) $x = 0.8$ (h) $x = 6.2$
(i) $x = 10.33$ (j) $x = 9.433$ (k) $x = 0.37$
(l) $x = 1$

9.A27 (a) $x = 3$ (b) $x = 3$ (c) $x = 3$ (d) $x = 8$
(e) $x = 6$ (f) $x = 2$ (g) $x = 0.15$ (h) $x = 2$

9.A28 (a) $x = 12$ (b) $x = 10$ (c) $x = 52$ (d) $x = 54$
(e) $x = 2.6$ (f) $x = 18$ (g) $x = 6$ (h) $p = 2.6$

9.A29 (a) $x = 4$ (b) $x = 3$ (c) $x = 8$ (d) $p = 2$
(e) $m = 12$ (f) $x = 10$ (g) $x = 8$ (h) $x = 0.25$

9.A30 (a) $x = 8$ (b) $x = 45$ (c) $x = 70$ (d) $m = 9$
(e) $k = 8$ (f) $t = 0.3$ (g) $x = 12$ (h) $x = 43.5$

9.A31 (a) $x = 14$ (b) $x = 12$ (c) $x = 14$ (d) $x = 13$
(e) $x = 2$ (f) $x = 10$ (g) $x = 25$ (h) $x = 0.014$

9.A32 (a) $x = 4$ (b) $x = 9$ (c) $x = 8$ (d) $x = 3$
(e) $x = 8$ (f) $x = 4$ (g) $x = 5$ (h) $x = 2.5$

9.A33 (a) $x = \dfrac{L}{rt}$ (b) $x = \dfrac{f-4}{g}$ (c) $x = \dfrac{S+3t}{p}$
(d) $x = \dfrac{f}{n}$ (e) $x = \dfrac{rp}{Q}$ (f) $x = \dfrac{yn}{m}$
(g) $x = \dfrac{a}{y} - 2$ (h) $x = \sqrt{\dfrac{A}{\pi}}$

9.A34 (a) $x = \dfrac{R}{my}$ (b) $m = \dfrac{g}{h}$ (c) $r = \dfrac{S}{\pi h}$ (d) $T = \dfrac{I}{PR}$
(e) $V = \dfrac{A}{\pi r^2}$ (f) $l = \dfrac{4V}{\pi D^2}$ (g) $m = \dfrac{n+q}{p}$

(h) $r = \sqrt{\dfrac{A}{\pi V}}$

9.A35 (a) $x = T(p+q)$ (b) $n = \dfrac{S}{p} + 2$ (c) $k = \dfrac{C}{100} + a$
(d) $x = \dfrac{y}{a} - b$ (e) $c = \dfrac{5(F-32)}{9}$ (f) $I = \sqrt{\dfrac{P}{R}}$
(g) $n = \dfrac{t-r}{s} + 1$ (h) $a = G(P+q)$
(i) $b = \dfrac{0.9}{x} - d$ (j) $r = \sqrt{\dfrac{A}{4\pi}}$

9.A36 (a) $2\,375.35\,\text{cm}^2$ (b) (i) $h = \dfrac{3V}{\pi r^2}$ (ii) $r = \sqrt{\dfrac{3V}{\pi h}}$

Exercises 9.B

9.B1 (a) $A = lw$ (b) $V = lbd$ (c) $V = \dfrac{\pi D^2}{4} \times l$
(d) $V = lbh$

9.B2 (a) $A = LB$ (b) $A = \dfrac{\pi D^2}{4}$ (c) $\dfrac{b}{2} \times h$ (d) xy
(e) $\dfrac{\pi r^2}{2}$ (f) $(2x \times 3y) + (x \times y)$ (g) $\dfrac{6lh}{2}$
(h) $(x \times 2r) + \left(\dfrac{\pi r^2}{2}\right)$

9.B3 (a) $H = 4a + 3b$ (b) $2c + d = W$ (c) $L = 2a + e$
(d) $V = HWL$

9.B4 (a) $H = 96\,\text{cm}$ (b) $W = 43\,\text{cm}$ (c) $L = 60\,\text{cm}$
(d) $247\,680\,\text{cm}^3$

9.B5 $x = 30W + 40A + 20P$

9.B6 £1 680

9.B7 (a) $x = 4\,800F + 5\,600V + 3\,700L$ (b) £139 400

9.B8 (a) $x = 3W + 2T + £15$ (b) £153

9.B9 (a) $x\,\text{in} = 2.54x\,\text{cm}$ (b) $x\,\text{miles} = 1.61x\,\text{km}$
(c) $x\,\text{ft}^3 = 0.028x\,\text{m}^3$

9.B10 (a) $A = bd + ac$ (b) $350\,\text{m}^2$

9.B11 $P = b + 2d + (b - a) + 2c + a$

9.B12 (a)(i) $P = \dfrac{100I}{RT}$ (ii) $R = \dfrac{100I}{PT}$ (iii) $T = \dfrac{100I}{PR}$
(b) £165

9.B13 £605

9.B14 $ab + \tfrac{7}{8}ac$

9.B15 40%

9.B16 (a)(i) £52x (ii) £52x + £52y (b)(i) x = £2
(ii) y = £4

9.B17 (a) $5x + 8$ (b) $x = 100$ (c) £103, £101, £100, £106, £98

9.B18 $5x^2 + 6x$

9.B19 (a) £3x + 40 (b) £62, £65, £63

9.B20 (a) $L = 52x\,\text{mm}$ (b) $L = x - 15b$ (c) $P = 12(x - y)$
(d) $x^3 + 28x^2 + 140x - 400$

Exercises 9.C

9.C1 (a) $x = y + 73$ mm (b) $x = 105 - y$ mm
 (c) $x = 115 - y$ mm (d) $x = 45$ m $+ 7y$
9.C2 (a) $x = 42 - y$ (b) $x = 2y - 24$ (c) $x = 230 - 3y$
 (d) $x = 1.5y = 87$ (e) $x = 4.8y$
 (f) $x = y + r - z$ (g) $x = 55 - \dfrac{D}{2}$
9.C3 $V = 200(ad + bc)$
9.C4 $A = lb - \dfrac{\pi d^2}{4}$
9.C5 $25\,863$ mm^2
9.C6 Rake angle = 90° − (Point angle + Clearance angle)
9.C7 Rake angle = 20°
9.C8 (a) $I = \dfrac{E}{(R_1 + R_2)}$ (b) 0.5 ohms
9.C9 (a) $V = \dfrac{\pi DN}{1\,000}$ (b) $\dfrac{1\,000\,V}{\pi D} = N$ (c) 341 r/min
9.C10 $x = 10m + 8f + 12w$
9.C11 £37
9.C12 2 000 W
9.C13 (a) $\dfrac{P}{E} = I$ (b) 0.5 A
9.C14 (a) $LB = A$ (b) $A = cd + ba$ (c) $A = 2xy + 2y^2$
 (d) $2ry + \dfrac{\pi r^2}{2}$ (e) $3x \times 3y - xy$ (f) $xy - ab$
9.C15 0.25 ohms
9.C16 1 652°F
9.C17 (a) $C = \tfrac{5}{9}(F - 32)$ (b)(i) 20°C (ii) 212°F
9.C18 (a) $L = A + \dfrac{\pi r}{2}$ (b) 519.94 mm
9.C19 50 V, 40 ohms, 1.5 A, 105 V, 50 ohms, 6 ohms, 0.2 A, 5 A, 20 ohms
9.C20 (a) $S = \dfrac{\pi dN}{1\,000}$ (b) 1 759.5 m/min
9.C21 (a) $E = V + IR$ (b) $I = \dfrac{E - V}{R}$ (c) $R = \dfrac{E - V}{I}$
9.C22 (a) $V = 2.3$ (b) 1.5 (c) 0.8 (d) 9.075

Exercises 9.D

9.D1 $2c\left(\dfrac{\pi d}{2} + \dfrac{\pi D}{2}\right)$
9.D2 4.524 m
9.D3 (a) $(l + 5)(w + 4)(t + 3)$ (b) $175\,000$ mm^2
9.D4 220 N
9.D5 $W = \dfrac{E}{0.02} - 50$
9.D6 (a) $2l + \pi r$ (b) $l + \dfrac{\pi R}{2} + \dfrac{\pi r}{2}$ (c) $l + \pi r$
 (d) $2l + \pi r + 200$
9.D7 (a) lbh (b) $4lh + 2hb$
9.D8 $A + 2B + 6C$
9.D9 $\dfrac{9(\pounds A + \pounds b)}{10}$
9.D10 (a) $\dfrac{\pi x^2}{4} \times y$ (b) xyz (c) $\tfrac{2}{3}\pi\left(\dfrac{x}{2}\right)^3 + \pi\left(\dfrac{x}{2}\right)^2 y$
 (d) x^3
9.D11 (a) $\dfrac{\pi D^2}{4} - \dfrac{\pi d^2}{4}$ (b) $at + (c - t)t$ (c) $a^2 - b^2$
 (d) $a\left(\dfrac{a}{2}\right) - d\left(\dfrac{b}{2}\right)$ (e) $c^2 + b\left(\dfrac{c}{2}\right)$ (f) $d2r + \dfrac{\pi r^2}{2}$
 (g) $ac + ab$ (h) ab
9.D12 (a) 439.88 mm^2 (b) 852 mm^2 (c) 1 980 mm^2
 (d) 945 mm^2 (e) 1 540 mm^2 (f) 2 793.9 mm^2
 (g) 2 544 mm^2 (h) 864 mm^2
9.D13 $\dfrac{\pi d^2 tm}{4} =$ mass in g
9.D14 1 382 480 g
9.D15 (a) $2t + s$ (b) $\dfrac{2t + s}{60}$
9.D16 (a) $xyz = V$ (b) 31.5 m^3

Exercises 9.E

9.E1 $L - 4x$
9.E2 (a) $636\,255$ mm^3 (b) $r = \sqrt{\dfrac{V}{\pi h}}$
9.E3 (a)(i) LW (ii) $2L + 2W$
 (b) Area 6 930 m^2; perimeter 346 m
9.E4 $\dfrac{mt\pi d^2}{4\,000}$
9.E5 16 514.35 g
9.E6 14 A
9.E7 (a) $abc = V$ (b) 2 m
9.E8 (a) $2b + M + N$ (b) $S - (2P + M + N)$
9.E9 99 m^3
9.E10 (a) $2b \times 3a$ (b) $3b \times 4b + a \times 4b$
 (c) $(2a \times 3b) + ab$ (d) $5a \times 5b - (a^2 + ac)$
 (e) $4a2b + 3b2a$ (f) $4a4b - \dfrac{\pi c^2}{4}$ (g) $2\pi b^2$
 (h) $(a + b)(a + 2b + c) - 2ba$
9.E11 (a) 78.72 m^2 (b) 254.2 m^2 (c) 91.84 m^2
 (d) 301.28 m^2 (e) 183.68 m^2 (f) 189.09 m^2
 (g) 105.63 m^2 (h) 94.58 m^2
9.E12 (a) $4b + 6a$ (b) $2(\sqrt{a^2 + (4b)^2}) + 6b + 2a$
 (c) $6a + 6b$ (d) $2(5b - a) + 12a$ (e) $10b + 8a$
 (f) $(4a - c) + (4b - c) + 4a + 4b + \dfrac{\pi c}{2}$
 (g) $\pi 2b + 4b$ (h) $5a + 7b + 2c$
9.E13 $N(2y + \pi 2x)$
9.E14 459.94 m

9.E15 (a)(i) $2x + 2y$ (ii) $xy\,m^2$ (b) $50\,m^2$

9.E16 (a)(i) $d = \dfrac{C}{\pi}$ (ii) $38.51\,m$

(b)(i) $d = \sqrt{\dfrac{4A}{\pi}}$ (ii) $1385.62\,cm^2$

Exercises 9.F

9.F1 (a) $2ba$ (b) $3bc - \dfrac{\pi a^2}{2}$ (c) $2b^2 + \dfrac{ab}{2}$

(d) $2b(c+a) - \dfrac{\pi b^2}{4}$ (e) $b(c+a) + b(c-b)$

(f) $(c-a)(c+a)$ (g) $6b^2 + 3ba$

(h) $\left(\dfrac{c+a}{2}\right)(c+b)$

9.F2 (a) $17\,m^2$ (b) $43.22\,m^2$ (c) $27.37\,m^2$
(d) $43.28\,m^2$ (e) $32.3\,m^2$ (f) $20.79\,m^2$
(g) $94.86\,m^2$ (h) $33.11\,m^2$

9.F3 $x - (3y + 2z)$

9.F4 πdx

9.F5 (a) £$(3x + 5y + 10z)$ (b) £$(2.5x + 1.3y + 4z)$
(c) $6x - 4z$ (d) $(8.5x + 3y) - (6y - 2z)$

9.F6 (a) £55.40 (b) £26.16 (c) £18.80 (d) £26.20

9.F7 (a) $x - y$ (b) $6x - 6y$

9.F8 $x + 32.5$

9.F9 £79.65

9.F10 (a) $76x\,mm$ (b) £$4x + 10 + \dfrac{6x + 8}{4}$

9.F11 (a) £72 (b) £76 (c) £81 (d) £84 (e) £79
(f) £70

9.F12 $(x + 75)(x + 15)(x - 10)$

9.F13 (a) $A = LW$ (b) $L = \dfrac{A}{W}$ (c) $V = LWD$

(d) $74705.4\,cm^3$

9.F14 (a) $a + b + c + d + e$ (b) $4b\,g$ (c) $a - e\,g$
(d) $835\,g$

9.F15 (a) $30x + 90y$ (b) £243

Exercises 10.A

10.A1 (a) 8.8 (b) 1.75 (c) 13.65 (d) 109
10.A2 (a) 15 (b) 0.5 (c) 14
10.A5 (a) £150 (b) £546 (c) £69 (d) £504
(e) £1120 (f) £480 (g) £688.50 (h) £150
10.A6 (a) 995 (b) 45 (c) 929 (d) 135 (e) 41
(f) 1036 (g) 408 (h) 966
10.A7 (a) 2h 55min (b) 2h 26min (c) 14.20
(d) 00.30 (e) 16.00 (f) 00.30 (g) 6
(h) 06.20
10.A8 (a)(i) £44.08 (ii) £66 (iii) £219.92 (iv) £11
(v) £351.92 (b)(i) £732 (ii) £888 (iii) £122.04
(iv) £33 (v) £1707.84 (c)(i) £20.33 (ii) £25.42
(iii) £18.50 (iv) £111 (v) £121.99

10.A12 (a) 100, 121, 144, 169, 196, 225, 256
(b)(i) 365 (ii) 144 (iii) 308 (iv) 87 (v) 217
10.A13 (a) 4, 5, 8, 11, 13, 15
(b)(i) 23 (ii) 33 (iii) 29 (iv) 2 (v) 79
10.A14 (a) 33.02 (b) 96.52 (c) 22.86 (d) 66.04
(e) 3.81 (f) 127 (g) 508 (h) 939.8
10.A15 (a) 3.686 (b) 58.2 (c) 6.589 (d) 2560.36
(e) 123763.24 (f) 554.13 (g) 0.0361
(h) 0.7259
10.A16 (a) 1.814 (b) 3.564 (c) 2.553 (d) 7.397
(e) 11.43 (f) 0.806 (g) 48.86 (h) 0.2966
10.A18 (a) 10 cm (b) 12 cm (c) 12 cm (d) 50 mm
(e) 16 mm (f) 24 cm
10.A19 (a) 18.03 cm (b) 47.88 mm (c) 5.657 cm
(d) 4.9299 cm (e) 6.268 cm (f) 8.74 cm
(g) 10.369 cm (h) 30.91 mm
10.A20 (a) 6.7 cm (b) 10.412 mm (c) 1.142 m
(d) 39.63 mm
10.A21 (a) 462.33 mm (b) 124.45 mm
10.A22 (a) 400.28 mm (b) 71.589 mm (c) 130.25 mm
(d) 93.94 mm
10.A30 (b)(i) 4.913 (ii) 175.616 (iii) 254.42
10.A31 (b)(i) 3.458 m (ii) 4.725 yd

Exercises 10.B

10.B1 (a) 23°
10.B2 (a) 43
10.B3 (a) 54 mm (b) 108 mm (c) 276 mm
10.B5 (a) 71060 (b) 556.2 (c) 236.47
10.B7 (a) £11.54 (b) £2.43 (c) £495.72
(d)(i) £36.24 (ii) £28.32 (iii) £50.76
(iv) £47.03 (v) £22.55 (vi) £80.1
10.B8 (a) £2127 (b) £2029 (c) £2127 (d) £1975
(e) £1959 (f) £10470 (g) £11265
(h) £10470 (i) £11553 (j) £11820
(k) £6528 (l) £8296
10.B9 (a) £600 (b) £536 (c) £1248 (d) £1900
(e) £1488 (f) £525 (g) £3550 (h) £35.4
(i) £1742 (j) £168.75 (k) £2673 (l) £152
(m) £54.56 (n) £1806 (o) £524.40
10.B12 (a) 12.7 m (b) 362.11 mm (c) 2.1 m
(d) 100.17 cm (e) 42.52 cm (f) 94.34 mm
(g) 34.12 cm (h) 76.48 cm
10.B13 255.98 mm
10.B17 (a) £85 (b)(i) £94 (ii) £115 (iii) £93 (iv) £98
(c) £2 (d) $6\tfrac{1}{2}$ h
10.B19 $A = 51.39\%$, $B = 15.28\%$, $C = 9.72\%$, $D = 8.33\%$, $E = 15.28\%$
10.B20 (a) Mon = £57000, Tue = £52000, Wed = £40000, Thur = £50000, Fri = £63000 (b) £262000
(c) £23000

Exercises 10.C

10.C1 (a) 11.1 mm (b) 18.15 mm (c) 22.93 mm
10.C2 (a) 32.32 (b) 18.60
10.C3 23.3 mm

10.C8 (a)(i) 0.42 and 6° (ii) 0.4 and 4° (iii) 1.5 and 10°
(iv) 5.83 and 30° (v) 1.42 and 10° (b)(i) 332
(ii) 2°
10.C9 141.254 mm
10.C10 130.11 mm
10.C11 339.705 mm
10.C12 (a) White (b) green (c) 155 tins
(d) 1 315 tins (e) 65.75 tins
10.C16 (b) 130 mA (c) 5.8 V
10.C17 (b) 358 r/mm (c) 64 mm

Exercises 10.D

10.D1 (a) 425 litres (b) 13 gal (c) 3.5 litres
10.D3 (a) £4 269.72 (b) £769.72 (c) £177.91
10.D5 (a) 52.92 m² (b) 43.68 m² (c) 67.2 m²
(d) 49.14 m² (e) 53.76 m² (f) 44.28 m²
10.D6 (a) 102.34 m (b) 4.37 m
10.D7 (a) 6.92 nautical miles (b) 3.2 m
10.D8 (a) Apr (b) Feb (c) 178 000 t (d) 47 333 t
(e) 24 000 t (f) 284 000 t
10.D10 (a) 454.66 t (b) 556 m (c) 7.6 m (d) 519.25 t
(e) 3.9 m (f) 419.67 t
10.D13 (a) Mon = 700, Tue = 640, Wed = 790,
Thur = 500, Fri = 580 (b) 642 (c) 240
10.D14 Material = £160, labour = £260, overheads = £180,
profit = £50, transport = £40, V.A.T. = £30
10.D15 (a)(i) 60p (ii) 60p (iii) 30p (iv) 40p (v) 70p
(b) 20p
10.D16 (b) 29 min (c) 121°C
10.D17 (a) 293°F (b) 215°C
10.D20 (a) B1 = 8, B2 = 11, B3 = 6, B4 = 14, B5 = 17,
B6 = 15, B7 = 13, B8 = 19
(b)(i) 9th (ii) 2nd (iii) 621 (iv) 69

Exercises 10.E

10.E1 (a) $2\frac{5}{8}$ (b) 94°F (c) ENE (d) 2 500 r/min
10.E2 (a) £126.14 (b) £241.50 (c) £101.56
(d) £108.12 (e) £193.20
10.E4 Notts. Forest = 52, Man. Utd. = 51,
West Ham = 50, QPR = 43
10.E7 (a) 111 m (b) 13.42 ft (c) 26.93 m
(d) 16.55 cm (e) 9.19 m (f) 26.4 m
10.E8 21.824 m²
10.E10 (a) 0.805 km (b) 10.78 km (c) 2.17 km
(d) 8.69 km (e) 2.74 km
10.E11 (b)(i) £26.25 (ii) £35 (iii) £29.75 (iv) £45.50
(v) £21 (vi) £10.50 (vii) £7
10.E12 (a)(i) 214 (ii) 384 (iii) 103 (iv) 277 (v) 200
(vi) 176 (vii) 299 (viii) 18 (b) 35.2
10.E17 (a) £14 268 (b) £253.90
10.E19 (a) 50 (b) 40 (c) 30
10.E20 (a)(i) 9p (ii) 6p (iii) 83p (b)(i) 5 (ii) 5
(c) £5.19
10.E21 (a) June (b) 6 100 m (c) March (d) 9 000 m
(e) 4 458.3 m (f) 26 750 m
10.E22 A = 18%, B = 6.95%, C = 16.67%, D = 36.16%,
E = 22.22%

Exercises 10.F

10.F1 (a) 122 (b)(i) 0°F (ii) −18°C
(c)(i) 11 kg (ii) 24 lb (d) 5.5 kg
10.F2 (a) £76.17 (b) £241.50 (c) £108.12
(d) £50.78
10.F4 (a) 24.57 m (b) 50.05 m (c) 32.76 m
(d) 20.02 m (e) 37.31 m
10.F6 (a) 5.45 kg (b) 12.71 kg (c) 6.36 kg
(d) 0.91 kg (e) 10.89 kg (f) 8.17 kg
(g) 0.11 kg (h) 5.45 kg (i) 2.04 kg
(j) 0.68 kg
10.F8 (a) 226.8 g (b) 1.7 litres (c) 11 lb
(d) 2.64 pints (e) 85.05 g (f) 426 ml
(g) 396.9 g (h) 1.14 litres
10.F10 (a) 400/425, 1 h 20 min (b) 375/400, 1 h 50 min
(c) 400/425, 1 h 15 min (d) 400/425, 2 h 5 min
(e) 375, 3 h 15 min (f) 400/425, 2 h
(g) 375/400, 1 h 20 min (h) 375, 4 h 15 min
10.F11 (a) 1 h 45 min (b) 25 min (c) 50 min
(d) 1 h 5 min (e) 25 min (f) 3 h 25 min
10.F12 (b) 4 h $7\frac{1}{2}$ min
10.F13 (a) 1972 (b) 1978 (c) 1971 and 1975
(d) 1975 (e) best
10.F14 (a) 1 065 (b) 4 575.8 (c) yes (d) 1 750
10.F16 1.93 m
10.F17 (a) 12.04 cm (b) 804 cm²
10.F18 (a)(i) 6 (ii) 4 (iii) £760 (iv) £254.44
(b) (ii) £990
10.F19 (b) 23°C (c) 6.7 min (d) 12.35 min
10.F20 (a) 1982 (b) 1975 (c) 2 300 (d) 23 900
(e) 3 983

Exercises 11

11.1 214 **11.2** 3 404.07 **11.3** 2.157 7
11.4 1 058.091 6 **11.5** 1.832 993 **11.6** 331 570.2
11.7 1 140.976 **11.8** 5.676 55 **11.9** 114 574.2
11.10 0.053 869 **11.11** 8.774 **11.12** 29 531.448
11.13 142.269 53 **11.14** 1 202.877
11.15 19 906.567 **11.16** 18 019.102
11.17 0.003 116 4 **11.18** 0.564 014 5
11.19 41.872 77 **11.20** 0.406 488 5 **11.21** 0.688 9
11.22 3 237.61 **11.23** 2.614 689 **11.24** 2.283 336
11.25 1.309 198 2 **11.26** 62.944 419
11.27 22.009 089 **11.28** 4.615 950 1
11.29 1.002 149 4 **11.30** 0.880 443 6
11.31 4.158 415 8 **11.32** 1.105 674
11.33 0.076 999 8 **11.34** 1.194 464 4
11.35 141.533 11 **11.36** 803.6 **11.37** 0.188
11.38 30.0, 43.9, 29.9, 46.5, 39.5, 22.6
11.39 46 784, 35 351, 179 334, 430 452, 225 317, 1 738 603,
3 511 745
11.40 550.0, 125.7, 443.0, 188.5, 926.8
11.41 I = 40.0, 0.214, 4.76, 1.71, 4.17; P = 800, 1.61, 238,
20.52, 417
11.42 (a) 95.15 (b) 5 320 000 (c) 41 600 (d) 5.588
(e) 38.86 (f) 2.122 (g) 0.334 4 (h) 891 000

	(i) 1.928 (j) 41.38 (k) 297.9 (l) 78.41
	(m) 981.9 (n) 2.610
11.43	£148.24
11.44	£226.46
11.45	£81.80
11.46	59.7
11.47	(a) 412.665 (b) 1.151 (c) 4606.624 (d) 50 531.197
11.48	(a) 209 820.3 (b) 1 699.7 (c) 76 853.9 (d) 547.2
11.49	(a) 216 266.6 (b) 28.417 25 (c) 0.322 538 3
11.50	(a) 15.38 (b) 89.47 (c) 66.67 (d) 36.94 (e) 32.43 (f) 20.69
11.54	0.252 057 7
11.55	21.050 592
11.56	12 763.52
11.71	22.342
11.73	152.995 55
11.86	21.213 333